Small and Short-Range Radar Systems

GREGORY L. CHARVAT SERIES ON PRACTICAL APPROACHES TO ELECTRICAL ENGINEERING

SERIES EDITOR

Gregory L. Charvat
Advisor, Co-Founder, Researcher,
Westbrook, Connecticut, USA

PUBLISHED TITLES

Small and Short-Range Radar Systems
Gregory L. Charvat

Small and Short-Range Radar Systems

Gregory L. Charvat

CRC Press is an imprint of the
Taylor & Francis Group, an informa business

CRC Press
Taylor & Francis Group
6000 Broken Sound Parkway NW, Suite 300
Boca Raton, FL 33487-2742

First issued in paperback 2017

ISBN-13: 978-1-4398-6599-6 (hbk)
ISBN-13: 978-1-138-07763-8 (pbk)

Library of Congress Cataloging-in-Publication Data

Charvat, Gregory L.
Small and short-range radar systems / author, Gregory L. Charvat.
pages cm -- (Gregory L. Charvat Series on Practical Approaches to Electrical Engineering ; 1)
Summary: "Coupling theory to reality by demonstrating the implementation of actual radar systems to enable those interested in developing small radar technology to move quickly from idea to proof of concept, Small and Short-Range Radar Systems analyzes and provides practical working examples of small and short-range radar systems. It supplies a derivation of the radar range equation, an explanation of applicable algorithms, and a case study for each of the several different types of small radar systems covered, including CW, ultrawideband (UWB) impulse, UWB linear FM, SAR, and phased array"-- Provided by publisher.
Includes bibliographical references and index.
ISBN 978-1-4398-6599-6 (hardback)
1. Radar. I. Title.

TK6575.C4746 2014
623.7'348--dc23 2013049551

Visit the Taylor & Francis Web site at
http://www.taylorandfrancis.com

and the CRC Press Web site at
http://www.crcpress.com

To my wife, Kathryn Phillips, for forcing me to complete this book rather than spend time repairing old radios and pocket watches. To my parents, Dave and Rita Charvat, who similarly encouraged me to write this book even though I recently joined a startup. To my daughter, may you find your passion in life.

Contents

List of Figures

List of Tables

Foreword

Technology development moves quickly the modern era. An engineer must "ball park" a design, rapidly create proof of concept, and analyze measured data to drive the technology's course. The purpose of this book is to enable those interested in developing small radar technology to move quickly from idea to proof of concept.

We can do it. Anyone with a background in electronics who is an undergraduate, professional, entrepreneur, researcher, or hobbyist can develop small radar sensors. Radar engineering need not be intimidating or full of theoretical derivations.

Learn by doing. The sooner a design is put to practice, the faster the designer will learn. Numerous examples are shown with sufficient detail to replicate or scale to your application.

Measured data is ugly. Measurements from examples are available to test your algorithms on real-world data before investing development time.

For your application. Applications are shown from automotive to through-wall imaging. Problems, solutions, and future directions of technology are presented.

Let's build something together. Skip to any chapter, section, or figure to learn what you need to get the job done. All material stands on its own.

About the Series

Gregory L. Charvat Series on Practical Approaches to Electrical Engineering

Learn, create, execute. Technology development moves quickly in the modern era. An engineer must "ballpark" a design, rapidly create proof of concept, and then measure results to drive product development. The purpose of this series is to enable those interested in moving from concept to practical implementation. Key concepts and background will be covered, with tangential discussion minimized. This series of texts will serve to bridge the gap between academia and implementation with an emphasis on practical approaches.

Preface

Inexpensive microwave components, micro-computing, and data acquisition have made possible widespread adaptation of small and short-range radar sensors. Initially, small radar sensors were continuous wave (CW) Doppler units for use as proximity fuses during the Second World War. These sensors were mounted on the top of an anti-aircraft shell, detonating the shell when it approached the body of an aircraft. Shortly after the war CW radar sensors found use in the law-enforcement application of measuring the speed of automobiles. During the late 1950s automotive radar sensors were developed but proved too costly. With the invention of the Gunn oscillator the small Doppler radar found use as a motion sensor for opening automatic doors. During the evolution of low-observable aircraft, short-range rail synthetic aperture radar (SAR) imaging systems were developed to measure radar cross section (RCS). Today, low-cost radar sensors are used in automatic cruise control, collision mitigation braking, blind spot detection, and as back-up aids in automobiles. Lower frequency portable and mobile sensors can penetrate concrete walls, providing a sensing capability for search and rescue missions. Compact SAR sensors are flown on unmanned air vehicles providing on-demand reconnaissance. Small radar sensors are utilized in autonomous vehicles (both air and ground) to sense the environment, detect hazards, attempt to classify objects, and enable autonomous decision making.

These radar systems are different in theory and significantly different in operation from conventional radar. This is due to the short-range geometry of target scenes. Short ranges require wide bandwidths to achieve useful range resolution and simultaneously transmit and receive because of practical hardware limitations, resulting in unusual radar architectures. These target scenes are full of clutter that requires coherent processing and detection algorithms. Furthermore, short-range radar systems operate in the near field, requiring special treatment of beamforming and imaging. To increase accuracy in automotive and unmanned vehicle applications data is often fused with other sensors to reduce false alarm rates.

In this book small and short-range radar systems will be explored with a high-level analysis and demonstrated with practical working examples. Several different types of small radar systems will be discussed, including CW, ultra-wideband (UWB) impulse, UWB linear FM, SAR, and phased array. For each type of radar system there will be a derivation of the radar range equation, an explanation of applicable algorithms, and a case study that will couple theory to reality by demonstrating the implementation of an actual radar system.

The application of small radar sensors is an evolving topic and therefore requires special treatment. The ubiquitous use of low-cost Doppler radar for use in law enforcement and motion sensing will be discussed. One chapter will focus on the topic of automotive radar and sensing, covering topics from physical design into vehicles to navigation and sensor fusion. Finally, a chapter on through-wall imaging will be presented.

This is the first text on the topic of small short-range radar systems and their applications. This text will capture the state of the art—exploring the theoretical, discussing the evolution of small radar and its application, and establishing a technical reference for the ongoing development of small radar.

The author can be reached at glcharvat.com for online book content including video demos and radar data, and MATLAB scripts can be found at glcharvat.com/shortrange

Acknowledgments

A special thank you to Chapter 9 authors, Shuqing Zeng and James N. Nickolaou, for hanging in there on this journey. Many thanks to the technical reviewers of this book, Alan Fenn and Raoul Ouedraogo, Massachusetts Institute of Technology (MIT) Lincoln Laboratory; Sam Piper and Greg Showman, Georgia Tech Research Institute (GTRI); Eli Brookner, Raytheon; Prof. Ed Rothwell, Michigan State University; Prof. Xiaoguang "Leo" Liu, University of California Davis; Robert Burkholder, Ohio State University; and Micha Feigin, MIT Media Lab. I really appreciated the editorial feedback from Mark A. Richards, Georgia Institute of Technology; Prof. Cary Rappaport, Northeastern University; Philip Erickson, MIT Haystack Observatory; Prof. Daniel Fleisch, Whittenberg University; and Prof. Xiaoguang "Leo" Liu, University of California Davis.

Over the years I've been very fortunate to work with numerous, amazing organizations, which served as funding sources, employers, or teachers, each of whom was an overwhelmingly positive influence on everything I've done. These organizations include MIT, MIT Lincoln Laboratory, the Naval Research Laboratory, the Air Force Research Laboratory, and Michigan State University.

I acknowledge one of my earliest role models, Kerry Pytel. To this day I like to share the story of how Kerry created a solar car team full of 7th and 8th graders. We designed and built from scratch a solar-powered electric vehicle—welding/cutting steel, designing the drive train, electronics, everything—which ran beautifully. This was a college-level project for middle schoolers! It was a life changing experience.

A special thanks to Ardis Herrold, who organized in my high school a group that designed and built from scratch a working radio telescope. Ardis provided us with the opportunity to compete in the science fair, learn basic astrophysics, and work as a team.

Thanks to Prof. Daniel Fleisch, from whom I learned at an early age that I must pursue my Ph.D.

Rather than admiring actors, musicians, or athletes, my childhood idols were (and continue to be) Thomas Edison and Henry Ford, whose visions and contributions to modern society are quite clear. On this thread, it is important to acknowledge the museums that have provided lifelong inspiration. Thanks to Henry Ford, from a very young age I realized that you can do anything if you put your mind to it, and that became clear while walking through Greenfield Village or touring the massive steam engines in the Henry Ford Museum. It was the Air Force Museum in Dayton, Ohio that showed me that humankind can do anything. It is difficult not to be inspired by admiring aircraft like the SR71 or the F117 and understanding the implications of this technology.

But, ultimately, if it were not for my wife, Kathryn Phillips, I would never have completed this book. She is the one who convinced me to push ahead and finish this project. I am extremely lucky to be raised by my parents, Dave and Rita Charvat, because they instilled independent thought, tenacity, and fostered this interest in electronics at a very early age, and they continue to do so.

About the Author

Gregory L. Charvat is co-founder of Butterfly Network Inc. and visiting research scientist at the Camera Culture Group at MIT Media Lab.

Dr. Charvat grew up in the metro Detroit area, where, at a young age, he would take apart old television sets and radios. Dr. Charvat eventually started making in high school amateur radio equipment and a radio telescope, and in college he learned to develop radar systems. He earned a PhD in electrical engineering in 2007, an MSEE in 2003, and a BSEE in 2002 from Michigan State University. He was a technical staff member at MIT Lincoln Laboratory from September 2007 to November 2011 where his work on through-wall radar won Best Paper at the 2010 MSS Tri-Services Radar Symposium and was an MIT Office of the Provost 2011 Research Highlight. Dr. Charvat has taught short radar courses at the Massachusetts Institute of Technology, where his Build a Small Radar Sensor... course was the top-ranked MIT professional education course in 2011.

Dr. Charvat authored or co-authored numerous journals, proceedings, magazine articles, and seminars on topics including: applied electromagnetics, synthetic aperture radar (SAR), phased array radar systems, RF and analog design, and project-based learning. He has developed numerous rail SAR imaging sensors, phased array radar systems, and impulse radar systems, and he holds a number of patents.

For fun, Dr. Charvat develops vacuum tube audio equipment, restores antique radios, watches, and clocks, designed his own amateur radio station from scratch, goes Lindy-hop dancing, and sails on the Long Island Sound.

Dr. Charvat is a senior member of the IEEE. He served on both the 2010 and 2013 IEEE Symposium on Phased Array Systems and Technology steering committees and the steering committee for the CMU 2012 Next Generation Medical Imaging Workshop. He served as chair of the IEEE AP-S Boston chapter from 2010-2011 and IEEE Boston Section member at large in 2012.

Further information on his work can be found on the front page of *MIT news* online, in Slashdot, *Popular Science* blog, MIT CSAIL news, ABC news, CNN blog, *Financial Times*, *Popular Mechanics* blog, *PC Magazine*, Fox news Boston, BBC news, *Wired UK*, Discovery news, *R & D Magazine*, MSNBC online, *MIT Alumni News*, *The State News*, *Wall Street Journal*, Make Magazine blog, *IEEE Spectrum Magazine*, *QST Magazine*, and others.

Chapter 9 Contributing Authors

Shuqing Zeng
General Motors Corporation
Detroit, Michigan

James N. Nickolaou
General Motors Corporation
Detroit, Michigan

1

Radio Direction and Ranging (RADAR)

RADAR is a World War II acronym that stands for radio direction and ranging. A radar system consists of a radio transmitter and a receiver, where pulses of electromagnetic fields are transmitted from a transmitter, scattered (reflected or echoed) off of a target, and returned to a receiver. Round-trip time is measured from the moment the transmitted pulse is radiated to the moment the scattered pulse is received (Fig. 1.1). Radio waves travel through air at approximately the speed of light. By measuring the time delay, the range to target can be determined.

To understand radar we must first describe how radio transmitters and receivers work (Sec. 1.1). With this understanding basic radar systems will be discussed (Sec. 1.2), where maximum range is estimated by using the radar range equation (Sec. 1.3). Finally, an introduction to small and short-range radar systems will be presented (Sec. 1.4).

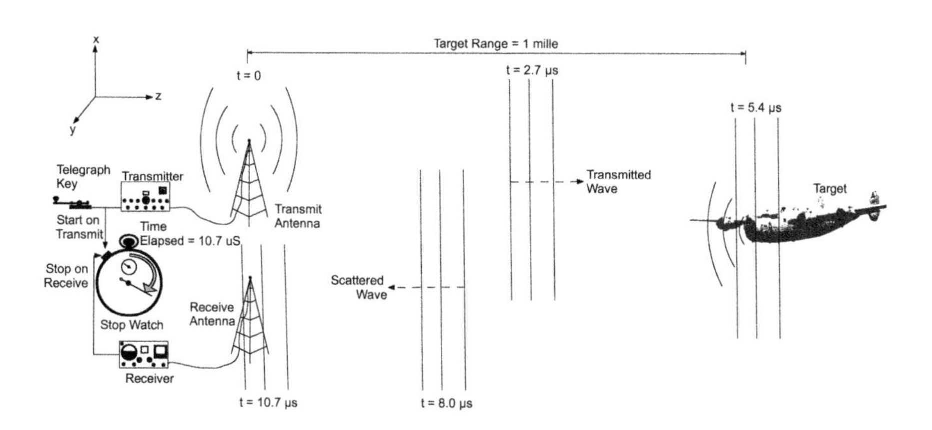

FIGURE 1.1
Electromagnetic fields travel at the speed of light in free space. In a radar system, range to target is measured by the round-trip time from when an electromagnetic pulse is transmitted, scattered off of a target, and when its reflection is received.

1.1 Radio Transmitters and Receivers

Radio transmitters radiate electromagnetic fields constrained to a specific wavelength that are modulated to carry information. Radio receivers collect electromagnetic fields at a certain wavelength and de-modulate to reproduce the transmitted information. A radar system is made up of a radio transmitter and receiver, where the modulation is designed to facilitate measurement (distance, velocity, imagery, etc.) of a target at a distance.

The basics of radio transmission and reception will be described starting with the generation of electromagnetic fields (Sec. 1.1.1). Electromagnetic fields are transferred between components inside of a transmitter or receiver using transmission lines (Sec. 1.1.2). Electromagnetic waves are radiated into free space by using antennas (Sec. 1.1.3). Maximum range between a radio transmitter and receiver can be estimated (Sec. 1.1.4).

1.1.1 Generating Electromagnetic Fields and Maxwell's Equations

The physics of electromagnetic fields is described by Maxwell's equations (Fig. 1.2).

Individual electrons are point charges and each point charge emits a static electric field. Electric conductors function because they contain free electrons that, when connected to a signal source like a battery or an AC generator, can be pushed down a wire to carry DC current or forced to accelerate back and forth to carry AC current.

An electromagnetic field is generated by connecting an oscillator to an antenna (Fig. 1.3). The oscillator accelerates electrons up and down the conductive antenna elements (wires) in the $\hat{x}$ direction at the frequency of oscillation f_c, continuously accelerating the current back and forth at the frequency of oscillation. We generate the electromagnetic field by forcing the electrons to accelerate we are changing the electric field, perturbing it at the rate of oscillation. Changes in the electric field are not observed instantly everywhere away from the antenna because these propagate through free space away from the antenna at the speed of light (approximately $c = 3 \cdot 10^8$ m/s in free space).

From Maxwell's equations we know that a change in electric field results in an induced magnetic field (Equation (1.6)) that is orthogonal to the electric field. This magnetic field is in phase with the electric field. As the electric field intensity rises so does the magnetic and as it falls the magnetic field falls as well. From this we now have an electromagnetic field, where the electric field lines are time varying and the magnetic field lines are time varying in an orthogonal direction. All of this is following the frequency of oscillation f_c and propagating outward from the antenna at velocity c when we are in free space.

Maxwell's equations [3], differential (instantaneous) form:

$$\nabla \times \mathscr{E} = -\mathscr{M}_i - \frac{\partial \mathscr{B}}{\partial t}, \tag{1.1}$$

$$\nabla \times \mathscr{H} = \mathscr{J}_i + \mathscr{J}_c + \frac{\partial \mathscr{D}}{\partial t}, \tag{1.2}$$

$$\nabla \cdot \mathscr{D} = \wp_{ev}, \tag{1.3}$$

$$\nabla \cdot \mathscr{B} = \wp_{mv}, \tag{1.4}$$

$$\nabla \cdot \mathscr{J}_{ic} = -\frac{\partial \wp_{ev}}{\partial t}. \tag{1.5}$$

where each vector is a function of space and time (for example $\mathscr{E} = \mathscr{E}(x, y, z, t)$) and is described:

$\mathscr{E}$ = electric field intensity (volts/m)
$\mathscr{M}_i$ = impressed magnetic current density (volts/m^2)
$\mathscr{B}$ = magnetic flux density (webers/m^2)
$\mathscr{H}$ = magnetic field intensity (amperes/m)
$\mathscr{J}_i$ = impressed electric current density (amperes/m^2)
$\mathscr{J}_c$ = conduction electric current density (amperes/m^2)
$\mathscr{D}$ = electric flux density (coulombs/m^2)
$\wp_{ev}$ = electric charge density (coulombs/m^3)
$\wp_{mv}$ = strictly theoretical magnetic charge density (webers/m^3)
$\mathscr{J}_{ic}$ = $\mathscr{J}_i + \mathscr{J}_c$

Maxwell's equations describe the behavior of electric fields, magnetic fields, and rates of change of charge as they relate to the generation of electromagnetic radiation. An electromagnetic field oscillating at a given frequency is time harmonic, meaning that it is continuously varying with time in a sinusoidal fashion. Applying this to the instantaneous form of Maxwell's equations above results in the time harmonic form of Maxwell's equations:

$$\nabla \times \mathbf{E} = -\mathbf{M} - j\omega \mathbf{B} \tag{1.6}$$

$$\nabla \times \mathbf{H} = \mathbf{J}_i + \mathbf{J}_c + j\omega \mathbf{D} \tag{1.7}$$

$$\nabla \cdot \mathbf{D} = q_{ev} \tag{1.8}$$

$$\nabla \cdot \mathbf{B} = q_{mv} \tag{1.9}$$

$$\nabla \cdot \mathbf{J}_{ic} = j\omega q_{ev} \tag{1.10}$$

where the relationship between each time harmonic vector field and the instantaneous vector fields follows $\mathscr{E}(x, y, z, t) = real\{\mathbf{E}(x, y, x)e^{-j\omega t}\}$.

FIGURE 1.2
Maxwell's equations.

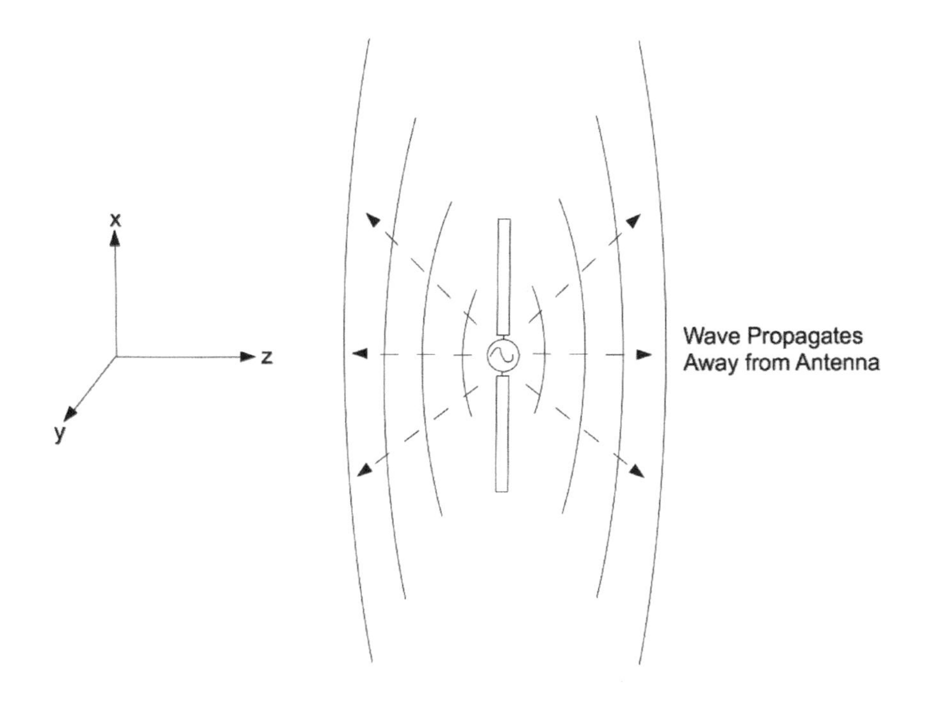

FIGURE 1.3
Generating electromagnetic fields using a dipole antenna.

Additionally, a time varying magnetic field induces an electric field. Each field component supports the other as it propagates through space. For this reason an electromagnetic wave is able to propagate and keep itself propagating as it travels away from the antenna.

1.1.1.1 Far Field and Near Field

Near the antenna the electric and magnetic field lines have a great deal of curvature due to the geometry of the antenna elements. This provides wavefront curvature as the field propagates away from the antenna in a spherical shape. As the electromagnetic field travels much further from the antenna the field lines become more planar, much like the surface of the earth seems flat at ground level. At approximately the distance $R = 2D^2/\lambda$ (where D is the largest dimension of the antenna, in this example D is the length of the antenna) the radiated field is considered to be a plane wave [2].

When observing an electromagnetic wave you are considered to be in the **far field** when $R \geq 2D^2/\lambda$ and the **near field** when $R < 2D^2/\lambda$.

In the far field, the electric field lines are time varying in $\hat{x}$ and the magnetic field lines are time varying in $\hat{y}$ to the frequency of oscillation f_c propagating outward in the $\hat{z}$ direction from the antenna at velocity c in free space.

1.1.1.2 Polarization

Linear antenna polarization is defined as being tangent to the electric field lines. In this example the horizon is in the $\hat{y}$ axis, the antenna element is vertical in the $\hat{x}$ axis, and wave propagation is in the $\hat{z}$ direction away from the antenna normal to the electric and magnetic field lines; therefore this dipole antenna is vertically polarized. Additional polarizations are possible including horizontal, circular, and elliptic. In this book all antennas will be linearly polarized, either vertical or horizontal.

1.1.1.3 Constitutive Parameters or the Medium in Which a Wave Propagates

The rate at which the electric and magnetic field lines form and propagate depends on the properties of the medium in which the fields are propagating. These properties are known as the constitutive parameters permittivity ϵ, permeability μ, and for a lossy medium loss tangent σ [3]. In this book most radars will be used in free space therefore these parameters are $\epsilon_o = 8.854 \cdot 10^{-12}$ (Farads/m), $\mu_o = 4 \cdot \pi \cdot 10^{-7}$ (Henries/m), and $\sigma \approx 0$ (siemens/m) and therefore wave velocity is $1/\sqrt{(\mu_o \epsilon_o)} = 3 \cdot 10^8 = c$ m/s.

1.1.1.4 Electric and Magnetic Antennas

Antennas can be designed to receive either the electric field component or the magnetic. A dipole acts on the electric field component. A loop antenna acts on the magnetic field component. Most antennas act on the electric field component. The use of magnetic loop antennas is the exception and these are typically used for applications where low frequency signals must be received but a small form factor is required at the expense of antenna efficiency (e.g., in AM broadcast or shortwave radio receivers).

1.1.1.5 Most-used Solution to the Wave Equation for Radar Systems

Solving Maxwell's equations for any geometry is challenging because they are coupled partial differential equations, requiring clever use of boundary conditions, vector calculus, and complex analysis. The least-difficult approach to solving most time harmonic problems is to derive the wave equation by re-arranging Equations (1.6)–(1.10) and applying boundary conditions. Such a derivation results in the wave equation for a source-free region [3],

$$\nabla^2 \mathbf{E} + \omega^2 \mu \epsilon \mathbf{E} = 0. \tag{1.11}$$

Source-free region means that this is a solution to Maxwell's equations for the electric field assuming there are no antennae or boundaries present.

Stated another way, it is assumed that the fields have already launched off of the antenna and have not yet scattered off of any targets just like the wave shown in Fig. 1.1 at $t = 2.7\mu s$. If we are in free space and the wave is traveling away from our vertical dipole antenna (Fig. 1.3) and we are in the far field of the antenna then the solution to the electric field of the wave equation along the z axis of propagation is

$$E_x = E_o e^{-j\omega z\sqrt{\mu_o \epsilon_o}}. \tag{1.12}$$

The magnetic field can be found by substituting the above into Equation (1.7), but since the magnetic field is proportional to the electric field we will focus only on solutions to the electric field. The free space wave number $k_o = \omega\sqrt{\mu_o \epsilon_o}$ is substituted into the above equations providing the more recognizable form:

$$E_x = E_o e^{-jk_o z}. \tag{1.13}$$

The electric field solution to the wave equation will be used throughout this book to model wave propagation from the radar to the target and back.

1.1.2 Transmission Lines

When generating electromagnetic fields it is often inconvenient to co-locate the transmitter (or oscillator in Fig. 1.3) directly on the antenna. For this reason a transmission line is used which provides a means of transmitting electromagnetic fields from the transmitter to the antenna some distance away.

A transmission line works in a similar fashion as an antenna, where an electromagnetic field is generated and propagated from a source (a transmitter or an oscillator). Unlike an antenna, the transmission line does not radiate electromagnetic field out and away from itself; instead it confines the electromagnetic field to travel down the path of two conductors, transmitting this field into the load of your choosing (either an antenna or another device). These two conductors can be a two-wire transmission line, a coaxial cable, or a microstrip line. These conductors can be curved to guide the transmission line up walls, around corners, out the door, and up your antenna tower (or in and around the chassis of your radar system).

An oscillator with an output impedance of Z_s is connected to a two-wire transmission line in Fig. 1.4. At the end of the transmission line is a load Z_l. The output of the oscillator is a sine wave as shown in Fig. 1.3 similar to the generation of electromagnetic fields described in Sec. 1.1.1. As the voltage output of the oscillator increases it pushes charge across the wires which form electric field lines between the wires. From Maxwell's equations we know that a change in electric field results in an induced magnetic field (Equation (1.6)) that is orthogonal to the electric field, thus an electromagnetic field between the two wires [4] is produced.

The electromagnetic field is contained between the wires as it travels down the wires away from the source toward the load. The transmission line has a

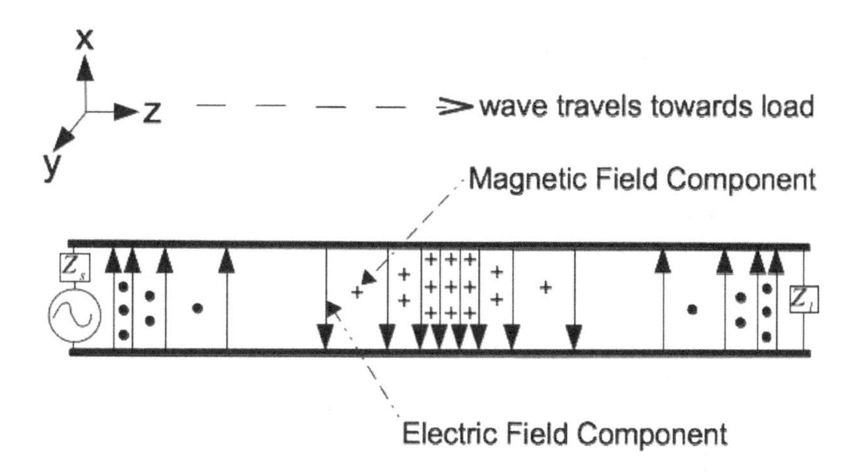

FIGURE 1.4
A two-wire transmission line which transmits electromagnetic fields between a source and a load, preventing radiation outside of the transmission line.

characteristic impedance Z_o, which is proportional to the spacing and the dielectric properties of the medium between the wires. Full power transfer occurs when $Z_s = Z_o = Z_l$. If this is not the case then the traveling wave will be reflected and bounced back toward the source (much like how a radar pulse bounces off of a target). The load for which the transmission line terminates to can be a resistor, another device (such as an amplifier input, filter, mixer, etc), or an antenna.

When we apply the wave equation (1.11) to the two-wire transmission line the solution for the electric field traveling in the z direction becomes

$$E_x = E_o e^{-jkz}, \tag{1.14}$$

where $k = \omega\sqrt{\mu\epsilon}$ is the wave number in the dielectric medium between the two wires. This solution is identical to the electromagnetic field traveling away from the antenna in the far field because both fields are traveling waves where the electric and magnetic field components are transverse to the direction of propagation, known as a transverse electromagnetic (TEM) field.

The more commonly known coaxial transmission lines behave in a similar fashion. Unless otherwise noted, all transmission lines in this book will be coaxial where $Z_o = 50$ ohms, output ports will have an impedance $Z_s = 50$ ohms, and input ports will have an impedance $Z_l = 50$ ohms.

Waveguides provide an alternative to transmission lines, where electromagnetic fields can be ducted through hollow pipes (circular or rectangular). Waveguides are typically used in applications where high power and low loss are required or as part of an antenna feed. This book is on the subject of small and short-range radar systems which do not use high power; therefore if a waveguide is used in this book it will be part of an antenna feed.

1.1.2.1 Scattering Parameters

All radar systems in this book will be shown at the modular RF level, where building blocks such as mixers, amplifiers, switches, circulators, and filters will be treated as modules with input and output ports that connect to transmission lines. It will be assumed that the inputs and outputs of each module are matched to 50 ohms. These modules will be considered 'black boxes' where we can abstract their functionality as the specifications of each module (e.g., an amplifier has a gain and a noise figure, a filter has bandpass, band stop frequency, and loss etc.).

In practice, most commercial connectorized RF modules are close to 50 ohms except for the LO ports of frequency mixers or the inputs and outputs of frequency multipliers. Similarly, most commercial antennas are 50 ohms at their specified frequency range.

If you must build your own RF module or antenna then it is important to understand scattering parameters (S-parameters). S-parameters provide a measurement of the reflected (scattered) and transmitted power over frequency and phase for a given two-port RF module (or device). The measurement of S-parameters facilitates matching the electrical impedance of a given device to 50 ohms (or any desired impedance). In-depth discussion of S-parameters for two-port RF amplifiers and filters is treated in [5] and for antennas in [2] and [6].

For the purposes of this book it is important to understand magnitude S-parameters of two-port devices (Fig. 1.5), where we are mostly concerned about S21 which is the gain (or loss) measurement of a given two-port device over a range of frequencies. For example, S21 of a crystal filter (ECS-10.7-30B) is measured where the filter is excited from port 1 and measured on port 2. The magnitude response of this crystal filter is shown (Fig. 1.6). From this measurement the filter passes frequencies with better than −3 dB of loss from approximately 10.685 to 10.715 MHz. This measurement showing the frequency response of the filter should look familiar to all electrical engineers because similar plots are used for low frequency circuit response using Bode analysis.

1.1.3 Characteristics of Antennas

Directional antennas focus the radiation (or reception) over a narrow field of view where most of the gain is in only one direction. Examples of directional antennas include satellite television dishes, Yagi antennas, and

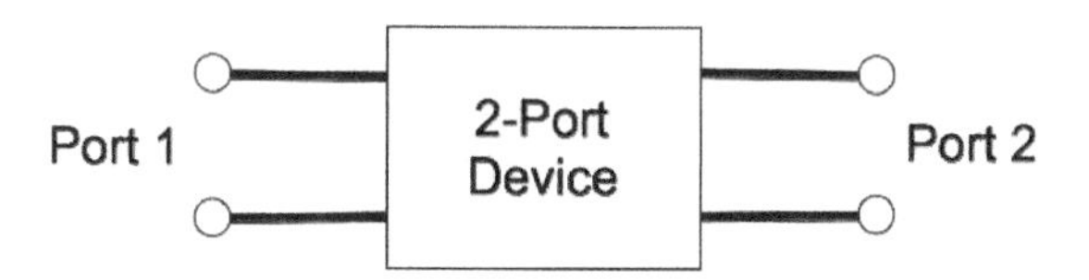

FIGURE 1.5
RF devices in this book will be treated as two-port devices, where details of the device's function are abstracted into its frequency response in gain (or loss) and noise figure.

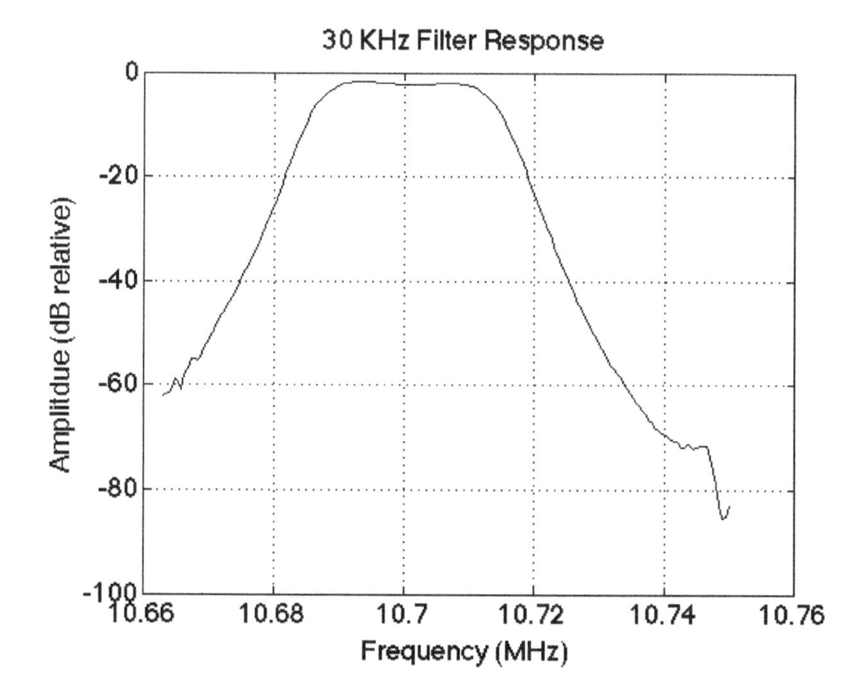

FIGURE 1.6
S21 magnitude response of a 30 KHz wide crystal filter centered at approximately 10.7 MHz.

FIGURE 1.7
Example of a directional antenna, a 10 GHz microwave horn with a WR-90 waveguide feed.

microwave horns (Fig. 1.7). Conversely, omni-directional antennas radiate (or receive from) a broad field of view which usually includes anything within the line of sight. Examples of non-directive (omni-directional) antennas include the dipole antenna in Fig. 1.3, AM/FM broadcast antenna on your car, cellular phone antennas etc.

We will use antennas to radiate, direct the transmission, and receive radar signals. In this book we are only concerned with directional antennas. There are many books on the topic of antennas ([2], [6] and others); in this book we will focus on the use of antennas for small radar systems, where we are primarily concerned with beamwidth, gain, and effective aperture.

Antenna performance of the 10 GHz microwave horn from Fig. 1.7 is shown, where a simulated pattern of the the main beam is directed toward the z axis and the antenna is located on the xy axis (Fig. 1.8). Gain of the antenna is plotted compared to an ideal isotropic radiator, meaning that this antenna radiates proportionally more or less power density than an isotropic antenna in the given directions shown.

The antenna beamwidth is defined as the point at which the main beam gain drops 3 dB from maximum gain. Taking the cross section of Fig. 1.8 along the yz plane shows the -3 dB points located at $\pm 10.1°$ which tells us that this antenna has a $20.2°$ beamwidth in the horizontal plane (Fig. 1.9).

Similarly, taking the cross section of Fig. 1.10 along the xz plane shows the -3 dB points located at $\pm 6.7°$ which tells us that this antenna has a $13.4°$ beamwidth in the vertical plane (Fig. 1.10).

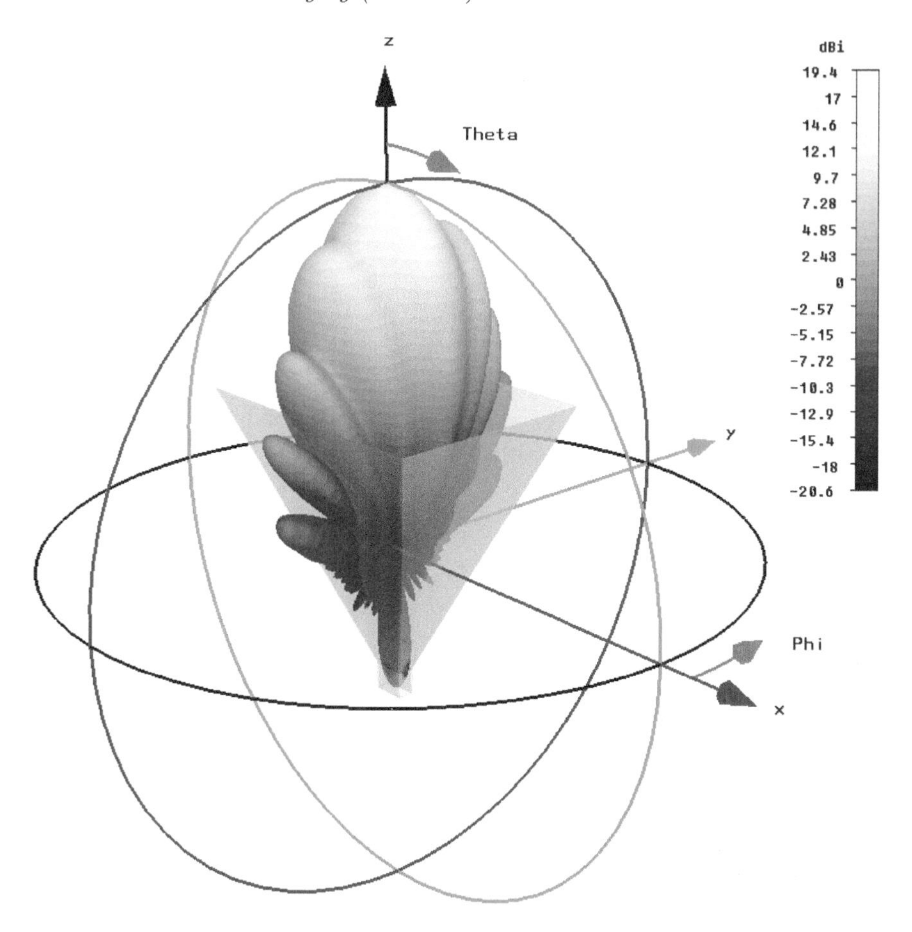

FIGURE 1.8
3D antenna pattern of a standard gain horn.

In all plots (Figs. 1.8 to 1.10) the antenna gain is the peak of the main lobe, which is 19.4 dBi. In this book the antenna gain accounts for antenna efficiency and impedance mismatch.

Given the half-power beamwidths, the antenna gain for a directional antenna can be estimated [2],

$$G_{dB} \simeq 10 \cdot \log\left(\rho \frac{4\pi}{\Theta_H \Theta_E}\right), \tag{1.15}$$

where ρ is the antenna efficiency (a typical efficiency is $\geqq 90\%$ for most well-matched metal antennas) and Θ_H and Θ_E are the half-power beamwidths in the magnetic and electric field planes. In this case the electric field component of this antenna is in the yz plane.

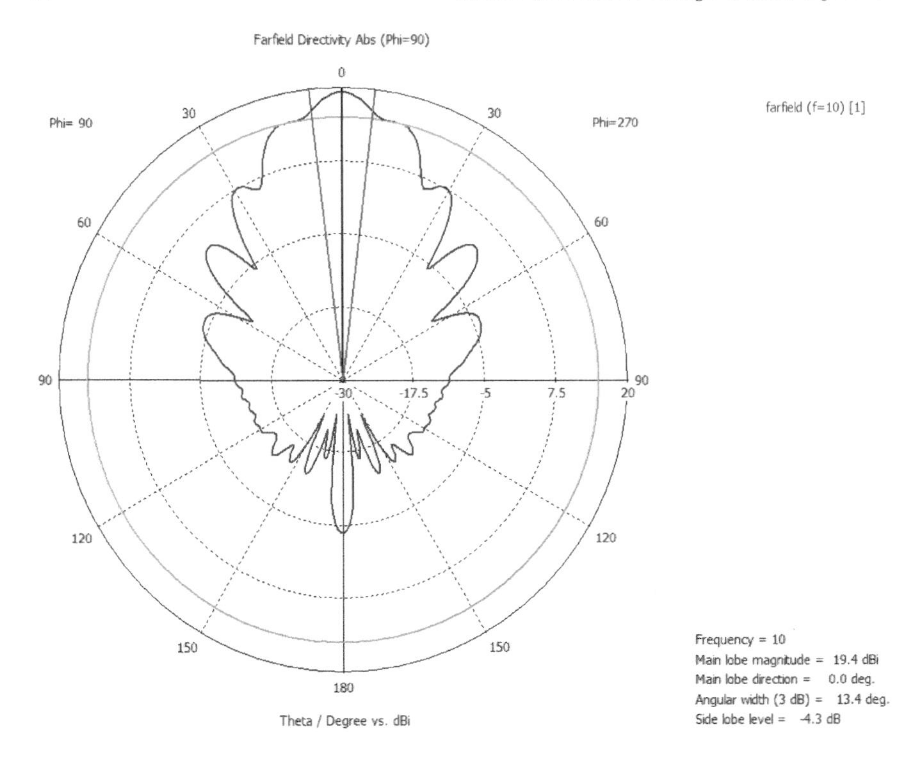

FIGURE 1.9
2D antenna pattern in the yz plane.

Another way of thinking about an antenna is to consider it as a device that intercepts a cross section of the traveling electromagnetic field. That cross section can be expressed in terms of aperture or area in m^2. Aperture and gain are proportional; the greater the gain, the larger the effective aperture. Antenna aperture does not always represent the physical area of an antenna. For the case of a dipole antenna, the effective aperture is larger than the physical aperture because the dipole elements are thin wires. For the case of a phased array radar system the aperture size of the array is typically equal to the effective antenna aperture. Effective antenna aperture and antenna gain are related [2],

$$A_{eff} = \frac{G\lambda^2}{4\pi}, \tag{1.16}$$

where $G = 10^{G_{dB}/10}$. Effective aperture is often used to model the performance of a radar system.

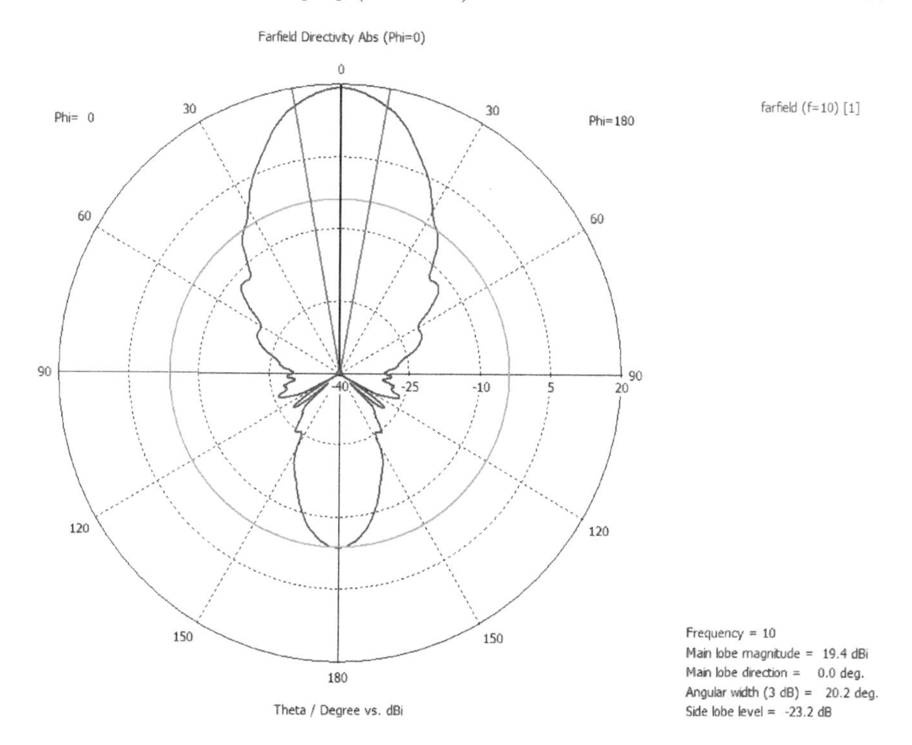

FIGURE 1.10
2D antenna pattern in the xz plane.

1.1.4 Friis Transmission Equation

The Friis transmission equation estimates the power transferred from a transmit antenna to a receive antenna for a given power input (Fig. 1.11). It is used to estimate the performance of radio communication and broadcast stations [2],

$$P_{rx} = P_{tx}\left(\frac{\lambda}{4\pi R}\right)^2 G_{tx}G_{rx}, \tag{1.17}$$

where, P_{tx} is the transmit power fed into the transmit antenna from the transmitter, P_{rx} is the receive power fed into the receiver from the receive antenna, G_{tx} and G_{rx} are the transmit and receive antenna gains, and R is the range or distance between the two antennas.

By estimating or measuring your receiver sensitivity [7]-[8] the minimum input power into your receiver required for reliable communications can be determined and substituted for P_{rx}. Knowing the gain of both transmit G_{tx} and G_{rx} receive antennas, the maximum range R of your communications link can be solved assuming both antennas are within line of sight (no obstructions).

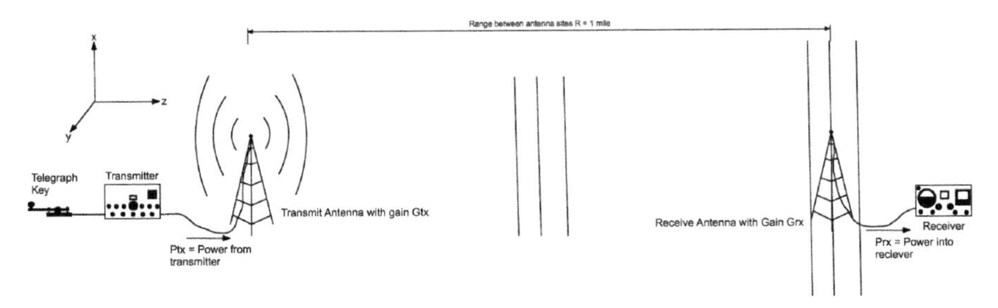

FIGURE 1.11
The Friis transmission equation estimates the power received at the antenna terminals given the power fed into the transmit antenna and the gain of both the transmit and receive antennas.

1.1.5 Radio Receivers

Radio receivers 'tune in,' amplify, and apply detection to a specific wavelength or range of wavelengths and exclude all others. For example, an AM broadcast radio receiver will tune in a range of frequencies from 580 KHz to 1.6 MHz using an envelope, or magnitude, detector where the output is fed into an audio amplifier and out a loudspeaker. Similarly, a frequency modulation (FM) broadcast radio receiver tunes in a range of frequencies from 88 MHz to 108 MHz using a FM detector. Modern wireless phones and digital television use analog-to-digital converters rather than detectors, where data is de-modulated after some degree of signal processing.

1.1.5.1 Tuned Radio Frequency (TRF) Receivers

Early receivers, known as tuned radio frequency (TRF), use an amplified resonant bandpass filter approach to tune in the desired wavelength (Fig. 1.12). The antenna ANT1 is fed into a resonant filter FL1 which is then amplified by AMP1 and fed into a second resonant bandpass filter FL2. The output of FL2 is fed into a second amplifier AMP2 followed by a third bandpass filter FL3. FL3 feeds AMP3 followed by FL4. FL1 through FL4 are tunable narrow bandwidth (approximately 10 KHz) bandpass filters that are mechanically coupled together on a common drive shaft so that when you rotate the tuning knob on your receiver it changes the center frequency of all filters. FL1 through FL4 can tune across the entire AM broadcast band. The output of FL4 is fed into an envelope detector DET1 and amplified by the audio frequency AF Signal Chain then out to a loudspeaker. An example of an early TRF receiver from 1929 that was restored by the author is shown in Fig. 1.13.

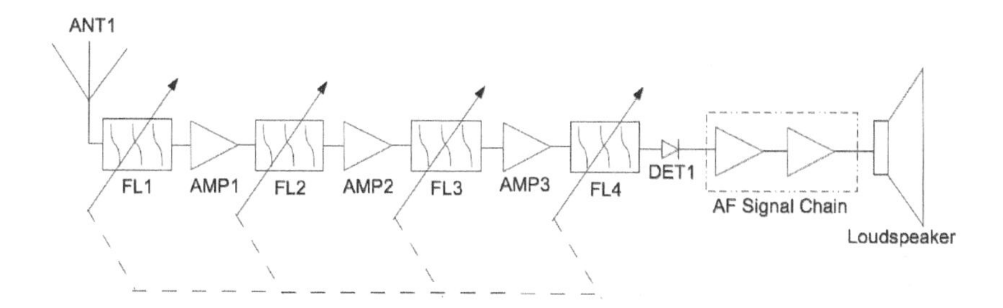

FIGURE 1.12
Block diagram of an early TRF radio receiver.

1.1.5.2 Heterodyne Receivers and the Frequency Mixer

Modern receivers, including all radar receivers in this book, use the heterodyne architecture developed by Edwin Armstrong which relies on frequency multiplication to shift the desired receive signal into the bandpass of a fixed narrow bandwidth filter rather than using a series of tunable bandpass filters. This architecture significantly reduced cost and improved performance, thereby democratizing access to radio receivers by the 1930's because single-frequency narrowband bandpass filters are less expensive and provide better characteristics than tunable filters. Tuning is accomplished by a variable frequency oscillator (VFO) and a frequency mixer, where the frequency of the VFO determines the wavelength to which the receiver is tuned.

To understand heterodyne receivers we must first consider the frequency mixer (Fig. 1.14), which has three ports the RF, LO, and IF. To the RF port we feed our RF signal that we are trying to detect at a frequency of f_{RF}. To the LO port we feed an oscillator of known frequency f_{LO}. The output of the mixer is the IF port which produces a pair of frequencies that are the result of a time domain multiplication of two sinusoidal signals [9],

$$f_{IF} = \begin{cases} f_{RF} - f_{LO} \\ f_{RF} + f_{LO} \end{cases} . \tag{1.18}$$

Practical mixers are implemented by the use of diodes or active devices which are inherently non-linear and do not provide perfect isolation between ports. For this reason additional spurious (e.g., unintentional mixer products) frequency components on the IF output, including some small portion of LO, the RF, and numerous products of harmonics. To mitigate spurious products from frequency mixers proper use of filtering allows them to function like ideal multipliers. This filtering includes filters on the RF port (or signal chain leading to the RF port) to reject unwanted images, on the IF port to reject unwanted frequency products, on the LO port to reject harmonics from the

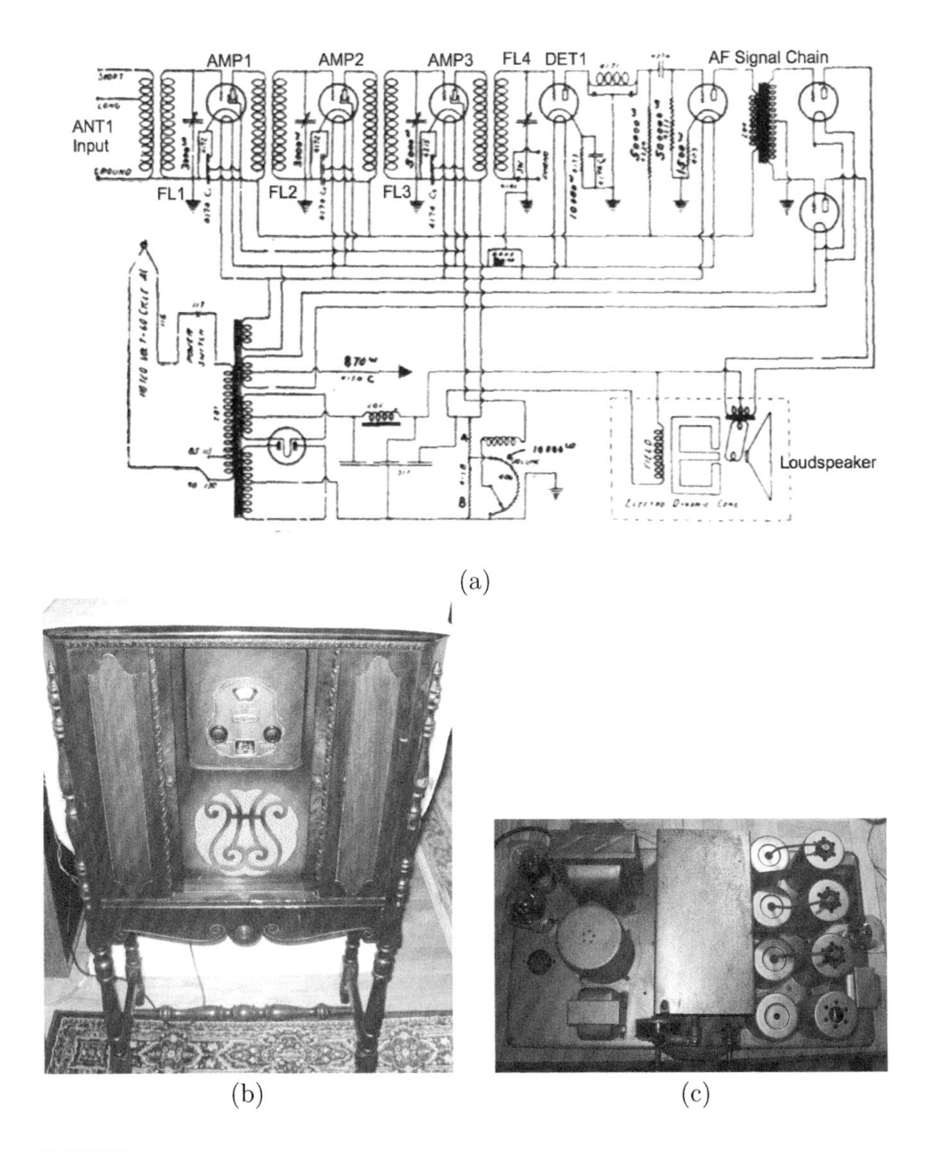

FIGURE 1.13
Example of a TRF receiver, the Colin B. Kennedy Model 20B built in 1929: schematic (a), front panel (b), electronics (c).

LO (if the LO source contains too many spurious emissions). Mitigation of spurious emissions from frequency mixers is a more acute problem for high performance radio receiver design than it is for wide-band radar design, but none the less the designer must be aware of these issues. Details on the use of practical frequency mixers used in this book are described (Fig. 1.15).

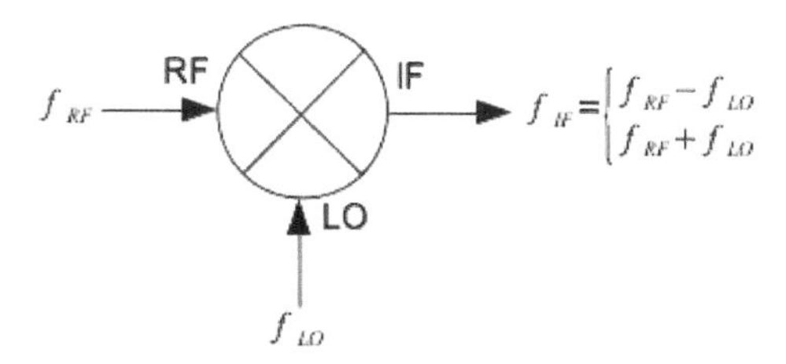

FIGURE 1.14
A frequency mixer performs a frequency domain multiply resulting in the mixing of a LO and an RF signal where the output is the sum and difference between the two.

The block diagram of a typical heterodyne AM broadcast receiver is shown in Fig. 1.16. The intermediate frequency (IF) signal chain in this radio consists of FL3, AMP2, and FL4. FL3 and FL4 are narrow bandwidth bandpass filters centered at $f_c = 455$ KHz with a bandwidth of 10 KHz. AMP2 provides gain between FL3 and FL4 sufficient to drive the envelope detector DET1 which then feeds an audio signal chain. The IF signal chain has a center frequency of $f_c = 455$ KHz; therefore we let $f_{IF} = f_c = 455$ KHz. Using the frequency mixer MXR1 we solve Equation (1.18) for f_{RF} which we know to be any frequency within the AM broadcast band of 520 to 1610 KHz. OSC1 is a VFO capable of oscillating anywhere from 975 to 2065 KHz.

To tune in am AM broadcast station at 850 KHz, OSC1 would be set to

Most mixers in this book are double-balanced mixers. For reliable operation you must drive it with enough power to cause the diodes to switch on and off. This is the LO drive power and will be clearly indicated in your mixer's data sheet. Most mixers in this book require 7 to 13 dBm of LO drive.
The **LO port of double-balanced mixers is often poorly matched to 50 ohms** and for this reason when using a double-balanced mixer at microwave frequencies it is good design practice to place an isolator between the LO port of the mixer and the source of its LO.

FIGURE 1.15
Notes on the use of practical frequency mixers.

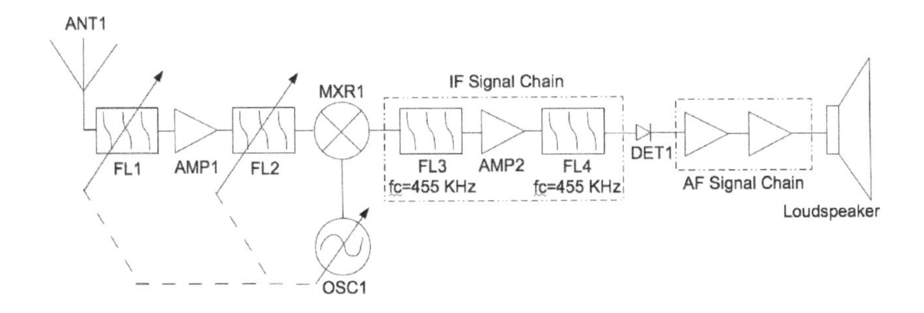

FIGURE 1.16
Block diagram of a modern heterodyne broadcast band receiver.

$850 + 455 = 1305$ KHz making the product on the IF output of the mixer 455 KHz because according to Equation (1.14) $850 - 1305 = -455$ KHz. A negative frequency simply manifests itself as a sine wave at 455 KHz with a 90° phase shift.

According to Equation (1.14) there also exists a mixer product at $1305 + 455 = 1760$ KHz where, a 1760 KHz radio station would also be tuned in when OSC1 is set to 1305 KHz. The purpose of FL1 and FL2 is to filter out this unwanted mixer product, known as an image. For any given LO frequency there exist two mixer images. Filtering at the RF port of the mixer rejects the unwanted mixer products, thereby feeding only one of the two images into the mixer. FL1 and FL2 are tunable bandpass filters that are coupled to OSC1, with relatively wide bandwidths, where the bandwidth is just narrow enough to reject the unwanted image.

An example of a heterodyne AM broadcast receiver is shown in Fig. 1.17. This receiver was built in 1946, restored by the author, and follows the modern heterodyne architecture shown in Fig. 1.16.

1.1.5.3 Single Sideband (SSB) Receivers

Single sideband receivers use a second mixer to multiply down (or down-convert) the IF to baseband where the RF is fed directly into the AF amplifiers and out of the loudspeaker (Fig. 1.18). SSB receivers do not use a detector. What is heard from the loudspeaker is the RF at the antenna down-converted to the audio frequency spectrum. Consequently the audio from a SSB receiver sounds very unusual when out of tune.

Electromagnetic fields are radiated into the antenna ANT1 (Fig. 1.19a). To reject unwanted images, FL1 is a bandpass filter that passes only the frequency spectrum of interest from 14 to 14.5 MHz. A low noise amplifier LNA1 amplifies this spectrum and feeds into FL2 which is a bandpass filter identical

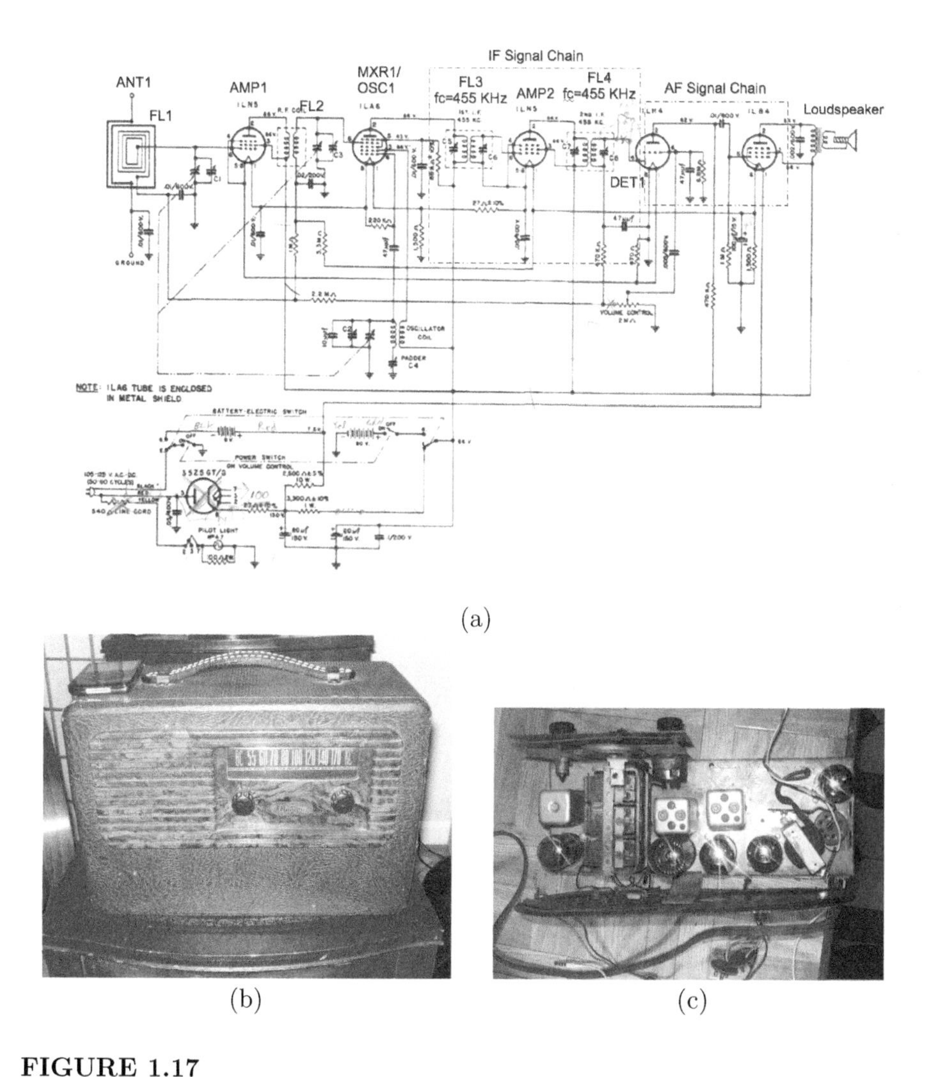

FIGURE 1.17
Example of a heterodyne receiver, the Olympic Model 6-606: schematic (a), front panel (b), electronics (c).

to FL1. All low noise amplifiers contribute some broadband noise to whatever is amplified as represented by the light grey in the spectrum plot on the output of LNA1. The noise contribution is characterized by an amplifier's noise figure (to be discussed in Sec. 1.1.5.4). The purpose of FL2 is to reject the broadband noise spectrum contribution of LNA1, thereby rejecting the noise that falls within the unwanted mixer image. In a modern LNA this noise spectrum expands well above and below the frequency spectrum of in-

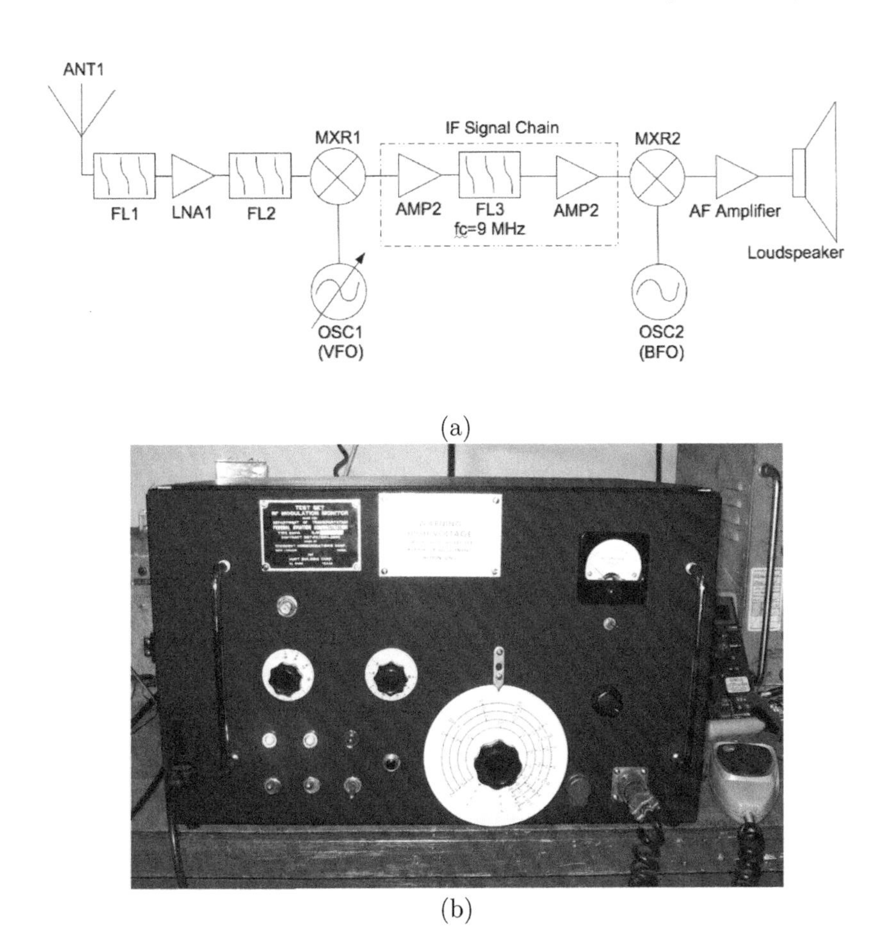

FIGURE 1.18
Example of a SSB transmitter and receiver for the 20 M amateur radio band (developed by the author): block diagram in receive mode (a), front panel (b).

terest. When mixed with the VFO OSC1, the noise image would fold over into the IF and effectively double the noise figure of the receiver, thereby reducing sensitivity. To prevent this, FL2 rejects the wide band noise spectrum produced by LNA1, preserving maximum possible receiver sensitivity.

The output of FL2 is fed into the frequency mixer MXR1. The LO port of MXR1 is fed by OSC1 which is a VFO capable of tuning $\pm$ 250 KHz around a center frequency of 5.25 MHz. When multiplied by the output of FL2 the mixer product provides two image spectra where each is 500 KHz wide and is a frequency shifted copy of the original 14 to 14.5 MHz input spectrum. One image spectra occupies 8.75 to 9.25 MHz and the other 19.25 to 19.75 MHz.

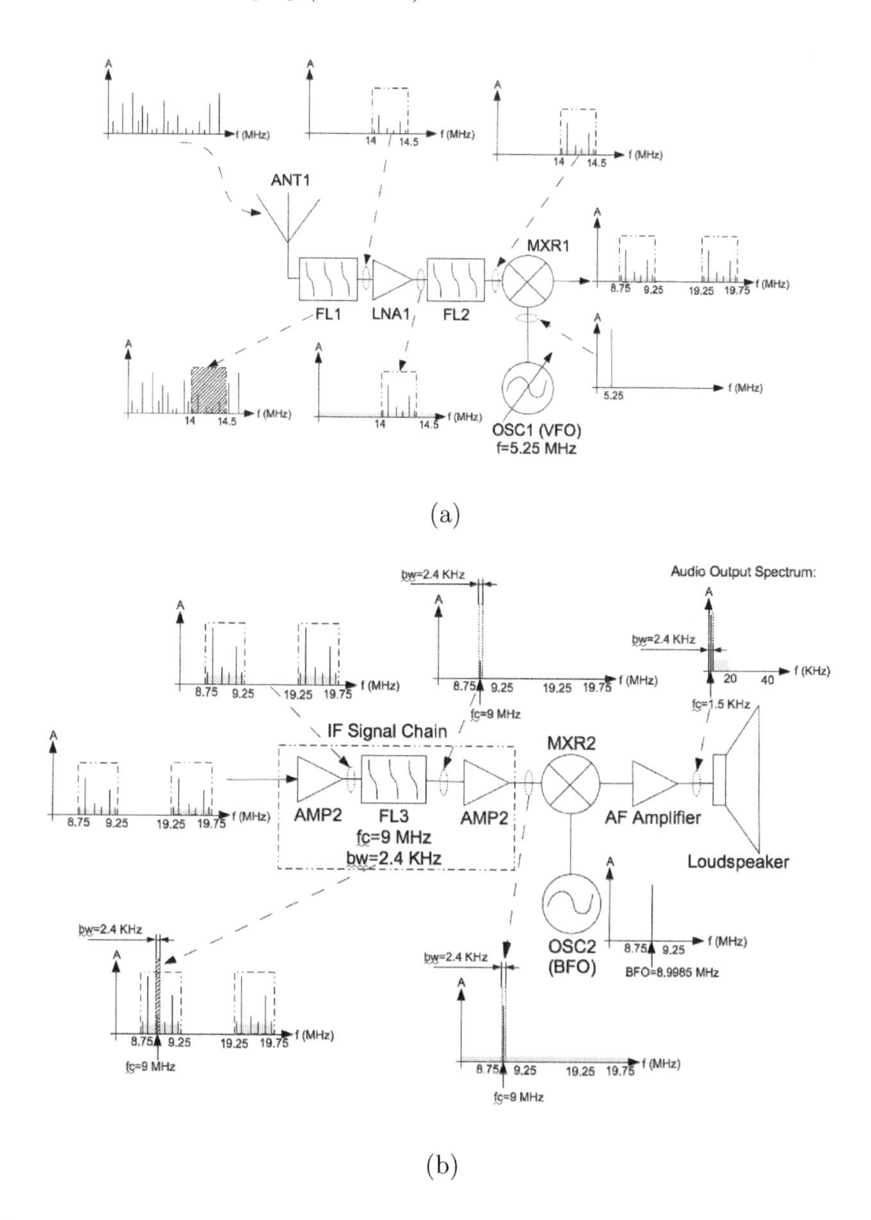

FIGURE 1.19
Explanation of SSB receiver: front end (a), IF and product detector (b).

The output of MXR2 is amplified by the IF amplifier AMP2 (Fig. 1.19b) to make up for the conversion loss from MXR1. Most frequency mixers have conversion loss. For a typical double balanced mixer, the expected conversion loss will range from 5 to 8 dB. For active mixers, where amplification is internal to the mixer, gain occurs during conversion (an example of an active mixer is the SA602).

The output of AMP2 is fed into FL3 which is a crystal bandpass filter with a center frequency of 9 MHz and a bandwidth of 2.4 KHz. This filter rejects all spectra outside of its pass band, thereby eliminating the 19.25 to 19.75 MHz image and all but 2.4 KHz of bandwidth centered at 9 MHz. FL3 is what sets the detection bandwidth of this receiver; it limits the spectrum which is eventually down-converted and fed out of the loudspeaker. By changing the frequency of OSC1 the center frequency that is presented to FL3 is changed, thereby tuning the radio to a different frequency.

The output of FL3 is amplified by AMP2 which is a high gain IF amplifier, often providing 40 to 60 dB or more of gain and adding some additional noise spectrum which is negligible compared to the noise originally added by the LNA. The output of AMP2 is fed into the frequency mixer MXR2.

The LO feeding MXR2 is a fixed frequency oscillator known as the beat frequency oscillator (BFO) which is set to 8.9985 MHz. MXR2 and OSC2 shift the output of of AMP2 from 9 MHz to the audio frequency spectrum where it can be heard by the radio operator. The output of MXR2 is amplified by audio frequency amplifiers which are not capable of amplifying and reproducing the second image product at the RF frequency of 17.9985 MHz; therefore all the radio operator hears is the down-converted spectrum.

This radio operates within the short wave band, defined as any frequency between 2 and 30 MHz. Long distance communication is possible on the short wave band to and from anywhere on the planet. This long distance propagation relies on the Earth's ionosphere, which reflects short wave signals radiated from a transmitter back down to Earth. Ionospheric conditions change in real time and cause signal fading; therefore an automatic gain control circuit is often applied to AMP2 to provide a relatively steady output to improve audibility of the received signal.

To determine a receiver's sensitivity the noise figure must be calculated and a signal-to-noise ratio (SNR) requirement for reliable reception is defined. Noise figure will be discussed (Section 1.1.5.4) followed by receiver sensitivity (Section 1.1.5.5).

1.1.5.4 Noise Figure

Noise figure is a measure of how much a given two-port device, such as an amplifier, degrades the SNR of a given input signal. Sources of noise include thermal noise, shot noise, and other effects due to the physics of various types of transistors, all of which are combined into the noise factor (F). F is the ratio of the SNR of an input signal (S_i/N_i) to the SNR of an output signal (S_o/N_o) [10] of a two-port device:

$$F = \frac{S_i/N_i}{S_o/N_o}. \tag{1.19}$$

The most commonly used specification for noise factor when specifying an LNA is the noise figure

$$NF = 10 \cdot \log_{10} F, \tag{1.20}$$

where NF is the noise figure in decibels (dB). In satellite communications or radio astronomy, it is more convenient to specify the noise figure in terms of equivalent noise temperature

$$T_e = T_o(F - 1), \tag{1.21}$$

where T_e is the equivalent noise temperature in degrees Kelvin (K) and T_o is the reference noise temperature of 290K, which is the noise temperature typically seen by terrestrial antennas directed toward the horizon.

For every component in a radio receiver signal chain there exist a gain and a noise figure. For LNA's and IF amplifiers these parameters are clearly stated but for attenuators, filters, mixers, and other devices the noise figure is equal to the insertion loss of the device. Each device in a signal chain contributes to the total receiver noise figure, known as the cascaded noise figure:

$$NF_{sys} = 10 \cdot \log_{10}\left(F_1 + \frac{F_2 - 1}{G_1} + \frac{F_3 - 1}{G_1G_2} + \frac{F_4 - 1}{G_1G_2G_3} \cdots \frac{F_n - 1}{G_1G_2G_3 \ldots G_{n-1}}\right), \tag{1.22}$$

where F_n and G_n are the noise factors and gains of each stage of a receiver signal chain.

For example, each component in the SSB receiver (Fig. 1.18) has the NF and gain specifications shown in Table 1.1. To calculate the cascaded noise figure, each NF and gain term must be linear, using

$$F = 10^{NF/10}, \tag{1.23}$$

and

$$G = 10^{G_{dB}/10}. \tag{1.24}$$

Applying Equation (1.22) to the linear gain and F terms results in a cascaded noise figure of 2.7 dB. To calculate the noise figure of your system use the MATLAB® script [11].

1.1.5.5 Receiver Sensitivity

The receiver sensitivity is defined as the minimum detectable signal (MDS) in dBm [8],

$$MDS = 10 \log_{10}\left(\frac{kT_o}{1mW}\right) + NF_{sys} + 10 \log BW + SNR_{out}, \tag{1.25}$$

where k is Boltzmann's constant ($1.38 \cdot 10^{-23}$ joules/Kelvin), BW is the receiver bandwidth (Hz), and SNR_{out} is the minimum required SNR for detection. Typical values for SNR_{out} are 10 dB and 13.4 dB for a SSB amateur radio receiver [12] and a long-range radar system [13].

TABLE 1.1
Gain and noise figure of each component from the SSB receiver in Fig. 1.18.

Component	NF (dB)	G_{db}
FL1	0.5	-0.5
LNA1	2	20
FL2	0.5	-0.5
MXR1	5	-5
AMP2	4	20
FL3	2	-2
AMP2	5	60
MXR2	5	-5
AF Amplifier	5	40

For example, the SSB receiver (Fig. 1.18) has a cascaded noise figure $NF_{sys} = 2.7$ dB, a bandwidth $BW = 2.4$ KHz, and a required $SNR_{out} = 10$ dB, resulting in a MDS of -127.5 dBm. To calculate the MDS of your system use the MATLAB script [14].

1.1.6 Radio Transmitters

The simplest radio transmitter is an oscillator connected directly to an antenna, as shown in Fig. 1.3, known as a continuous wave (CW) transmitter. Turn on and off the power to this oscillator according to the international Morse code to modulate this transmitter so that its transmission contains information (examples include early spark gap transmitters and later tube-type CW transmitters).

To transmit voice you can vary the oscillator's power output amplitude accordingly, providing an amplitude modulated (AM) transmitter.

Modern radio transmitters that transmit SSB voice and phase modulated data are heterodyne receivers in reverse, directly up-converting data or voice spectra to IF then RF frequencies by use of frequency mixers in the same way that a SSB receiver downconverts RF to baseband. These upconverted signals are amplified and transmitted out of the antenna.

The SSB receiver shown in Fig. 1.18 is capable of transmitting when the direction of its amplifiers are reversed to that shown in Fig. 1.20. The audio spectrum of a voice is collected by the microphone which is amplified by the AF amplifier. The audio frequency (AF) amplifier amplifies this audio spectrum and multiplies it with OSC2, the BFO, up to the IF frequency of 9 MHz. AMP1 amplifies this product and feeds it into the IF crystal filter FL3 which is centered at 9 MHz and has a bandwidth of 2.4 KHz, thereby passing one mixer product (sideband) and rejecting the other. AMP2 is bypassed because the signal strength is already large enough to be fed into MXR1, where the IF is multiplied by the VFO OSC1. The VFO tunes from 5 to 5.5 MHz resulting

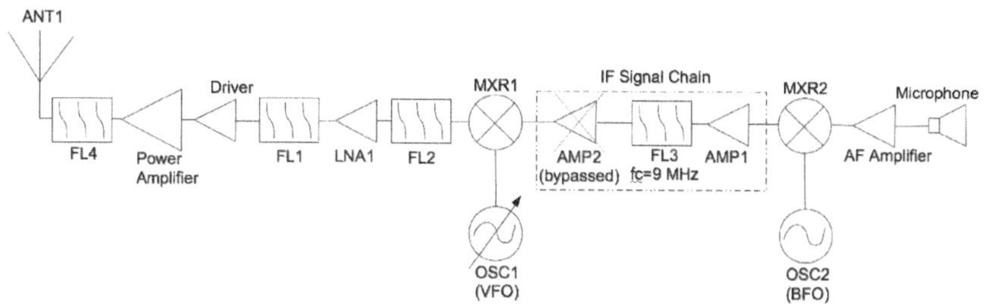

FIGURE 1.20
Block diagram of SSB transmitter; it is simply a SSB receiver in reverse.

in an output from MXR1 of two images, one at the desired band of 14 to 14.5 MHz and the other at 3.5 to 4 MHz. FL2 is a bandpass filter that rejects everything but the 14 to 14.5 MHz frequency range. LNA1 is the pre-driver, increasing the weak mixer output signal. FL1 is another bandpass filter which passes only 14 to 14.5 MHz, thereby greatly attenuating the undesired image. The driver amplifies the desired signal to the point of driving a high power transistor represented by the power amplifier. FL4 is a harmonic filter which rejects the spurious harmonics generated by the power amplifier because power amplifiers are often driven close to or above saturation in communications and radar systems.

Noise figure is typically not considered when developing a transmitter because the SNR of the initial audio (or phase modulated data) spectrum is very high. The exception to this is in the case of more complicated communication systems with co-located transmitters and receivers.

1.2 Radio Direction and Ranging (Radar)

Aircraft performance was increasing steadily between the First and Second World Wars. Every year aircraft, including military bombers, were flying at ever increasing speeds. Existing detection technology included field glasses and large directional microphones but these proved ineffective at providing a timely warning when high speed bombing formations approached. Radar was the solution, capable of locating enemy aircraft out to hundreds of kilometers and providing enough time for air defenses to be mobilized [15].

Radar measures the time of flight for a transmitted radio signal to propagate to and from a target (Fig. 1.1). Radio waves propagate at the speed of light in free space (c); therefore this timing measurement is not easy.

Numerous types of military radar systems are used for ground-to-air, ship-to-air, or air-to-air defense. These are usually pulsed radar systems, where a very large peak-power transmitter emits a relatively short duration pulse of electromagnetic energy. Round-trip propagation time is measured and a reflected (or scattered) signal is used to determine target characteristics.

Examples of radar systems include those used in air traffic control (Fig. 1.21a), detecting and tracing ballistic missiles (Fig. 1.21b), providing the 3D location of aircraft in a military theater of operation (Fig. 1.21c), and long-range instrumentation and tracking (Fig. 1.21d).

Phase coherent radar systems are those where the local oscillators and data acquisition clocks are phase-locked to each other. A special mode of ground mapping, known as synthetic aperture radar (SAR) uses phase coherent radar to create a photo-like image of what is below and away from an aircraft [16] and [17]. Some military and civilian radar systems are also phase coherent so that moving targets are more easily identified.

The basics of pulsed radar will be discussed (Sec. 1.2.1) followed by an overview of phase coherent radar (Sec. 1.2.2).

1.2.1 Pulsed Radar System

The most common radar is the pulsed radar system, where a high power but short-time duration microwave pulse is generated, fed through a duplexer circuit, out to a transmit antenna, radiated toward the target scene, scattered off of the target, collected by the same antenna, and fed into a receiver. Scattered returns are typically shown on a plan position indicator (PPI), which is a polar display showing range and bearing (angle) to target. PPI displays were adopted during the Second World War because they allowed commanders to more easily make battle plans based on the situation displayed on the radar screen.

A block diagram of this radar system is shown (Fig. 1.22). A pulse of relatively short duration τ (low duty cycle) is generated in the high voltage pulse generator PULSE1. PULSE1 powers the cavity magnetron (MAG1 which is a special vacuum tube designed to output high peak-power microwave pulses) for a short period of time. The output of MAG1 is fed into what is often referred to as the duplexer circuit consisting of microwave circulator CIRC1 and receiver protection switch SW1. CIRC1 is a magnetic circulator which allows microwave signals to pass in a circle from port 1 to 2, 2 to 3, and 3 to 1 but not in reverse order. The output from port 2 is fed to the antenna ANT1, which is a feed horn at the focal point on a parabolic dish.

The feed line for ANT1 passes through a rotary joint, which allows the feed line to be rotated freely as the antenna rotates. ANT1 and its parabolic dish are rotated by the motor. A drive shaft couples the motor to an angle encoder and the rotary joint. The rotary joint couples motion of the motor drive shaft to the antenna. Angular position of the parabolic dish (θ) is fed to the PPI from the encoder.

FIGURE 1.21
Examples of radar systems and their antennas: the ASR-9 air traffic control radar (a), the PAVE-PAWS ballistic missile defense phased array radar (b), the TPS-59 3D forward-deployed ground-to-air surveillance radar (c), and the ALTAIR long-range instrumentation and tracking radar (d).

The parabolic dish directs the microwave pulses from ANT1 in a focused beam which sweeps the horizon as the dish rotates. If a target is present, one or more microwave pulses will scatter off of the target back into the parabolic dish. This scattered energy is fed into port 2 of CIRC1 which will then be

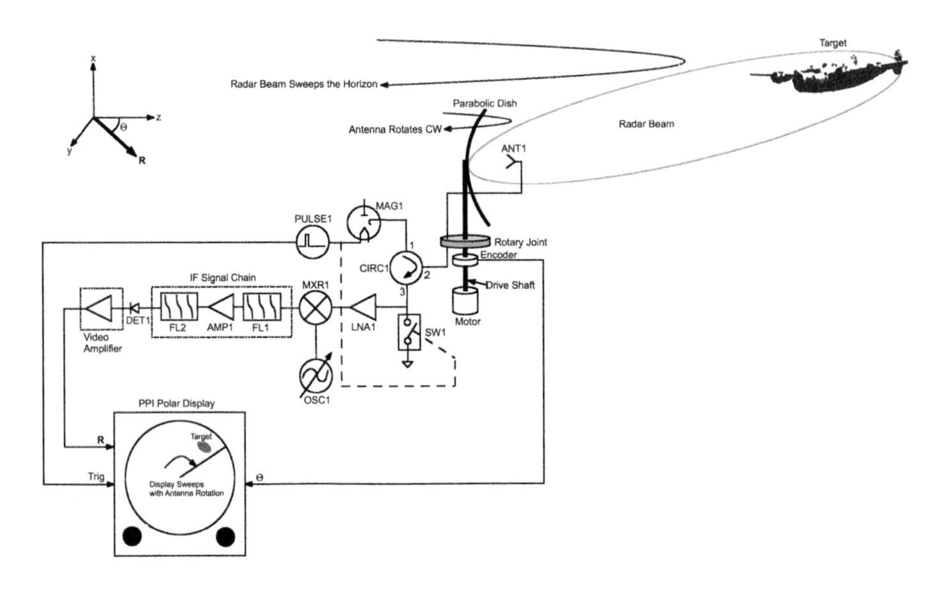

FIGURE 1.22
Block diagram of a common pulsed radar system.

circulated over to port 3. The output of port 3 is fed into the low noise amplifier LNA, where SW1 is a receiver protection switch used to protect LNA1 from the transmit pulse. SW1 is usually wired into PULSE1. SW1 can be a solid-state PiN diode switch, a gas discharge tube, or similar high-power high-speed switch.

The output of LNA1 is fed into the frequency mixer MXR1, where it is mixed with a microwave local oscillator OSC1. OSC1 is tunable, where there is degree of automatic frequency control (AFC) between the output of the magnetron and the OSC1, so that OSC1 tracks MAG1 with some frequency offset equal to the IF signal chain's center frequency.

The output of MXR1 is fed into the IF signal chain, consisting of FL1, AMP1, and FL2. FL1 and FL2 are band-limiting filters with a bandwidth $B \geq 1/\tau$. AMP1 is often variable and has a large amount of gain. The gain of AMP1 can be controlled by an automatic gain control (AGC) circuit or can be scaled in time (less gain after less time, more gain as time passes by over the length of a pulse interval) by the use of a constant false alarm rate (CFAR) circuit.

The output of FL2 is fed into a detector DET1, which can be a simple envelope detector, a square law detector, or a logarithmic detector. The output of DET1 is amplified by the video amplifier and fed into the PPI polar display.

The range R, bearing angle θ, and the magnitude of the scattered return from the target are plotted on the PPI. Range is derived from the time difference between the rising edge of the trig signal and the output of the video amplifier. The target bearing angle θ is given by the encoder. This is how a conventional pulse radar system works.

1.2.2 Phase Coherent Radar System

Phase coherent radar systems use the same oscillator for both the transmitter and receiver or the transmitter and receiver's oscillators are phase-locked to each other. By doing this, coherent radar systems are capable of measuring the scattered phase of the target relative to their own transmitter. Measured phase information enables advanced radar modes such as velocity measurement, detecting only moving targets, target characterization, imaging, and beamforming using antenna arrays [13]–[17].

1.2.2.1 A Simple Phase Coherent Radar System

A simple coherent radar system is shown (Fig. 1.23), where OSC1 is an unmodulated oscillator generating a sine wave at a frequency of 2.4 GHz ($= f_c$). The output of OSC1 is amplified to a sufficient power level by AMP1. The output of AMP1 is fed into a two-way power splitter SPLTR1, which sends half of the power from AMP1 to ANT1 and the other half to the LO port of MXR1. The input of ANT1 is a 2.4 GHz sine wave. An electromagnetic field is radiated out of ANT1 in the direction of the target, represented by the solid lines (a direct measurement of this spherical wavefront is shown in the video demonstration [18]).

The transmitted electromagnetic field is scattered off the target, represented by the dashed lines, and back to the receive antenna ANT2 with a phase delay that is proportional to the range to target

$$\phi_R = \frac{2\pi f_c R}{c}. \tag{1.26}$$

The signal from ANT2 is amplified by LNA1. The output of LNA1 is fed into the RF port of MXR1 where it is multiplied with the output of SPLTR1, thereby making this a coherent radar system. In other words, the transmitted waveform is multiplied by the scattered receive waveform. The output of MXR1 is amplified and low-pass-filtered by the video amplifier.

For this example, the two time domain signals are multiplied in MXR1 and because each signal is at the same frequency of 2.4 GHz the mixer product is a direct current (DC) potential which is proportional to the phase difference between the scattered and the incident fields (an amplitude term is also present, but ignored here to keep this example simple). In effect, the output of this mixer is the round-trip phase response of the target at 2.4 GHz.

To make this radar more interesting, a quadrature mixer could be substituted for MXR1 to provide two mixer products, both in-phase and 90° out-of-phase products resulting in an effective real and imaginary scattered

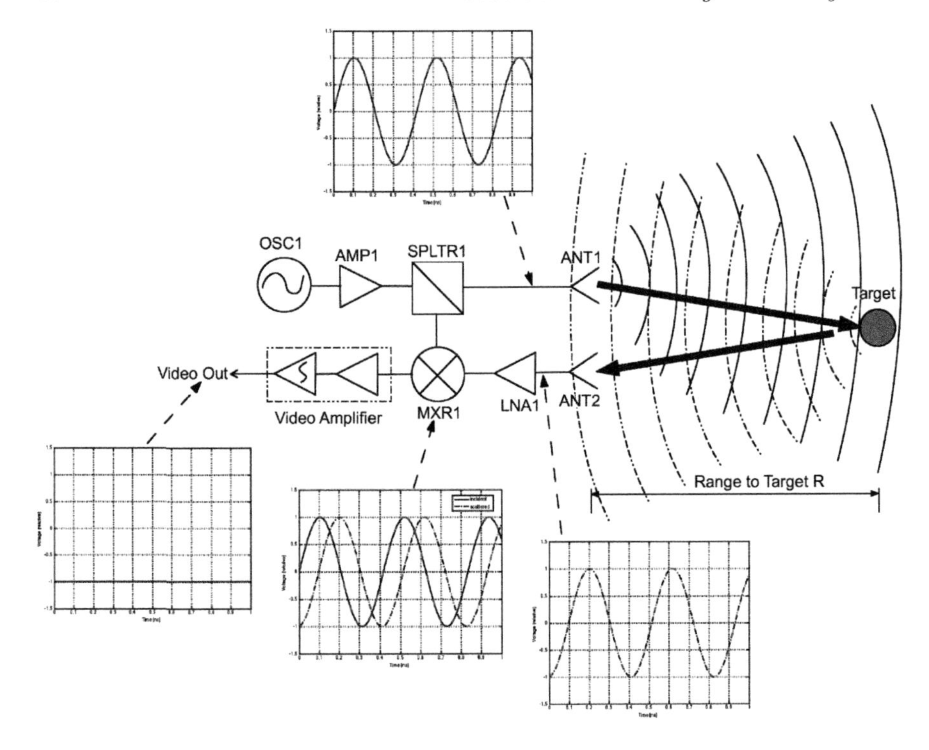

FIGURE 1.23
A phase coherent radar system.

field measurement which would provide both magnitude and phase information. To measure range, OSC1 could be FM modulated, thereby providing a spatial frequency product at the output of MXR1. Many other modulation methods and waveforms are possible and depend on the application.

1.2.2.2 Pulsed Phase Coherent Radar Systems

Coherent radar architectures can also be applied to long range pulsed radar systems, thereby providing a means to detect moving targets by coherently comparing the current scattered signal to the previous ones and applying various Doppler filtering techniques. This architecture is achieved by making the waveform generation, analog-to-digital converter clocks (ADC), and local oscillators phase coherent. Phase coherence can be be achieved by sharing LO signals or by the use of phase-lock-loop (PLL) synthesizers for these various clocks, where a master clock is used as a coherent reference to all oscillators, digitizers, and waveform generators throughout the system (Fig. 1.24).

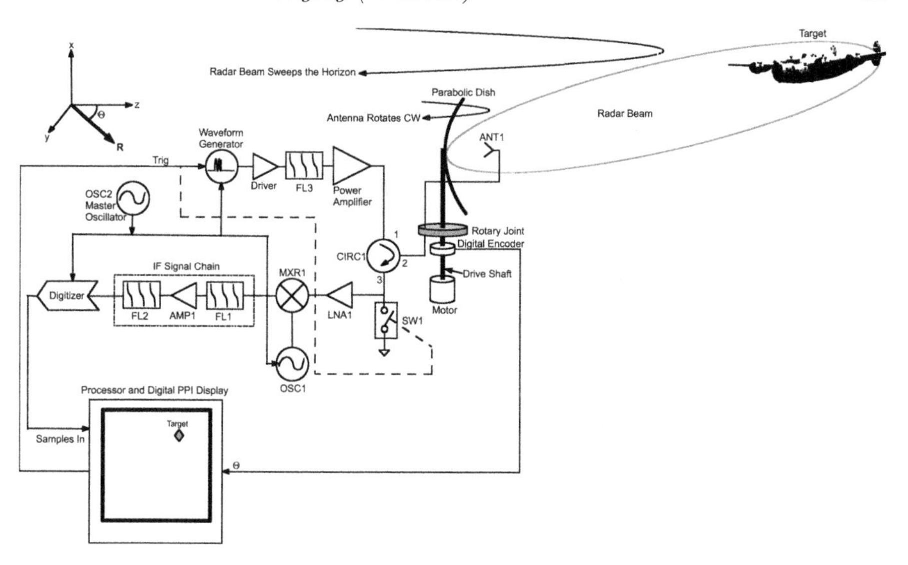

FIGURE 1.24
Block diagram of a phase coherent pulsed radar system, where all oscillators, digitizers, and waveform generators are phase-locked to OSC2, which is typically a high-stability reference oscillator.

1.3 Estimating Radar Performance Using the Radar Range Equation

The radar range equation provides a method for estimating system performance. The radar range equation combines antenna gain and aperture (Sec. 1.1.3), the Friis transmission equation (Sec. 1.1.4), cascaded noise figure (Sec. 1.1.5.4), receiver sensitivity (Sec. 1.1.5.5), and assumes a pulsed radar architecture (Sec. 1.2.1).

The most widely accepted radar range equation is solved for maximum range [13],

$$R_{max}^4 = \frac{P_{ave} G_{tx} A_{rx} \rho_{rx} \sigma n E_i(n) e^{(2\alpha R_{max})}}{(4\pi)^2 k T_o F_n B_n \tau F_r (SNR)_1 L_s}, \tag{1.27}$$

where:

$R_{max} =$ maximum range of radar system (m)
$P_{ave} =$ average transmit power (watts)
$G_{tx} =$ transmit antenna gain
$A_{rx} =$ receive antenna effective aperture (m^2)
$\rho_{rx} =$ receiver antenna efficiency
$\sigma =$ radar cross section (m^2) for target of interest
$n =$ number of received pulses integrated
$E_i(n) =$ integration efficiency
$L_s =$ miscellaneous system losses
$\alpha =$ attenuation constant of propagation medium
$F_n =$ receiver noise figure (derived from procedure outlined in Sec 1.1.5.4)
$k =$ $1.38 \cdot 10^{-23}$ (joul/deg) Boltzmann's constant
$T_o =$ 290^oK standard temperature
$B_n =$ system noise bandwidth (Hz)
$\tau =$ pulse width (s)
$f_r =$ pulse repetition frequency (Hz)
$(SNR)_1 =$ single-pulse signal-to-noise ratio requirement

In this equation, receive aperture (A_{rx}) and efficiency (ρ) are related to receiver antenna gain (G_{rx}) by

$$G_{rx} = \frac{4\pi A_{rx}\rho}{\lambda_c^2}, \tag{1.28}$$

where λ_c is the center frequency wavelength of the radar system. Antenna efficiency (ρ) includes, ohmic losses in the antenna, impedance mismatch losses, and any other additional losses not otherwise accounted for.

Average transmit power (P_{ave}) is related to the root-mean-square power of a single pulse (P_t), the pulse length (τ) and pulse repetition frequency (f_r) by

$$\frac{P_{ave}}{P_t} = \tau f_r = \text{duty cycle of the pulsing transmitter.} \tag{1.29}$$

The radar range equation can also be written in terms of peak power $P_t = P_{ave}/(\tau F_r)$.

The attenuation constant (α) is typically 0 or close to 0 for most short-range radar systems used in free space (unless they are operating at extremely high microwave frequencies or are imaging through dielectric media), thereby making the term $e^{(2\alpha R_{max})} = 1$.

The single-pulse signal-to-noise ratio $(SNR)_1$ depends on the specified probability of detection (P_d) and probability of false alarm (P_{fa}) and can be found using the figure in [19]. A typical specification is to have a P_d of 95% and a P_{fa} of 10^{-6}, thereby requiring an $(SNR)_1 = 13.4$ dB.

Radar cross section σ is similar in principle to antenna effective aperture A, where a plane wave intersects a radar target and some square area (m^2) of the field is scattered off of the target. Typical radar cross sections (RCS) are shown (Table 1.2).

TABLE 1.2
Typical RCS values of targets.

Target	σ (m^2)	Definitions
Sphere	$\sigma = \pi a^2$	a = radius [20]
Flat plate	$\sigma = \frac{4\pi A^2}{\lambda_c^2}$	A = area of plate (m^2) [20]
Cylinder	$\sigma = \frac{a\lambda cos(\theta) sin^2(kLsin(\theta))}{2\pi sin^2(\theta)}$	a = radius, L = length, θ = angle of incidence [20]
Bird	0.01	at f_c = 10 GHz [21]
Man	1	at f_c = 10 GHz [21]
Cabin cruiser	10	at f_c = 10 GHz [21]
Automobile	100	at f_c = 10 GHz [21]
Truck	200	at f_c = 10 GHz [21]

1.4 Small and Short-Range Radars

Small and short-range radar systems are different in theory and significantly different in operation from conventional radar. This is due to the short-range geometry of target scenes. Short ranges require wide bandwidths to achieve useful range resolution and simultaneously transmit and receive because of practical hardware limitations resulting in unusual radar architectures. These target scenes are full of clutter that requires coherent processing and detection algorithms. Furthermore, short-range radar systems operate in the near field, requiring special treatment of beamforming and imaging. To increase accuracy in automotive and unmanned vehicle applications, data is often fused with other sensors to reduce false alarm rates.

In this book, several different types of small radar systems will be discussed, including; CW, ultrawideband (UWB) impulse, UWB linear FM, SAR, and phased array. For each type of radar system there will be a special derivation of the radar range equation, an explanation of applicable algorithms, and a case study that will couple theory to reality by demonstrating the implementation of an actual radar system. Finally, additional chapters will be devoted to real-world applications of small radar sensors.

Bibliography

[1] S. A. Schelkunoff, H. T. Friis, *Antenna Theory and Practice,* John Wiley & Sons, New York, NY, 1952.

[2] C.A. Balanis, *Antenna Theory Analysis and Design,* John Wiley & Sons, New York, NY, 1997.

[3] C.A. Balanis, *Advanced Engineering Electromagnetics,* John Wiley & Sons, New York, NY, 1989.

[4] E. C. Jordan, K. G. Bulmain, *Electromagnetic Waves and Radiating Systems, 2nd ed.,* Prentice Hall Inc., Englewood Cliffs NJ, 1968, pp. 187–188.

[5] C. Bowick, *RF Circuit Design,* Newnes, Boston, MA, 1982.

[6] J. D. Kraus, *Antennas, 2nd ed.,* McGraw Hill, Boston, MA, 1988.

[7] P. Vizmuller, *RF Design Guide Systems, Circuits, and Equations,* Artech House, Norwood, MA, 1995.

[8] U. L. Rhode, J. Whitaker, T. T. N. Bucher, *Communications Receivers, 2nd Ed.,* McGraw Hill, New York, NY, 1996.

[9] R. E. Ziemer, W. H. Tranter, *Principles of Communications, Systems, Modulation, and Noise* John Wiley & Sons, New York, NY, 1995.

[10] "Fundamentals of RF and microwave noise figure measurements," Agilent Application Note 57-1, August 5, 2010.

[11] cascaded_noise_figure.m, `http://glcharvat.com/shortrange/radar/`

[12] *The ARRL Handbook, 87th ed.,* The American Radio Relay League, Inc., Newington, CT, 2010.

[13] M. I. Skolnik, *Introduction to Radar Systems,* McGraw Hill, New York, NY, 1962.

[14] minimum_detectable_signal.m, `http://glcharvat.com/shortrange/radar/`

[15] L. Brown, *A Radar History of World War II,* Carnegie Institution of Washington, Washington, DC, 1999.

[16] G.W. Stimson, *Introduction to Airborne Radar,* Hughes Aircraft Company, El Segundo, California, 1983.

[17] W.G. Carrara, R.S. Goodman, and R.M. Majewski, *Spotlight Synthetic Aperture Radar Signal Processing Algorithms,* Artech House, Boston, MA, 1995.

[18] Time Harmonic Electromagnetic Wave Propagation Demo, `http://glcharvat.com/shortrange/radar/`

[19] M. I. Skolnik, *Introduction to Radar Systems,* McGraw Hill, New York, NY, 1962, Fig. 2.7, p. 34.

[20] M. I. Skolnik, *Introduction to Radar Systems,* McGraw Hill, New York, NY, 1962, Table 2.2, p. 43.

[21] M. I. Skolnik, *Introduction to Radar Systems 2nd ed.*, McGraw Hill, New York, NY, 1980, p. 44.

Part I

Short-Range Radar Systems and Implementations

2

Continuous Wave (CW) Radar

Small CW Doppler radar sets are used as motion sensors to open the doors at large retail stores and by law enforcement for measuring the speed of automobiles. Purpose-built instrumentation CW Doppler radar units can be used to observe physical phenomena, such as the Doppler spectra of moving targets or observation of nature in general. Discussion of CW radar architectures will be presented (Sec. 2.1), followed by the radar range equation applied to small CW Doppler radar (Sec. 2.3). Signal processing will be discussed (Sec. 2.2), and working examples of small CW Doppler radar sets will be detailed (Sec. 2.4).

2.1 CW Radar Architecture

Most small CW Doppler radar systems follow the coherent radar architecture shown (Fig. 1.23), where a CW carrier is radiated toward and scattered from a target except for in this case, the target is moving at velocity $\vec{v}$ (Fig. 2.1), where $\vec{v}$ is the vector projection of target velocity normal to (in the direction of) the radar's antenna beam.

From Sec. 1.2.2, we understand that the range to target for a CW coherent radar system can be described in terms of phase, where

$$\phi_R = \frac{2\pi f_c R}{c}. \tag{2.1}$$

If the target is moving at a velocity from an initial position R_i then its position with respect to velocity and time can be described

$$R = vt + R_i. \tag{2.2}$$

Therefore the phase of the scattered field from the moving target is changing with respect to time. By substituting Equation (2.2) into (2.1) we have a changing scattered phase as a function of time

$$\phi(t)_R = \frac{2\pi f_c (vt + R_i)}{c}. \tag{2.3}$$

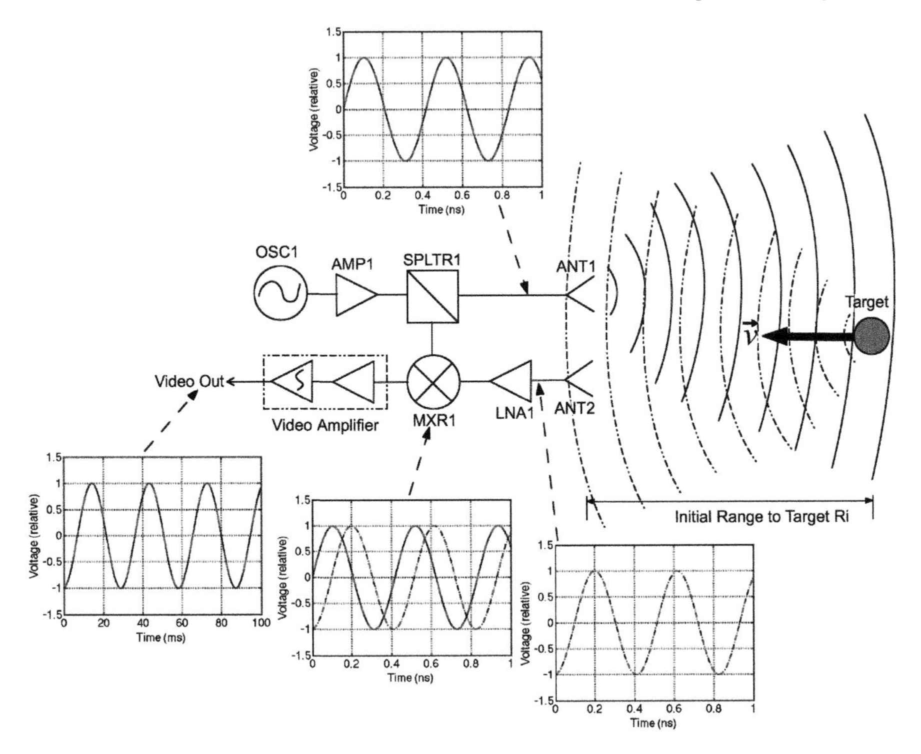

FIGURE 2.1
A coherent CW Doppler radar sensor.

Doppler is the time rate change of scattered phase. A time rate change of phase is a frequency shift. This frequency shift is described by the Doppler equation which can be derived by re-arranging terms from above, substituting $f_c = c/\lambda_c$, and ignoring the static phase due to initial target position R_i, where Δf_D is the frequency shift in Hz [1]

$$\Delta f_D = \frac{2v}{\lambda}. \tag{2.4}$$

Consider a CW doppler radar radiating at a frequency of $f_c = 2.4$ GHz. If the target is moving directly at the radar sensor at a velocity of 60 MPH, then when the transmit carrier is multiplied by the scattered field in MXR1, the output product would be a sine wave at the Doppler frequency Δf_D of 34 Hz (plus the static phase of OSC1 and any additionally static phase due to static delays within the radar device).

From Equation (2.4), it is apparent that both the incoming and outgoing velocity ($\pm v$) can be determined and would manifest as $\pm\Delta f_D$. Unfortunately, we know from Sec. 1.1.5.2 that a negative frequency manifests

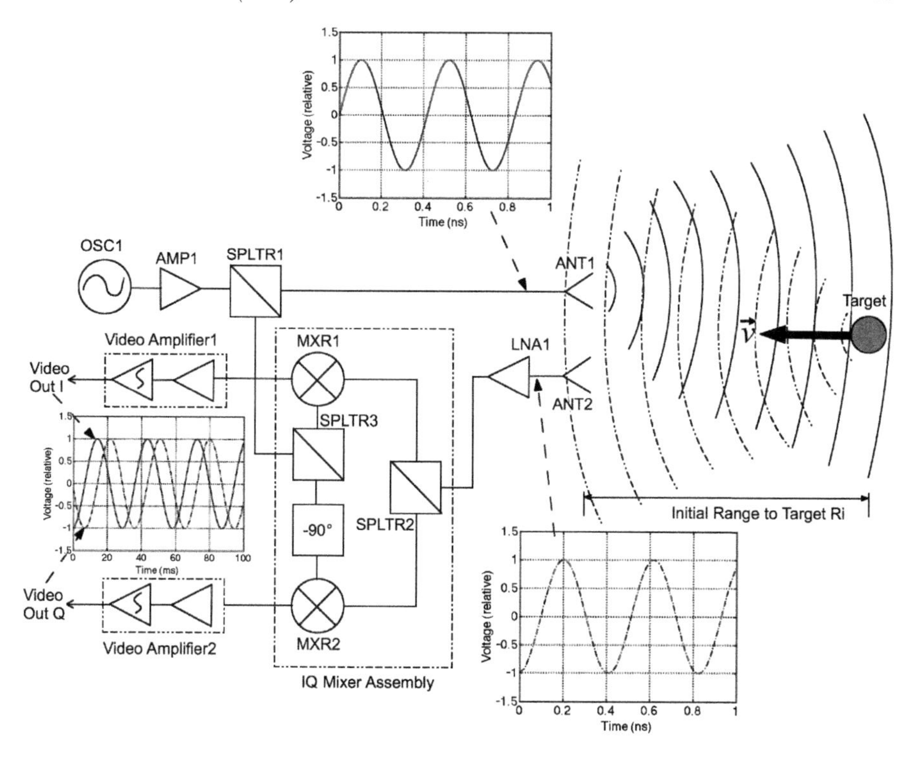

FIGURE 2.2
A CW Doppler radar using quadrature mixing.

itself as 90° phase shift but is otherwise indistinguishable from its in-phase positive-frequency variant. Therefore, the architecture shown in Fig. 2.1 is incapable of differentiating $\pm\Delta f_D$ unless the initial velocity of the target is known.

To measure $\pm\Delta f_D$ without a priori target information we must be able to measure the phase. This can be done by using two separate mixers where one is fed with an LO that is 90° out of phase from the other (Fig. 2.2). As before, OSC1 is a CW oscillator that is amplified by AMP1 to a sufficient level for transmission and to drive the two mixers MXR1 and 2. One output from the power splitter (or directional coupler) SPLTR1 is fed to the antenna where it is radiated out toward the target, the other is fed to SPLTR3, which is part of the IQ mixer assembly (also known as a quadrature mixer or an image-reject mixer). A microwave field is radiated out of ANT1.

Scattered fields from the moving target are collected by ANT2 and amplified by the low noise amplifier LNA1. The output of LNA1 is fed to SPLTR2 inside the IQ mixer assembly which equally splits the power between the RF input ports of MXR1 and MXR2.

One output from SPLTR3 is fed to the LO port of MXR1. MXR1 is considered in phase with OSC1. The second output port from SPLTR3 is fed through a 90° phase shifter, thereby lagging the phase of OSC1 by 90°, to the LO port of MXR2. The IF output of MXR2 is now 90° out of phase but not delayed in time with respect to its RF input. By lagging the LO in phase we can shift the phase of the IF through a mixer without affecting the time of signal arrival through the RF signal chain.

The IF output of MXR1 is amplified and filtered by Video Amplifier1 providing the in-phase (I) channel output. Similarly, the IF output of MXR2 is amplified and filtered by Video Amplifier2 providing the quadrature (Q) channel output.

When the target velocity (v) is positive the I channel will lead the Q channel, but when the target velocity is negative the Q channel leads the I channel, thereby properly measuring the phase of an incoming or outgoing target. When the I and Q channels are digitized the I channel can be considered real and the Q channel imaginary. Many other interesting features are possible when digitizing both I and Q channels, including instantaneous amplitude, phase, and a reduction of individual digitizer bandwidths.

IQ mixing can also be implemented in software but data must be digitized at an IF frequency that is shifted at least twice the necessary bandwidth to capture positive and negative Doppler velocities. In the example (Fig. 2.3), a CW carrier from OSC1 is amplified by AMP1 and radiated toward the moving target by ANT1. OSC1 is phase locked to OSC2 which is also phase locked to OSC3, thereby making OSC1 and OSC2 phase coherent. Scattered fields from the moving target are collected by ANT2, amplified by LNA1, and downconverted by MXR1 to an IF frequency (of 10.7 MHz in this case, but can be whatever is most useful) where it is amplified by AMP2 and band limited by FL1. FL1 is a bandpass filter, such as a ceramic filter, surface acoustic wave (SAW), or a ladder network made up of quartz crystals (crystal filter). The output of FL1 is digitized by the high frequency digitizer. The resulting spectrum from the digitizer shows either a positive or negative Doppler frequency centered around 10.7 MHz, thereby preserving the phase information needed to determine incoming or outgoing velocities.

Additional CW Doppler radar architectures can also be used to measure Doppler velocities, but when the signal equation is derived these methods will result in the same Δf_D.

2.2 Signal Processing for CW Doppler Radar

Three types of signal processing will be discussed: the simple frequency counter, frequency-to-voltage converter, and the discrete Fourier transform.

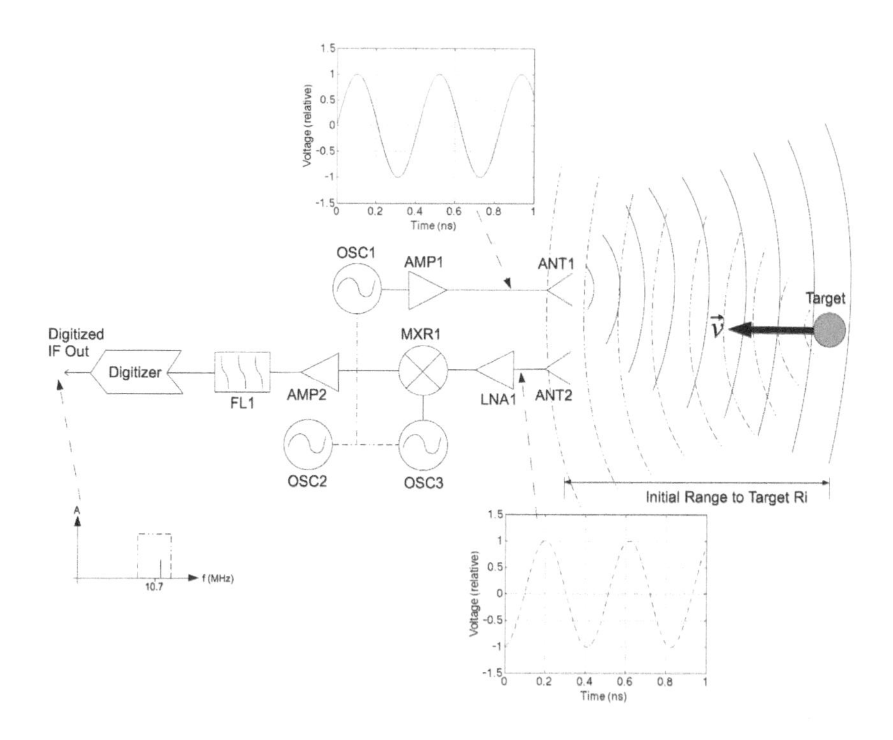

FIGURE 2.3
A CW Doppler radar using IF digitization.

2.2.1 Frequency Counter

Frequency counting is the least complicated method of processing CW Doppler radar data, where a digital frequency counter can be plugged into the video output of the radar shown in Fig. 2.1, thereby displaying Δf_D.

Unfortunately this method only works when a single target is present. When multiple targets are present the strongest target will be counted because the frequency counter works by counting zero crossings, where the counter counts the number of times the voltage output from the video amplifier crosses 0 V. The largest signal will dominate zero crossings.

2.2.2 Frequency-to-Voltage Converter

Before low-cost digital electronics, one method to measure frequency was to build a frequency-to-voltage converter. These converter circuits suffer from the same problems that a digital counter would, where they output a voltage proportional only to the largest magnitude Doppler frequency.

2.2.3 Discrete Fourier Transform

When multiple moving targets are present there exists a plurality of Doppler frequencies Δf_D at the video output. This Doppler spectrum can be analyzed by first digitizing then applying the discrete Fourier transform (DFT), which converts a discretely sampled (digital) time domain signal into a frequency spectrum by plotting the amplitude and phase compared to frequency. This analysis technique has a further benefit where it reduces the effective noise bandwidth of the radar by the duration over which the DFT is applied ($B_n = 1/\tau$ where τ is the length in time over which the DFT is applied).

The frequency domain representation S_ω of a discretely sampled signal s_n (from the Doppler radar) with N total number of samples is computed using the DFT [2]

$$S_\omega = \frac{1}{N} \sum_{n=-\frac{N}{2}+1}^{\frac{N}{2}} s_n e^{-j\omega n/N}, \tag{2.5}$$

where $j = \sqrt{-1}$, $\omega = 2\pi f$, f is the relative frequency, and n is the discrete sample number in the discretely sampled data set being analyzed. The discretely sampled data s_n can be real (as would be acquired from the radar in Fig. 2.1), it can be complex (as would be the acquired from the radar in Fig. 2.2), or it can be over-sampled and digitally converted to complex data (as would be the case for the radar in Fig. 2.3). The DFT is a standard function in MATLAB and other software packages.

When using the DFT, the important issue to understand is that the resulting calculation S_ω provides only N points, the same number of data points that were sampled by the radar system's digitizer. To back-out the resulting frequencies represented by the points in S_ω one must know the total sample time from start to finish t_{sample}, from the digitizer's first sample to its last sample. This can be calculated by $t_{sample} = N/f_s$, where f_s is the sample rate of the digitizer. Therefore, the frequency step between each point in S_ω is $\Delta f = 1/t_{sample}$ for complex valued s_n and $\Delta f = 2/t_{sample}$ for real-valued s_n [3].

There are only up to N number of useful data points in a DFT but it is often the case that humans prefer smooth signals when viewing radar data sets. For this reason there is a demand to interpolate between the resulting data S_ω and this can be achieved by arbitrarily adding 0's to the end or beginning of the sampled signal s_n in software (known as zero padding or up-sampling), thereby calculating the DFT over a larger number of points. The result is a smoothed-over DFT plot that looks nice for briefings to your sponsor or data displayed to the radar operator.

There exist computationally faster versions of the DFT known as the fast Fourier transform (FFT) [4]. These are often automatically selected when using software packages such as MATLAB®, where if the data set s_n contains a number of samples that is equal to a power of 2 then an FFT will automatically be selected. Other fast methods of the Fourier transform exist. In this book, the DFT will be used to convert digitally sampled data from the time domain into the frequency domain.

2.3 The Radar Range Equation for CW Doppler Radar

The radar range equation for CW Doppler radar is different from a conventional radar system because there are no pulses. This requires some simplification of the general radar range equation (1.27), where the terms relating to pulse integration (n and $E_i(n)$) are dropped (set equal to 1),

$$R_{max}^4 = \frac{P_{ave} G_{tx} A_{rx} \rho_{rx} \sigma e^{(2\alpha R_{max})}}{(4\pi)^2 k T_o F_n B_n \tau F_r (SNR)_1 L_s}, \tag{2.6}$$

where:

R_{max} = maximum range of radar system (m)
P_{ave} = average transmit power (watts)
G_{tx} = transmit antenna gain
A_{rx} = receive antenna effective aperture (m^2)
ρ_{rx} = receiver antenna efficiency
σ = radar cross section (m^2) for target of interest
L_s = miscellaneous system losses
α = attenuation constant of propagation medium
F_n = receiver noise figure (derived from procedure outlined in Sec 1.1.5.4)
k = $1.38 \cdot 10^{-23}$ (joul/deg) Boltzmann's constant
T_o = 290°K standard temperature
B_n = system noise bandwidth (Hz)
τF_r = 1, the duty cycle for a CW radar
$(SNR)_1$ = single-pulse signal-to-noise ratio requirement

For an image-reject CW Doppler radar using DFT processing, the noise bandwidth is inversely proportional to the discrete sample length $B_n = 1/t_{sample}$. For a direct conversion CW Doppler radar using DFT processing the noise bandwidth is twice as wide as a radar following an image rejection architecture where $B_n = 2/t_{sample}$. If a simple frequency counter is used for processing with a direct conversion architecture, then $B_n = 2f_{-3dB}$ where f_{-3dB} is the -3 dB cut-off frequency of a band-limiting filter at base band.

The radar range equation for CW Doppler radar will be applied to several working examples in Sec. 2.4.

2.4 Examples of CW Radar Systems

Working examples of small CW Doppler radar sets will be shown and demonstrated, including the MIT 'coffee can' radar (Sec. 2.4.1) and an X-band laboratory radar (Sec. 2.4.2).

2.4.1 The MIT Independent Activities Period (IAP) Radar in Doppler Mode

The author developed a short radar course at MIT for both IAP and professional education, where the students learned about radar by building their own and using it in a series of field experiments [5]-[7]. This course was the top ranked MIT professional education course in 2011, showing the value of project-based learning.

The radar kit developed for this course costs about $360 USD and was fairly sophisticated, capable of Doppler, ranging, and forming SAR imagery. The kit is built from coffee cans, wood, runs on AA batteries, and data is digitized by the stereo audio input port of any computer (Fig. 2.4). The author encourages readers to build this kit by following the step-by-step instructions [7]. In this section, the Doppler mode will be described.

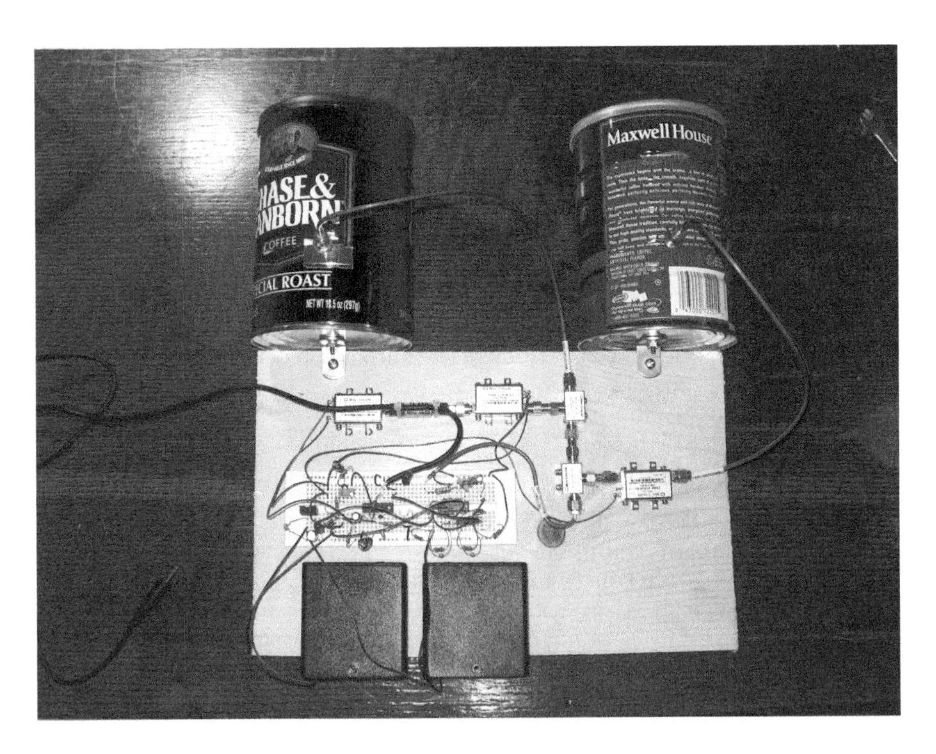

FIGURE 2.4
Photo of the MIT 'coffee can' radar built from coffee cans, wood, some coaxial microwave components, analog prototyping board, that runs on eight AA batteries.

Block diagram and call-outs are shown (Fig. 2.5) and a list of RF components is provided (Table 2.1). OSC1 is a voltage-controlled oscillator (VCO) biased to a single frequency at the upper edge of the 2.4 GHz industrial, scientific, and medical (ISM) band. The output of OSC1 is attenuated slightly by ATT1 then amplified by PA1, which is a low-noise amplifier capable of 18.5 dBm output power. ATT1 simply prevents PA1 from being saturated. The output of PA1 is fed into the −3 dB splitter, SPLTR1. Half of the output of SPLTR1 is fed to the transmit antenna, Cantenna1. The other half of the output of SPLTR1 is fed into the double balanced mixer MXR1.

Cantenna1 is a low-cost 2.4 GHz horn antenna which is in effect an open-ended circular waveguide (Fig. 2.6). This antenna is built from a small coffee can and it is capable of radiating efficiently from just below 2.3 GHz to 2.58 GHz with a measured gain of 7.2 dBi and−3 dB beam width of 72° [7] and [6].

Cantenna1 radiates toward the target scene and scattered energy is collected by Cantenna2. LNA1 is a low-noise amplifier that amplifies the signals collected by Cantenna1 and feeds them into the RF port of MXR1. The LO port of MXR1 is fed by SPLTR1, multiplying the CW carrier by what is received, thereby making this radar phase coherent.

The IF output of MXR1 is amplified by Video Amp1, which includes an amplification stage followed by a 15 KHz 4th order active low pass filter (details of Video Amp1 shown in Fig. 2.7). Video Amp1 is based on a MAX414 or other equivalent quad op-amp; it amplifies and provides an anti-alias filter. The output of Video Amp1 is fed into the right audio channel of a 3.5 mm audio connector, where the product is digitized by a computer's audio input port. Co-located with Video Amp1 are linear voltage regulators and the DC bias to select the CW carrier frequency.

2.4.1.1 Expected Performance of the MIT 'Coffee Can' Radar in Doppler Mode

Substituting the performance parameters for the coffee can radar in CW Doppler mode (Table 2.2) into the radar range equation (2.6), the performance of the MIT coffee can radar is estimated to be 1200 m for a 10 dBsm target using the MATLAB script [8].

Antenna gains and transmit power were measured. System noise figure is based on the front-end LNA noise figure of 1.2 dB because we know from Sec. 1.1.5.4 that the first amplifier effectively sets the noise of the entire system if it has sufficient gain. Due to the inherently short-range nature of this radar, we can assume the attenuation constant of propagation $\alpha = 0$. We have chosen to use $(SNR)_1 = 13.4\ dB$ because we want to achieve the gold standard of radar performance providing a probability of detection 95% and a probability of false alarm 10^{-6}. We will apply the DFT over $t_{sample} = 100$ ms intervals of data and this radar architecture uses a direct conversion receiver as previously shown (Fig. 2.1), therefore the noise bandwidth $B_n = 2/t_{sample}$. The RCS $\alpha = 10$ dBsm was chosen because it is the approximate RCS of an automobile at 10 GHz (Table 1.2).

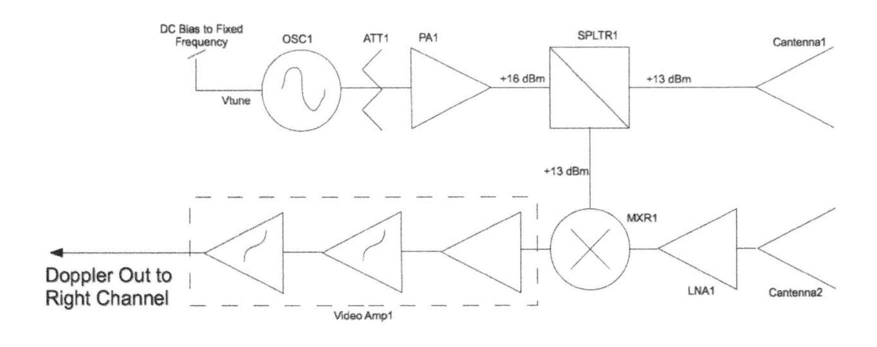

(a)

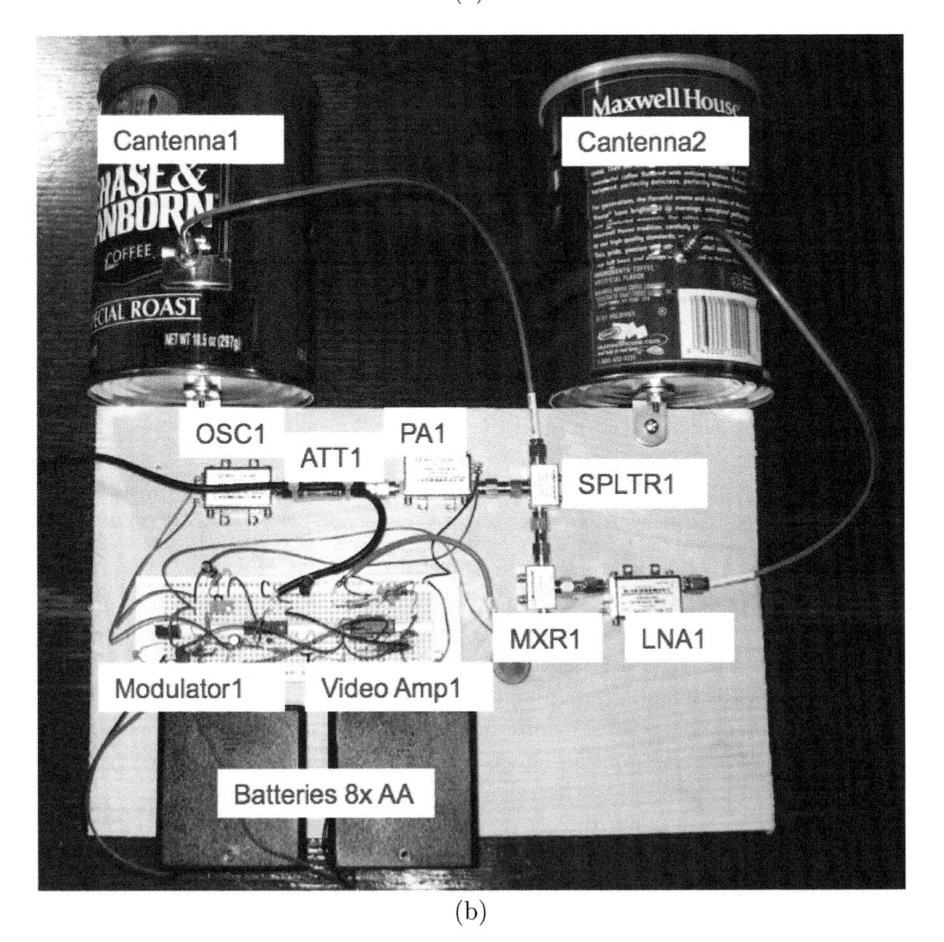

(b)

FIGURE 2.5
Block diagram (a) and callouts (b) of the MIT 'coffee can' radar configured for Doppler mode.

TABLE 2.1
MIT ‘coffee can’ radar RF modular component list.

Component	Description
OSC1	Mini-circuits ZX95-2536C+, 2315 to 2536 MHz VCO, +6 dBm out
ATT1	Mini-circuits VAT-3+, 3 dB attenuator
PA1	Mini-circuits ZX60-272LN-S+ LNA, gain = 14 dB, NF = 1.2 dB, IP1 = 18.5 dBm
ANT1	Figure 2.6
ANT2	Figure 2.6
LNA1	Mini-circuits ZX60-272LN-S+ LNA, gain = 14 dB, NF = 1.2 dB, IP1 = 18.5 dBm
MXR1	Mini-circuits ZX05-43MH-S+, 13 dBm LO, RF-LO loss = 6.1 dB, IP1 = 9.1 dBm
Video amplifier	Figure 2.7

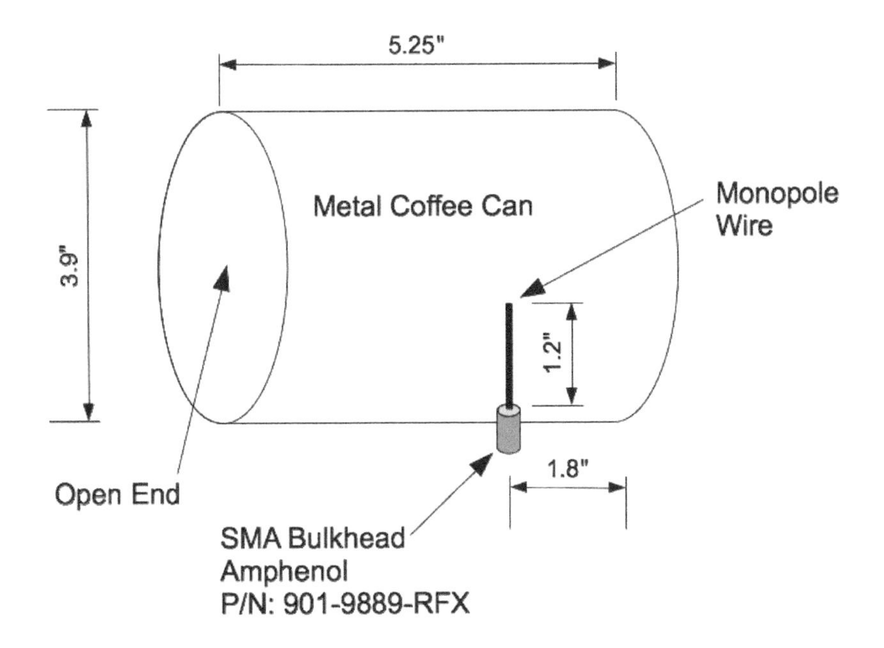

FIGURE 2.6
Diagram of the antenna used in the MIT 'coffee can' instructional radar system.

The maximum range of 1200 m is reasonable given the transmit power, antenna gains, and long DFT processing length. In general, system models are not precise and only work in an ideal world without clutter, radio frequency interference, and other obscuring objects. For this reason a loss margin was added to this calculation in the form of miscellaneous system losses $L_s = 6$ dB, providing 6 dB of performance margin.

2.4.1.2 Working Example of the MIT 'Coffee Can' Radar in Doppler Mode

The 'coffee can' radar works well for measuring Doppler velocities of moving vehicles. To demonstrate, this radar was held outside of the window of the author's car, directed toward Park Street, while he was stopped on Tremont street near Newton Corner outside of Boston (Fig. 2.8).

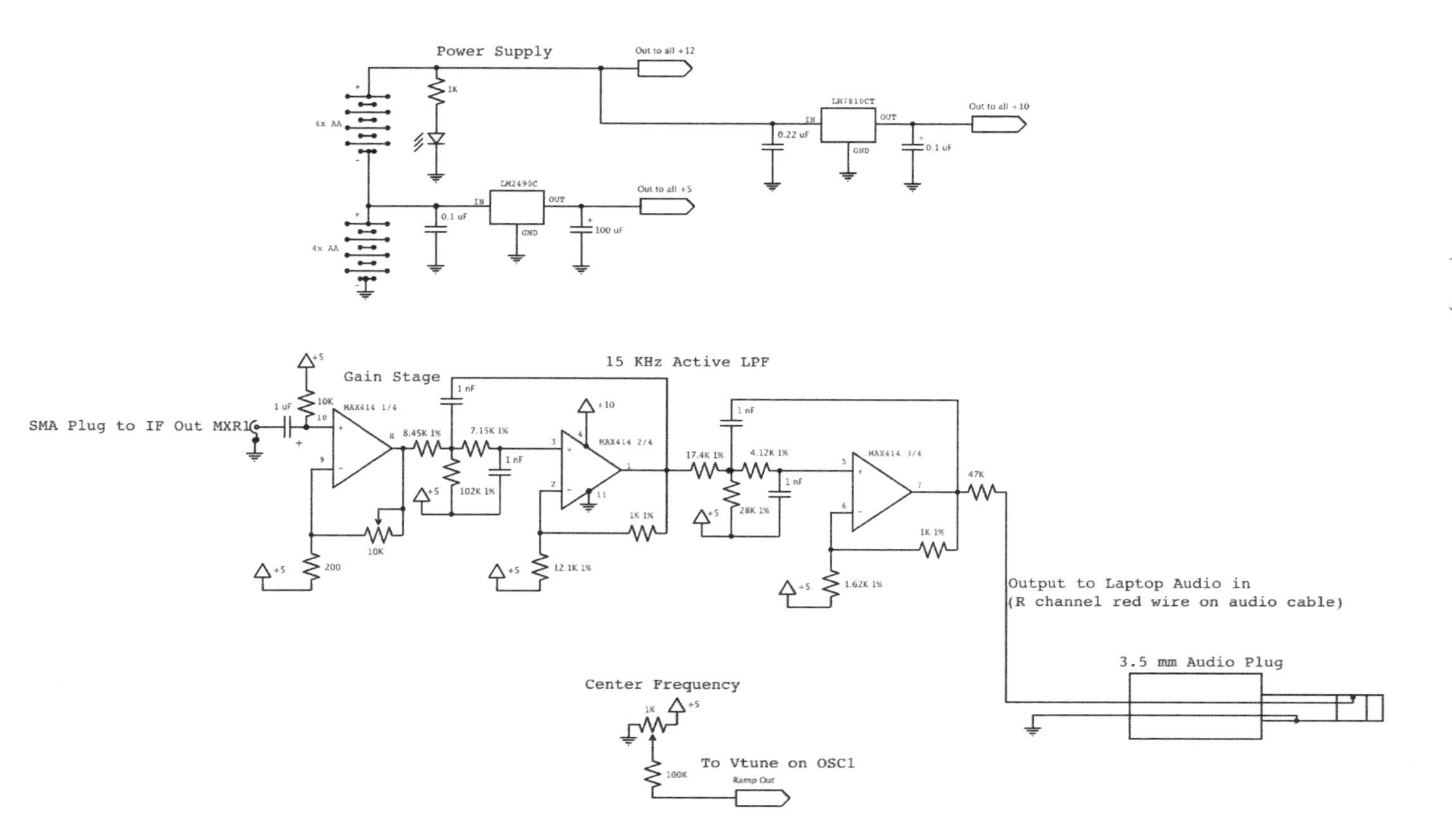

FIGURE 2.7
Schematic of the MIT 'coffee can' radar analog circuitry.

TABLE 2.2
MIT 'coffee can' radar in CW Doppler mode range equation parameters.

$P_{ave} = 10^{-3}$ (watts)
$G_{tx} =$ 7.2 dBi measured antenna gain
$G_{rx} =$ 7.2 dBi measured antenna gain
$A_{rx} = G_{rx}\lambda_c^2/(4\pi)$ (m^2)
$\lambda_c = c/f_c$ (m) wavelength of carrier frequency
$f_c =$ 2.4 GHz center frequency of radar
$\rho_{rx} =$ 1 because antenna efficiency is accounted for in measured antenna gain
$\sigma =$ 10 (m^2) for target of interest
$L_s =$ 6 dB miscellaneous system losses
$\alpha =$ 0 attenuation constant of propagation medium
$F_n =$ 1.2 dB receiver noise figure
$B_n = 2/t_{sample}$ system noise bandwidth (Hz) where $t_{sample} = 100$ ms
$(SNR)_1 =$ 13.4 dB

Data was acquired with a laptop audio input port for over 70 seconds. To process, the data was broken up into $t_{sample} = 100$ ms blocks and the DFT applied to each. The logarithmic relative power was calculated ($20 \log_{10}|Sw|$) and plotted for each 100 ms block to produce a Doppler velocity versus time plot (Fig. 2.9).

Looking at the last 15 seconds of data, five automobiles can be seen traveling at five different velocities from approximately 6 m/s to 13 m/s. The first vehicle starts at approximately 6 m/s then slows to about 2 m/s at about $t = 62$ s and speeds up again out of the antenna beam. The second vehicle at $t = 60$ s is initially traveling at approximately 6 m/s, speeds up to 8 m/s, then passes the author's vehicle, thereby dropping out of view as shown by the sudden streak of magnitude that reaches down to 0 m/s. The other vehicles are shown at nearly constant speeds until they pass the author's vehicle. To process this data yourself, please download it from the MIT Open Courseware site [9]. To view a video demo of the 'coffee can radar' measuring the Doppler velocities of passing vehicles, please watch the video [10].

2.4.2 An X-band CW Radar System

An X-band coherent radar system was developed for the purposes of learning the principles of coherent radar, high resolution ranging, and high resolution imaging (Fig. 2.10). Due to the expense of X-band microwave devices, the surplus market (mostly from amateur radio hamfests) was leveraged to supply all coaxial components and cables. This effort resulted in a wide band phase coherent radar system for laboratory experimentation capable of Doppler, high resolution ranging, and high resolution SAR imaging [11]–[14].

In this section the X-band laboratory radar will be configured to operate in a Doppler mode similar to the MIT coffee can radar in Sec. 2.4.1. A block

FIGURE 2.8
Experimental setup: measuring the velocities of passing traffic at Newton corner outside of Boston, Massachusetts.

diagram is shown (Fig. 2.11) with a complete bill of material (Table 2.3) and call-outs (Fig. 2.12). OSC1 is an yttrium iron garnet (YIG) oscillator set to output a single CW frequency of 10 GHz. The output of OSC1 is fed through the directional coupler CLPR1. CLPR1 feeds almost all power through and couples some power off at the ratio of −10 dB compared to its input. Power fed through CLPR1 goes to the input of circulator CIRC1. CIRC1 is a ferromagnetic circulator which passes traveling microwaves from one port through to the next in a clock-wise circle but prevents traveling microwaves to move counter-clock-wise. In this case, power is fed from CLPR1 through CRC1 to ANT1. Any reflected power from ANT1 is fed through CIRC1 into a load.

ANT1 radiates the microwave carrier toward the target scene. Scattered energy is collected by ANT2 and amplified by LNA1. There exists some degree of electrical coupling from ANT1 to ANT2; therefore part of the signal received at ANT2 is always directly from ANT1. For this reason, the output of LNA1

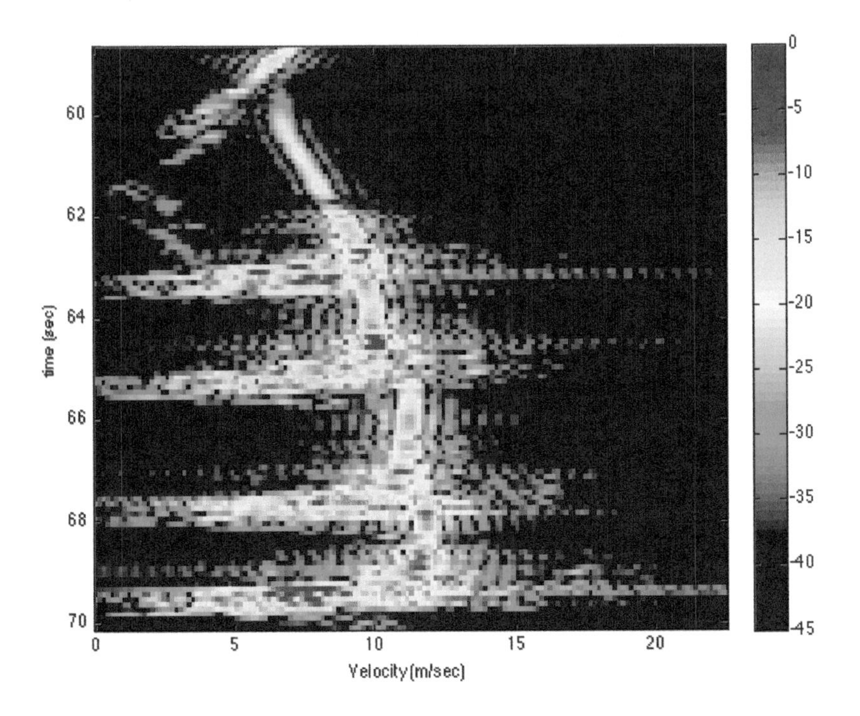

FIGURE 2.9
Doppler versus time plot of automobiles rounding the corner.

is attenuated by 5 dB because LNA1 has too much gain and would otherwise saturate MXR1 with parasitically (unwanted) coupled energy from ANT1.

The LO port of MXR1 is fed by the coupled signal from CLPR1 through CIRC2. The input impedance of the LO port of most double-balanced mixers is very poor relative to 50 ohms and for this reason there is a good amount of reflected power. This reflected power is absorbed by CIRC1, thereby reducing multi-bounce reflections on the LO feed line of MXR1.

In effect, MXR1 multiplies the carrier from OSC1 by what is being collected by ANT2. The output product of MXR1 is amplified by the video amplifier which includes a gain stage and a fourth order active low-pass filter with an 80 KHz −3dB cut-off frequency (Fig. 2.13).

2.4.2.1 Expected Performance of the X-band CW Radar System

Substituting the performance parameters for the X-band Doppler radar (Table 2.4) into the radar range equation (2.6), the maximum range is estimated to be 2100 m for a 10 dBsm automobile target using the MATLAB script [8].

The antenna gains G_{rx} and G_{tx} were from the data sheets and the noise figure is an educated guess based on the apparent age of the surplus microwave

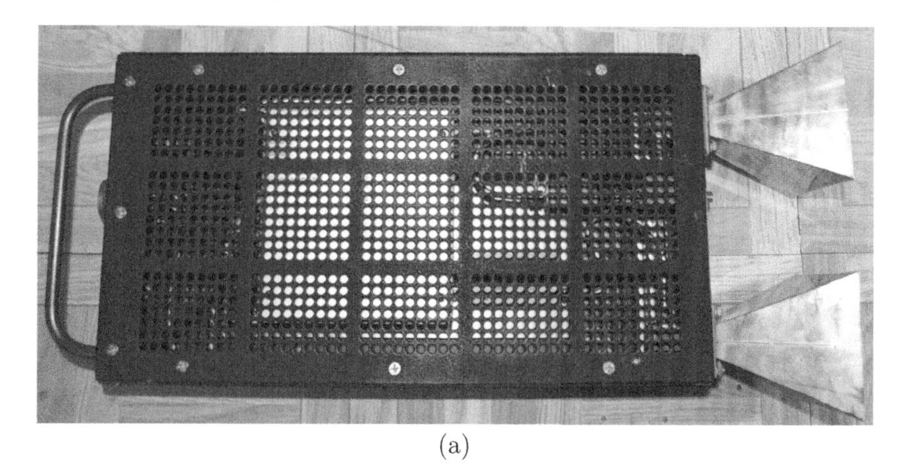

(a)

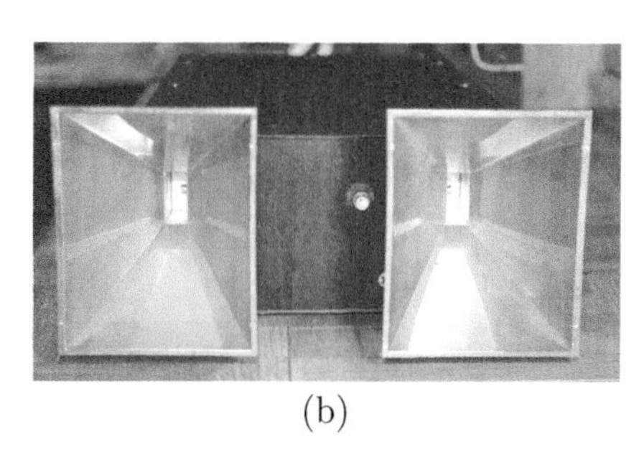

(b)

(c)

FIGURE 2.10
Example of a coherent X-band CW radar system (a), horn antennas for receive and transmit (b), inside view of electronics (c).

amplifier LNA1. With these parameters and the fact that the DFT is applied to 100 ms intervals of data, the maximum range of 2100 m is reasonable.

2.4.2.2 Working Example of the X-band CW Radar System

This radar system was deployed in the author's backyard, where it was used to settle a bet as to who is the fastest runner in a 30 yard sprint. The radar was directed downrange. The author's friend, Alex, ran first as shown by the first Doppler signature running outbound away from the radar and the second Doppler signature running back inbound toward the radar. The author was running outbound on the third Doppler signature and inbound on the fourth Doppler signature. A video of this experiment is shown [15] and a DTI plot was made of this race (Fig. 2.14).

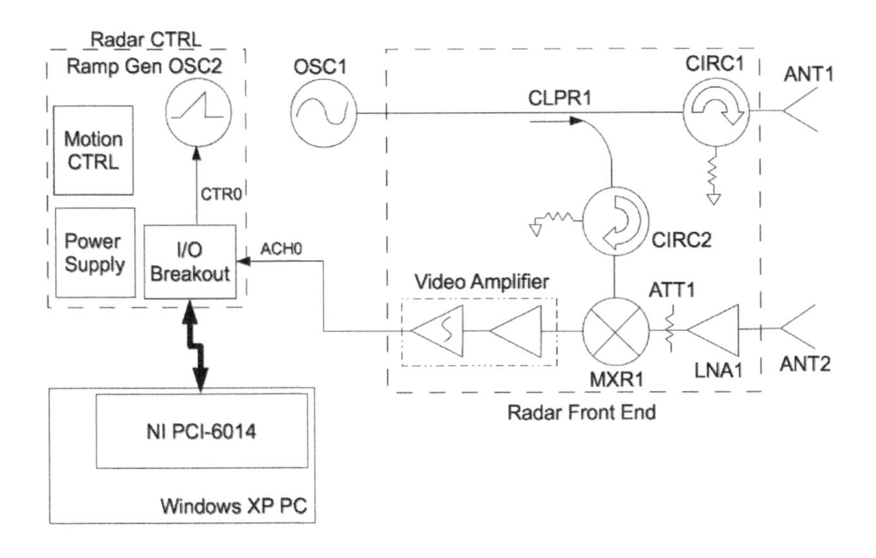

FIGURE 2.11
Block diagram of the X-band CW radar system.

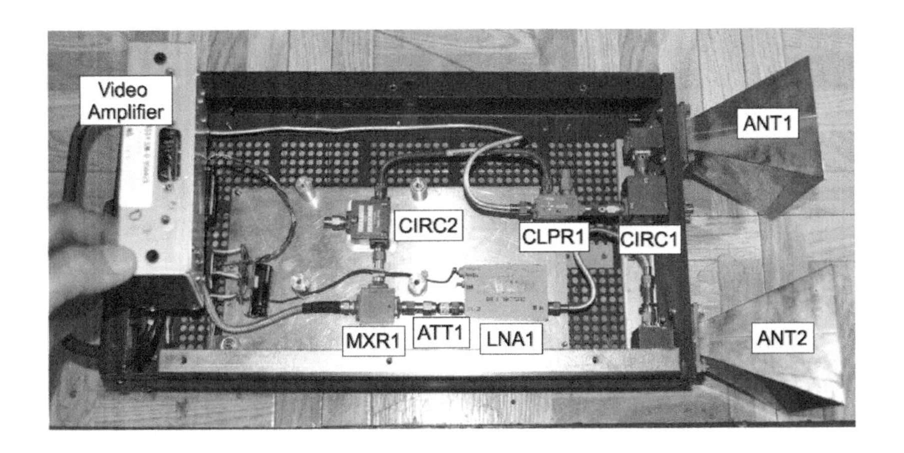

FIGURE 2.12
Call-out diagram of the X-band CW radar system.

TABLE 2.3
X-band CW radar system RF modular component list.

Component	Description
OSC1	Weinschel Engineering 430A Sweep Oscillator with 434A RF Unit
CLPR1	Narda Model 4015C-10, −10 dB, 7 to 12.4 GHz
CIRC1	Raytron Model 300049 Circulator
CIRC2	Raytron Model 300049 Circulator
ANT1	M/A-COM MA86551, 17 dBi gain 25 deg E and H plane beamwidths
ANT2	M/A-COM MA86551, 17 dBi gain 25 deg E and H plane beamwidths
LNA1	Dexcel, Inc, 58985, NF = 4 dB (educated guess), gain = 30 dB
ATT1	Midwest Microwave 5 dB attenuator
MXR1	X-band mixer of unknown manufacture
Video amplifier	Figure 2.13

TABLE 2.4
X-band CW radar system radar range equation parameters.

$$
\begin{aligned}
P_{ave} &= 31 \cdot 10^{-3} \text{ (watts)} \\
G_{tx} &= 17 \text{ dBi antenna gain} \\
G_{rx} &= 17 \text{ dBi antenna gain} \\
A_{rx} &= G_{rx}\lambda_c^2/(4\pi) \text{ (m}^2\text{)} \\
\lambda_c &= c/f_c \text{ (m) wavelength of carrier frequency} \\
f_c &= 10 \text{ GHz center frequency of radar} \\
\rho_{rx} &= 1 \text{ because antenna efficiency is accounted for in antenna gain} \\
\sigma &= 10 \text{ (m}^2\text{) for automobile at 10 GHz} \\
L_s &= 6 \text{ dB miscellaneous system losses} \\
\alpha &= 0 \text{ attenuation constant of propagation medium} \\
F_n &= 4 \text{ dB receiver noise figure} \\
B_n &= 2/t_{sample} \text{ system noise bandwidth (Hz) where } t_{sample} = 100 \text{ ms} \\
(SNR)_1 &= 13.4 \text{ dB}
\end{aligned}
$$

Upon closer examination of the data it is clear that Alex was the fastest with a maximum velocity of approximately 4.5 m/s (Fig. 2.15).

Data from this experiment and a MATLAB script are available [16].

2.5 Harmonic Radar

Harmonic radar is used for finding people stuck in snow avalanches, tracking the location of insects, and non-destructive testing. Generally speaking, harmonic radar can be used for applications where a specific target must be tracked at a stand-off distance in a high clutter environment and where that target is sufficiently cooperative so that it can be 'tagged' with a harmonic transponder. When this harmonic transponder is radiated at a fundamental frequency f_c it scatters (or re-transmits) a weak signal at $2f_c$. Only targets with this non-linear response will be detectable using a harmonic radar system.

2.5.1 CW Harmonic Radar System at 917 MHz

In this section, a harmonic radar system will be described for non-destructive testing and evaluation of metal pipes, structures, and storage tanks, where planar harmonic radar tags are placed into an outdoor industrial setting and located [17]–[20].

2.5.1.1 Implementation

This harmonic radar system consists of a CW transmitter at 917 MHz (f_c) and a receiver with a logarithmic amplifier detector at 1834 MHz ($2f_c$). The

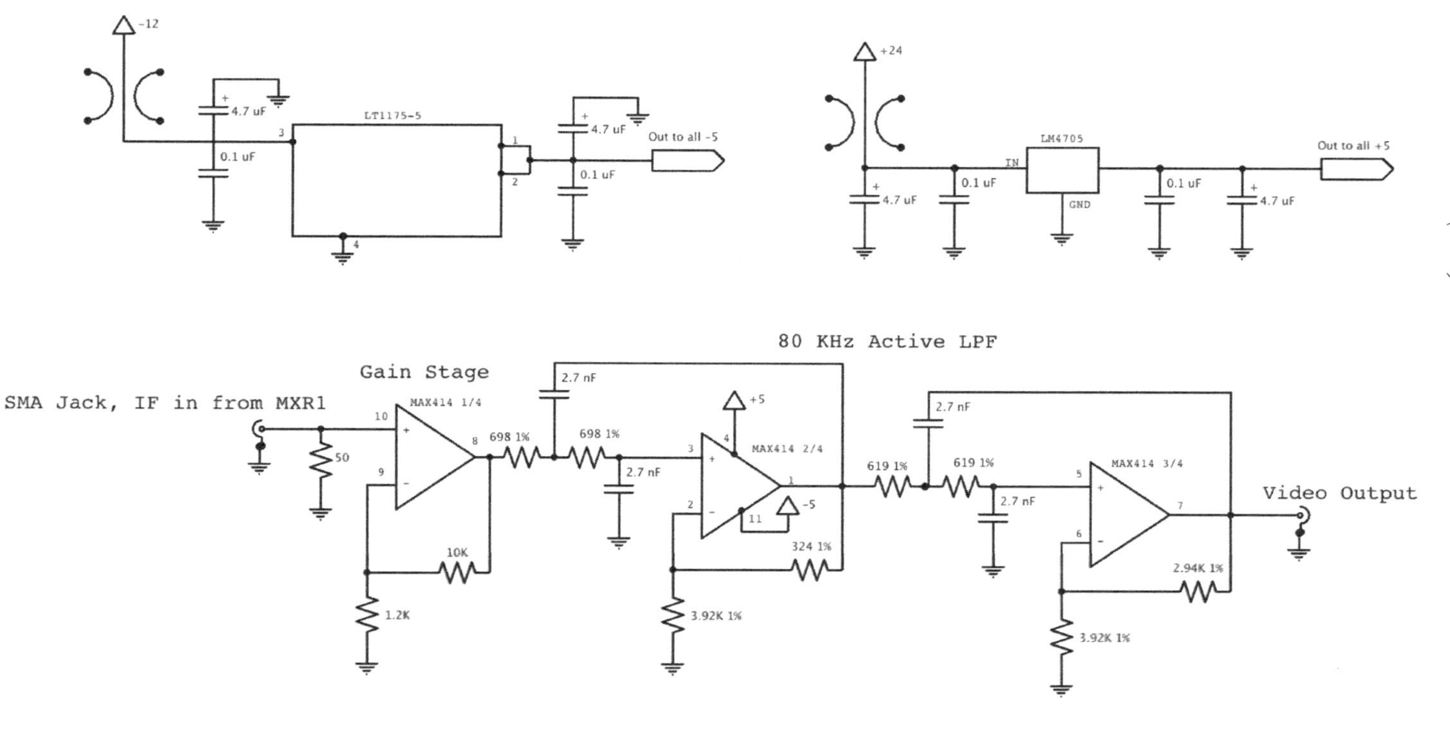

FIGURE 2.13
Schematic of the video amplifier for the X-band Doppler radar.

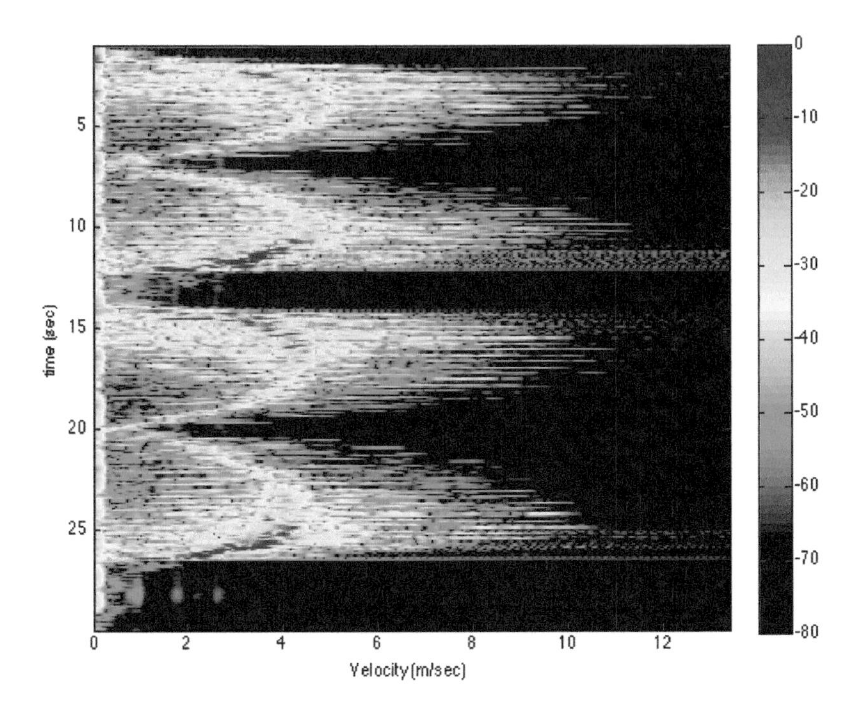

FIGURE 2.14
An X-band DTI plot of a race between the author and his friend.

user directs the transmitter toward a target scene. If a harmonic radar tag is present and within range of the radar then a signal strength reading is displayed on an analog signal meter, units in dB relative.

A photo of the radar system is shown (Fig. 2.16) and its transmitter consists of a CW oscillator fed into a 2.5 watt power amplifier. The output of the power amplifier feeds a Yagi transmit antenna which radiates the target scene at 917 MHz. Additional filtering is placed in series with the Yagi to reject the transmitter's second harmonic.

Scattered field from the radar tag (if present within the traget scene) at 1834 MHz is collected by the receive antenna, which is a second Yagi tuned to 1834 MHz. The receiver is a dual conversion heterodyne receiver similar to the SSB receiver described previously (Sec. 1.1.5.3) except that at its IF it uses a logarithmic amplifier that outputs relative receive power in decibels to an analog signal meter.

2.5.1.2 Harmonic Radar Tags

Three types of tags were developed, a dipole tag and two different planar harmonic radar tags.

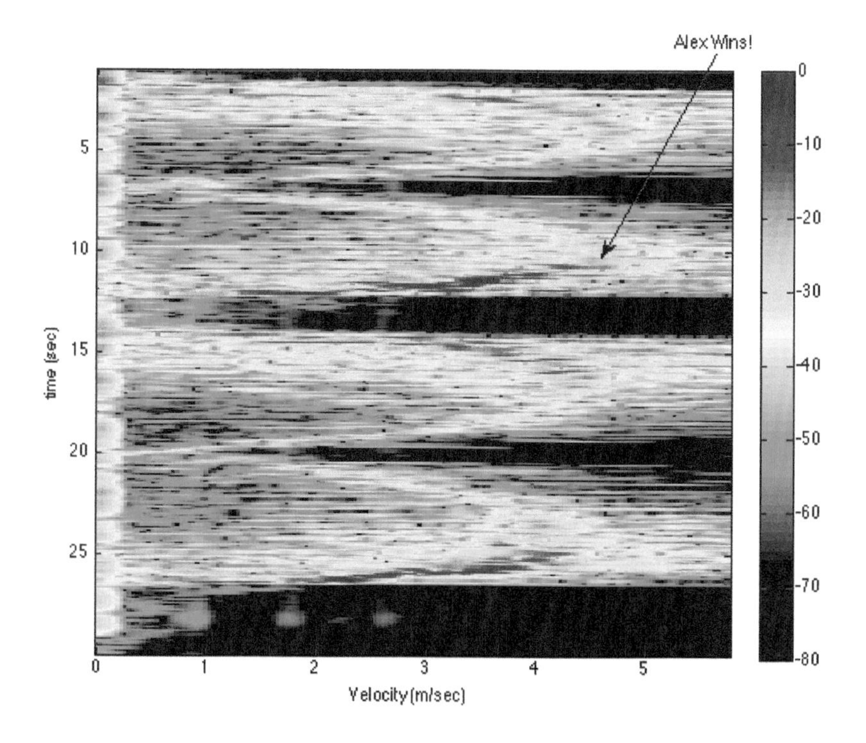

FIGURE 2.15
After closer examination it is clear that the author's friend, Alex, was the fastest runner.

The dipole radar tag is simply a high frequency diode connected to a dipole antenna (Fig. 2.17). When the transmit frequency f_c is incident on this tag it re-transmits a weak signal at $2f_c$. The re-transmit amplitude depends on how well the dipole resonates with the fundamental. Dipole radar tags are often used for tracking insects because they are light weight objects.

Unfortunately, a dipole tag cannot be placed against a metal industrial pipe or storage tank because it would be de-tuned (unless it was located approximately one quarter wavelength away from the metal). For this reason, a planar harmonic radar tag was developed so that it could be affixed to metal industrial tanks and pipes. This tag was developed using patch antennas because the patch antenna relies on its own ground plane on the opposite side of a printed circuit board, where the circuit board material and distance to ground plane are controlled. With its own ground plane, the patch antenna is not de-tuned when placed on metal structures. A modular design approach was used, where a separate receive patch antenna was built at 917 MHz, re-transmit antenna at 1834 MHz, and a microstrip frequency doubler circuit (Fig. 2.18).

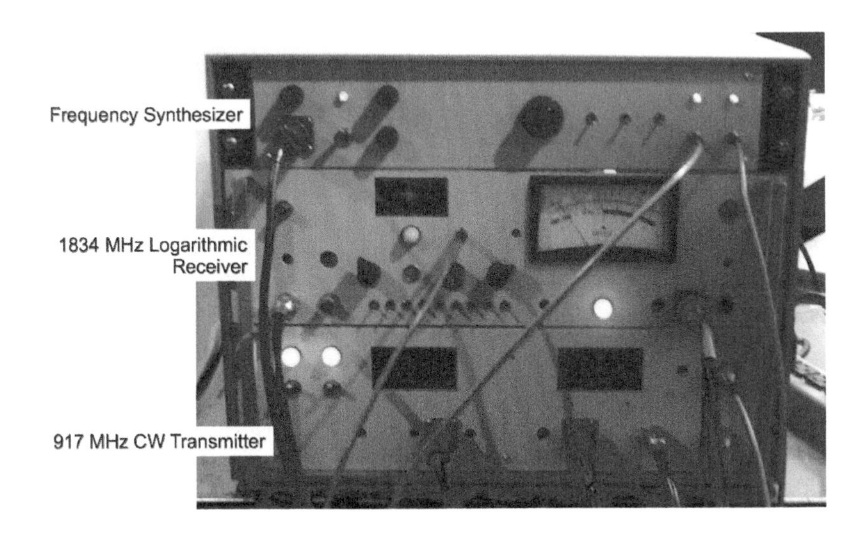

(a)

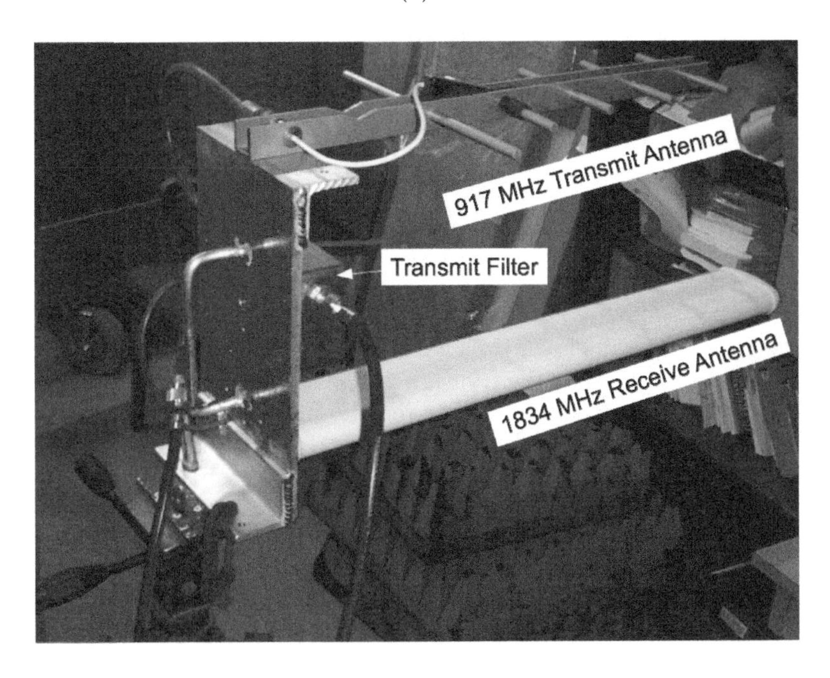

(b)

FIGURE 2.16
Harmonic radar (a) and antenna assembly (b).

FIGURE 2.17
A dipole harmonic radar tag is simply a diode soldered to two elements of a dipole antenna.

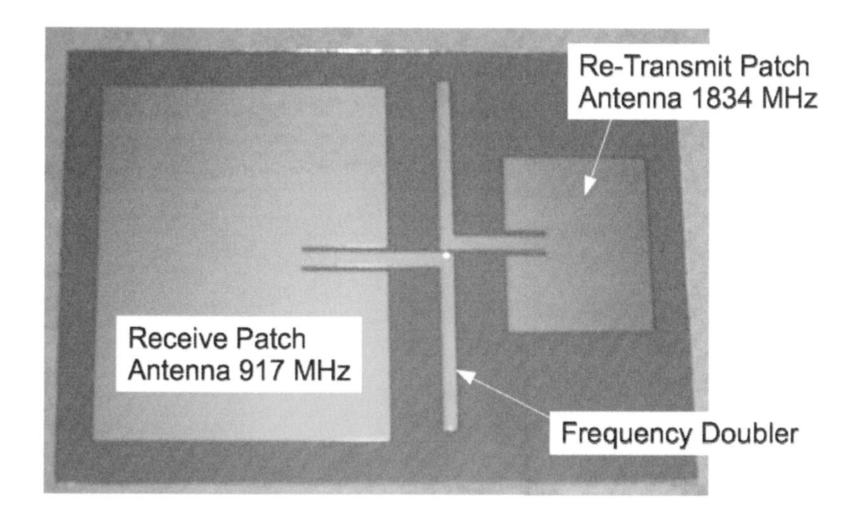

(a)

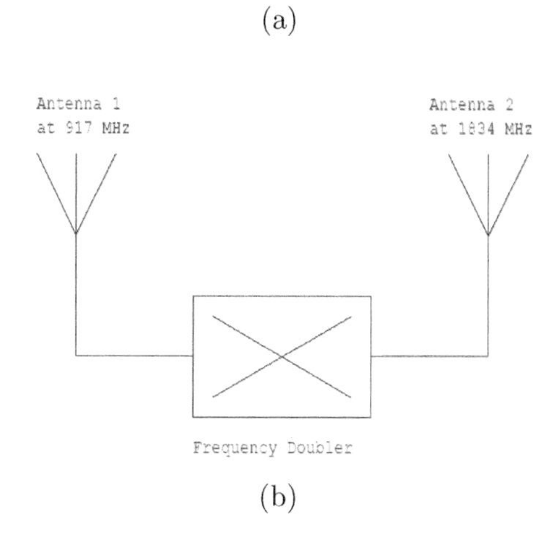

(b)

FIGURE 2.18
Planar harmonic radar tag on microwave substrate (a) with a receive antenna at the fundamental frequency f_c feeding a frequency doubler circuit followed by a re-transmit antenna at twice the fundamental $2f_c$ (b).

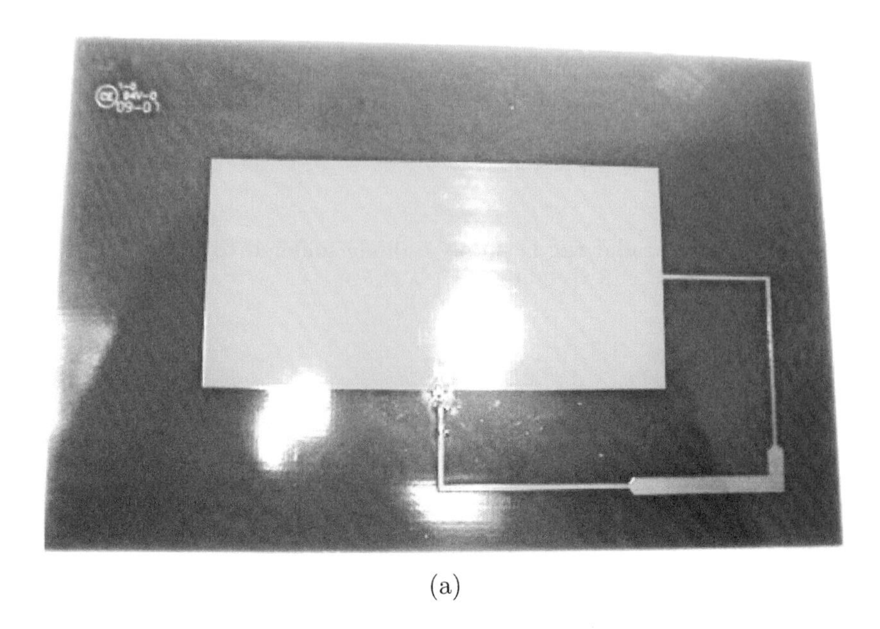

(a)

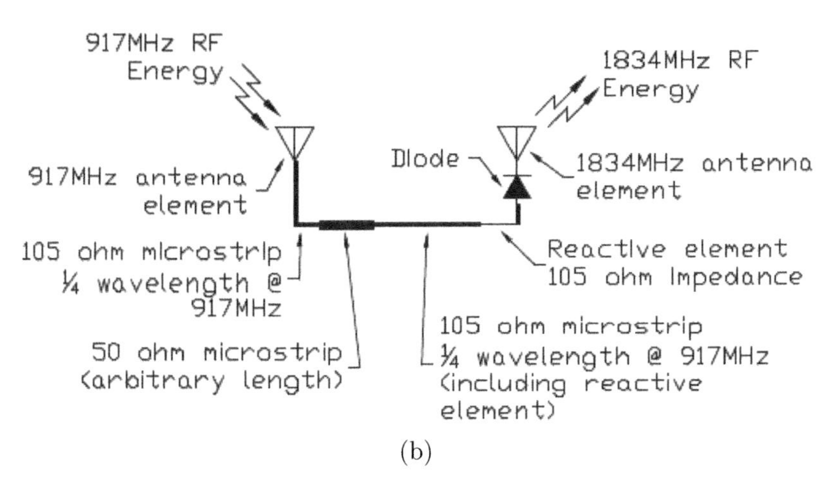

(b)

FIGURE 2.19
Smaller form factor planar harmonic radar tag (a) and block diagram (b).

A much smaller form factor was developed by using a dual-band patch antenna and impedance matching to and from the diode (Fig. 2.19). With its smaller size, this tag is a more flexible design that can be more easily placed on a greater variety of targets.

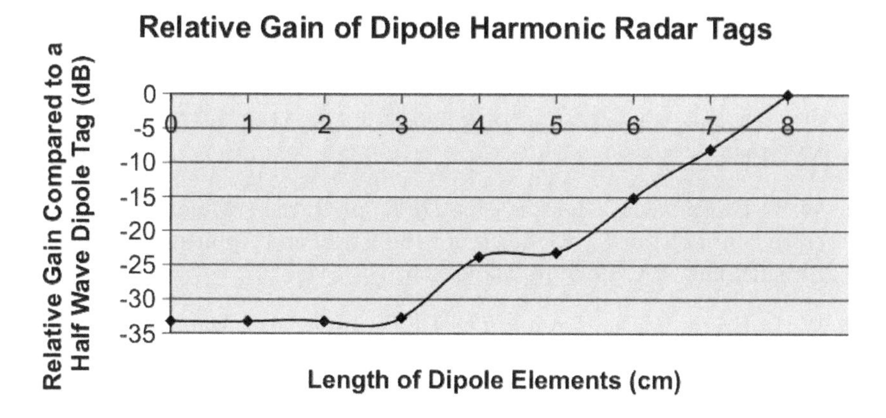

FIGURE 2.20
Relative return of dipole harmonic radar tags at $f_c = 917$ MHz.

2.5.1.3 Results

A variety of dipole tags were cut and measured at a fixed range. The relative magnitude of these tags is plotted (Fig. 2.20), showing that the half-wave dipole tag at the fundamental f_c was the most effective.

When testing planar harmonic radar tags it was realized that maximum range for both types was measured to be approximately 30 ft before the scattered harmonic was less than the receiver's noise floor.

Harmonic radar has potential. It is also possible to implement pulse ranging, a coherent CW radar for Doppler, or for FMCW operation using harmonic radar. This would require an architecture to support frequency doubling on the receiver's front end. It should also be possible to perform SAR imaging or other beamforming techniques.

2.6 Summary

Doppler CW radar architectures were shown with maximum range model developed and signal processing discussed. Two examples of CW Doppler radar were used to measure the velocity of moving vehicles and Doppler signatures of various targets including automobiles and people running. CW harmonic radar was also discussed for use in non-destructive testing and evaluation applications. In the next chapter, using the same architecture the CW oscillator will be frequency modulated so that range-to-target information can be computed.

Bibliography

[1] M. I. Skolnik, *Introduction to Radar Systems,* McGraw Hill, New York, NY, 1962, p. 3.

[2] W. L. Briggs, V. E. Henson, *The DFT, an Owner's Manual for the Discrete Fourier Transform,* Society for Industrial and Applied Mathematics, Philadelphia, PA, 1995, p. 23, eq. 2.6.

[3] W. L. Briggs, V. E. Henson, *The DFT, an Owner's Manual for the Discrete Fourier Transform,* Society for Industrial and Applied Mathematics, Philadelphia, PA, 1995, p. 22, eq. 2.5.

[4] W. L. Briggs, V. E. Henson, *The DFT, an Owner's Manual for the Discrete Fourier Transform,* Society for Industrial and Applied Mathematics, Philadelphia, PA, 1995.

[5] G. L. Charvat, J. H. Williams, A. J. Fenn, S. M. Kogon, J. S. Herd, "The MIT IAP 2011 radar course: build a small radar system capable of sensing range, Doppler, and SAR," The Boston Chapters of the IEEE Life Members, AP-S, AES, and GRSS, May 24, 2011. IEEE Boston Section, Education Society and Women in Engineering, September 13, 2011.

[6] G. L. Charvat, A. J. Fenn, B. T. Perry, "The MIT IAP radar course: build a small radar system capable of sensing range, doppler and synthetic aperture radar (SAR) imaging," Atlanta, GA: IEEE Radar Conference, May 2012.

[7] G. L. Charvat, J. H. Williams, A. J. Fenn, S. M. Kogon, J. S. Herd, "RES.LL-003 Build a Small Radar System Capable of Sensing Range, Doppler, and Synthetic Aperture Radar Imaging, January IAP 2011," (Massachusetts Institute of Technology: MIT OpenCourseWare), http://ocw.mit.edu (Accessed 02 Sep, 2012). License: Creative Commons BY-NC-SA.

[8] radar_range_eq_doppler.m, `http://glcharvat.com/shortrange/continuous-wave-cw-radar/`

[9] G. L. Charvat, J. H. Williams, A. J. Fenn, S. M. Kogon, J. S. Herd, "RES.LL-003 Build a Small Radar System Capable of Sensing Range, Doppler, and Synthetic Aperture Radar Imaging, January IAP 2011," (Massachusetts Institute of Technology: MIT OpenCourseWare), http://ocw.mit.edu (Accessed 02 Sep, 2012). License: Creative Commons BY-NC-SA, goto Experiment 1: Doppler vs. Time.

[10] Doppler Radar Explanation and Demo using the coffee can radar, `http://glcharvat.com/shortrange/continuous-wave-cw-radar/`

[11] G. L. Charvat, L. C. Kempel. "Low-Cost, High Resolution X-Band Laboratory Radar System for Synthetic Aperture Radar Applications," East Lansing, MI, IEEE Electro/Information Technology Conference, May 2006.

[12] G. L. Charvat, "Low-Cost, High Resolution X-Band Laboratory Radar System for Synthetic Aperture Radar Applications," Antenna Measurement Techniques Association Conference, Austin, Texas, October 2006.

[13] G. L. Charvat. "Build a high resolution synthetic aperture radar imaging system in your backyard," MIT Haystack Observatory, May 12, 2010.

[14] T. S. Ralston, G. L. Charvat, S. G. Adie, B. J. Davis, S. Carney, S. A. Boppart. "Interferometric synthetic aperture microscopy, microscopic laser radar," Optics and Photonics News, June 2010, Vol. 21, No. 6, pp. 32-38.

[15] X-Band CW Doppler Radar Experiment, `http://glcharvat.com/shortrange/continuous-wave-cw-radar/`

[16] CW_xband_doppler_radar.zip, `http://glcharvat.com/shortrange/continuous-wave-cw-radar/`

[17] G. L. Charvat, L. C. Kempel, "Harmonic radar tag measurement and characterization," IEEE Antennas and Propagation Conference, Ottawa Ontario, June 2003.

[18] G. L. Charvat, L. C. Kempel, E. L. Liening, T. E. Miller, M. W. Warren, "Harmonic wireless transponder sensor and method," US Patent 7145453, December 6, 2006.

[19] M. A. Volz, G. L. Charvat, L. C. Kempel, M. Warren, E. Liening, "Harmonic radar planar antenna miniaturization," North American Radio Science Meeting, Ottawa Ontario, July 2007.

[20] M. A. Volz, B. Crowgey, G. L. Charvat, E. Rothwell, L. C. Kempel, E. Liening, M. Warren, "Recent developments in miniaturized planar harmonic radar antennas," Antenna Measurement and Techniques Association Conference, Boston MA, November 2008.

3

Frequency Modulated Continuous Wave (FMCW) Radar

Doppler measurements were demonstrated using coherent CW radar discussed in Chapter 2, where a CW carrier was radiated toward a target scene and the scattered carrier was multiplied by a copy of the original, resulting in the measurement of phase over time, thereby providing a Doppler measurement. In this chapter ranging will be achieved by FM modulating the CW oscillator in such a way that its frequency changes linearly with time. By multiplying this transmitted waveform by what is scattered off of the target, range-to-target distance can be determined.

FMCW radar was originally developed for radar altimeters for aircraft in the mid 1930's. Today, FMCW is useful in applications where wide-band high-resolution time-of-flight measurements must be made with low-power transmitters [1]–[5]. Applications include automotive radar, short-range imaging, and many others.

Frequency modulated continuous wave radar (FMCW) architecture and processing will be described (Sec. 3.1), the radar range equation will be applied to FMCW radar (Sec. 3.2.1), and examples of FMCW radar systems will be shown (Sec. 3.3).

3.1 FMCW Architecture and Signal Processing

FMCW radar offers an elegant solution to wide-band short-range radar design. The radar does not need to pulse, instead it transmits and receives continuously. Targets at or near zero range can be detected. Fourier analysis of received data greatly increases sensitivity. FMCW is inexpensive, requiring only a frequency modulated (FM) oscillator and a frequency mixer. FMCW radar is an ideal mode for short-range radar systems where low cost, wide bandwidth, and high sensitivity are requirements that must not be compromised.

Architecture and signal processing are closely coupled in FMCW radar systems. The radar provides analog information in the spatial frequency domain which must then be processed to determine range to target(s). FMCW

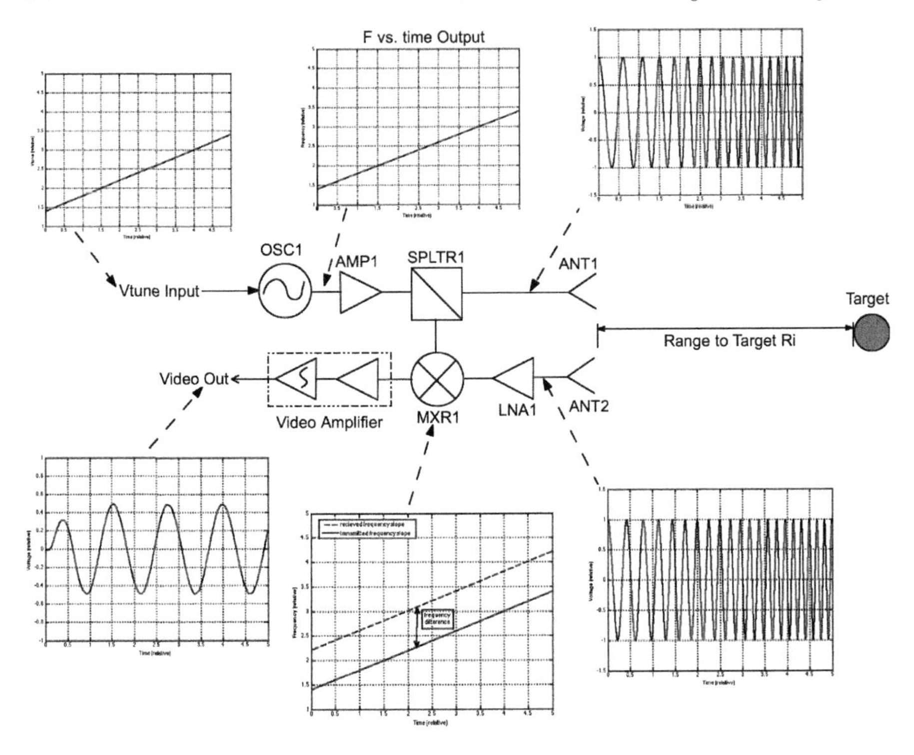

FIGURE 3.1
Simplified block diagram of a coherent linear FMCW radar sensor.

architecture will be discussed (Sec. 3.1.1) followed by signal processing (Sec. 3.1.3).

3.1.1 FMCW Architecture

A simplified block diagram of an FMCW radar is shown (Fig. 3.1), where OSC1 is a voltage controlled oscillator (VCO) that outputs a frequency in linear proportion to its input control voltage (Vtune). FM is achieved by changing OSC1's Vtune over time. In this case we modulate Vtune with a linear up-ramp. The output of OSC1 is a sinusoidal waveform that is changing frequency over time.

This waveform is amplified by AMP1 and fed into power splitter SPLTR1, where half of it is radiated out of ANT1 and the rest is fed into the LO port of MXR1.

What is radiated out of ANT1 looks like an 'accordion' waveform, where the early portion of the waveform is at a lower frequency than the later portion. This waveform propagates through space, scatters off the target, and propagates back toward the radar where a portion of it is collected by ANT2.

The waveform collected by ANT2 is a delayed version of the original accordion-like waveform.

The signal from ANT2 is amplified by the low-noise amplifier LNA1 and fed into the RF port of MXR1. Within MXR1, the delayed version of the accordion-like scattered waveform is multiplied by the transmitted waveform.

Consider the instantaneous frequency versus time of each waveform plotted on top of another. There is a constant frequency offset between what was transmitted and what is received. This constant frequency offset is directly proportional to range-to-target R_i and the greater this frequency offset, the further the range to target.

When the transmit waveform is multiplied by the delayed receive waveform within MXR1 the product difference (amplified and low-pass filtered by the video amplifier) is this constant frequency offset known as the beat frequency.

If multiple targets are present then multiple beat frequencies will be superimposed on each other in the video output, each uniquely representing target range, phase, and scattered amplitude. Using Fourier analysis, range to target can be determined for each.

Another way of considering the range to target in terms of beat frequencies is to think of each as a spatial frequency, where the further a target is in space the greater the spatial frequency.

3.1.2 Mathematics of FMCW Radar

FMCW radar uses a linearly modulated VCO (or a linear modulated waveform from an arbitrary waveform generator). This waveform is amplified and radiated out toward a target scene and is represented by

$$TX(t) = \cos\left(2\pi(f_{osc} + c_r t)t\right), \tag{3.1}$$

where:

$f_{osc} =$ start frequency of linear ramp modulated VCO
$c_r =$ radar chirp rate

The transmitted signal is scattered off the target and collected by the receiver antenna ANT2. The round trip time from ANT1 to the target and back to ANT2 is t_{delay}. What is collected by ANT2 is a time-delayed copy of the transmit signal ($TX(t - t_{delay})$) that is amplified by LNA1 and fed into the RF port of MXR1.

A portion of the power from AMP1 is split off to the LO port of MXR1, thereby providing a copy of the transmit waveform to MXR1 and making the transmitter and receiver phase coherent so that $TX(t)$ is multiplied by $TX(t - t_{delay})$. The resulting product is amplified and filtered by the video amplifier, represented by (disregarding amplitude coefficients)

$$Video(t) = TX(t) \cdot TX(t - t_{delay}),$$

$$Video(t) = \cos\left(2\pi(f_{osc} + c_r t)t\right) \cdot \cos\left(2\pi(f_{osc} + c_r t)(t - t_{delay})\right).$$

The higher frequency term resulting from this multiplication is ignored because the video amplifier usually also includes a low-pass filter circuit (in addition to this, the IF port of a typical microwave mixer could not produce the resulting microwave frequencies). The resulting video signal is at a very low frequency which is proportional to the chirp rate c_r and round-trip time delay t_{delay} plus a DC phase term:

$$Video(t) = \cos\left(2\pi f_{osc} t_{delay} + c_r t t_{delay}\right). \tag{3.2}$$

The further the target, the higher the beat frequency sampled at the video output. If there are a number of targets down range then the signal $Video(t)$ will be a superposition of beat frequencies at various amplitudes representing each target.

3.1.3 Signal Processing for FMCW Radar and the Inverse Discrete Fourier Transform

Two methods of processing range-to-target information from FMCW beat frequencies will be presented and a method of processing range-to-target data for only targets that have moved will be shown.

3.1.3.1 Frequency Counter and Frequency-to-Voltage Converters

Range to target measurement for one target can be achieved by connecting a frequency counter or a voltage-to-frequency converter to the video output, thereby counting zero crossings. This was done for radar altimeters in use since the mid 1930s [6]–[9]. This method of calculating range for an FMCW radar is only effective when one large target is being measured, such as the ground return for the case of a radar altimeter or the level of a liquid storage tank for industrial applications.

3.1.3.2 Inverse Discrete Fourier Transform

To process range to multiple targets and to retrieve phase and amplitude information, the inverse discrete Fourier transform (IDFT) is used to convert from the spatial frequency domain to the time domain [10]. This analysis technique has a further benefit where it reduces the effective noise bandwidth of the radar by the duration over which the IDFT is applied where $B_n = 1/\tau$ where τ is the length in time over which the DFT is applied.

The time domain representation s_n of a discretely sampled signal S_ω (from the FMCW radar) with N total number of samples is computed using

$$s_n = \sum_{\omega=-\frac{N}{2}+1}^{\frac{N}{2}} S_\omega e^{j\omega n/N}, \tag{3.3}$$

where $j = \sqrt{-1}$, $\omega = 2\pi f$, f is the relative frequency, and n is the discrete sample number in the discretely sampled data set being analyzed. The discretely sampled data S_ω can be real (as would be acquired from the radar in Fig. 3.1), it can be complex (as would be the acquired from the radar in Fig. 2.2), or it can be over-sampled and digitally converted to complex data (as would be the case for the radar in Fig. 2.3). The IDFT is a standard function in MATLAB® and other software packages.

When using the IDFT, it is important to understand that the resulting calculation S_ω provides only N points, the same number of data points that were sampled by the radar system's digitizer. To back out the resulting time-domain signal represented by the points in s_n one must know the total frequency bandwidth (BW) spanned from start (f_{start}) to stop frequency (f_{stop}) and the number of samples acquired N, where $BW = f_{stop} - f_{start}$. With this information, the time between calculated samples s_n (in seconds) becomes $\Delta n = 1/BW$ [11] where there are N total number of samples for a complex valued signal S_ω and $N/2$ total number of samples for a real-valued signal S_ω.

There are only up to N number of useful data points resulting from an IDFT but it is often the case that humans prefer smooth signals when viewing radar data sets. For this reason there is a demand to interpolate between the the resulting data s_n and this can be achieved by arbitrarily adding 0's to the end or beginning of the sampled signal S_ω in software (known as zero padding or up-sampling), thereby calculating the IDFT over a larger number of points. The result is a smoothed-over IDFT plot that looks nice for briefings or to the radar operator.

There exist computationally faster versions of the IDFT known as the fast Fourier transform (FFT). These are automatically selected when using software packages such as MATLAB, where if the data set S_ω contains a number of samples that is equal to a power of 2, then an FFT will automatically be selected. In this book, the IDFT will be used as a black box to convert data from the time domain into the frequency domain.

3.1.3.3 Coherent Change Detection (CCD)

By leveraging the fact that this is a coherent radar and that most VCO's (or waveform generators) are fairly stable from chirp-to-chirp, coherent change detection can be used to subtract the non-moving targets from those that are moving, thereby revealing the moving targets. This is done on a sweep-by-sweep (or pulse-by-pulse depending on how you want to define it) basis, where the previous de-chirp data ($S_\omega(t-1)$) is subtracted from the current ($S_\omega(t)$), providing de-chirp data for only those targets that have moved or changed since the previous pulse

$$S_\omega(changed) = S_\omega(t) - S_\omega(t-1). \tag{3.4}$$

Range to targets that have changed or moved can be found by plotting IDFT($S_\omega(changed)$).

Terminology for radar devices is as diverse as the authors who write about it. This process for CCD as described is also known as two-pulse clutter cancellation or moving target indication (MTI).

3.2 FMCW Performance

Two key performance metrics will be discussed, maximum range (Sec. 3.2.1) and range resolution (Sec. 3.2.2).

3.2.1 The Radar Range Equation for FMCW Radar

Similar to the Doppler CW radar when using Fourier analysis, the FMCW radar's effective noise bandwidth is inversely proportional to its chirp time. The longer the chirp time, the lower the effective noise bandwidth and therefore the greater the sensitivity. This relationship provides high performance with a low peak power transmitter, either longer ranges for a given target size or smaller target sizes detectable for a given range.

An estimate for the maximum range for an FMCW radar system is given by

$$R_{max}^4 = \frac{P_{ave} G_{tx} A_{rx} \rho_{rx} \sigma e^{(2\alpha R_{max})}}{(4\pi)^2 k T_o F_n B_n \tau F_r (SNR)_1 L_s}, \tag{3.5}$$

where:

$R_{max} =$ maximum range of radar system (m)
$P_{ave} =$ average transmit power (watts)
$G_{tx} =$ transmit antenna gain
$A_{rx} =$ receive antenna effective aperture (m^2)
$\rho_{rx} =$ receiver antenna efficiency
$\sigma =$ radar cross section (m^2) for target of interest
$L_s =$ miscellaneous system losses
$\alpha =$ attenuation constant of propagation medium
$F_n =$ receiver noise figure (derived from procedure outlined in Sec 1.1.5.4)
$k =$ $1.38 \cdot 10^{-23}$ (joul/deg) Boltzmann's constant
$T_o =$ 290°K standard temperature
$B_n =$ system noise bandwidth (Hz)
$\tau F_r =$ 1, the duty cycle for a CW radar
$(SNR)_1 =$ single-pulse signal-to-noise ratio requirement

The noise bandwidth is inversely proportional to the discrete sample length $B_n = 1/t_{sample}$. Similar to the Doppler radar, for a direct conversion FMCW radar (this applies to all FMCW radar examples in this book) using IDFT processing, the noise bandwidth is twice as wide as a radar following an image rejection architecture where $B_n = 2/t_{sample}$. If a simple frequency counter is used for processing with a direct conversion architecture, then $B_n = 2f_{-3dB}$

TABLE 3.1
Excess bandwidth factor K to account for several weighting functions, additional K factors shown [15].

Weighting	K
Uniform rectangular	0.89
Uniform circular	1.03
Hanning	1.43

where f_{-3dB} is the -3 dB cut-off frequency of a band-limiting filter at base band that would be placed in front of the counter.

The basic radar range equation as shown is a good first-order approximation of system performance. It is important to note that clutter and additional system noise effects (beyond thermal noise as modeled here), such as 1/F noise or oscillator phase noise, may reduce the maximum range of a practical radar system. Additional information on modeling FMCW radar performance is well documented [12] and [13].

3.2.2 Range Resolution

Range resolution is a metric of how well a radar system can differentiate between two targets in range. This is the actual minimum distance between two targets before they can no longer be differentiated in range. The expected range resolution, defined as the -3 dB points below peak target response, for a linear FM radar system depends on the chirp bandwidth calculated [14]

$$\rho_r = \frac{cK_r}{2BW}, \tag{3.6}$$

where c is the speed of light in free space, K_r is the down range excess bandwidth factor which depends on the weighting function applied before the IDFT (Table 3.1), and BW is the chirp bandwidth.

This estimate for range resolution does not take into account higher order effects, such as the linearity of the FM chirp source, overlap of transmit and receive pulses, and receiver processing bandwidth. This estimate assumes perfectly linear source, good overlap of transmit and receive pulses, and sufficiently wide receiver processing bandwidth to capture the full pulse. These assumptions are sufficiently sound to 'ball park' radar performances of devices discussed in this book but should be considered for your application if you have a very rapid chirp rate or a low-cost chirp source with linearity issues.

3.3 Examples of FMCW Radar Systems

Practical examples of FMCW radar systems will be shown, including X-band ultrawideband (UBW) radar (Sec. 3.3.1), the MIT coffee can radar system in ranging mode (Sec. 3.3.2), and two examples (S and X-band) of analog range-gated FMCW radar (Sec. 3.3.3).

3.3.1 X-Band UWB FMCW Radar System

UWB X-band FMCW radar is useful for precise ranging of targets down to zero range, high sensitivity and wide bandwidth range profiles of targets at moderate ranges, and for high resolution imaging [16]–[19].

This conventional FMCW radar system will be described, where OSC1 from the X-band CW Doppler radar (Sec. 2.4.2) is frequency modulated with a linear ramp from 7.835 GHz (f_{start}) to 12.817 GHz (f_{stop}). This radar system contains four major subsystems including a Windows XP computer with a National Instruments PCI6014 to control the radar, OSC1, and the front end (Fig. 3.2).

The computer runs a Labview virtual instrument (VI) computer program to control the radar. The VI controls the radar through an NI PCI-6014 data acquisition card. The I/O from this card is fed through the Radar CTRL box which also contains power supplies, signal conditioning, and motion control for a linear rail stage (Fig. 3.4). The radar front-end plugs into the radar CTRL box. A complete block diagram (Fig. 3.3) and bill of material (BOM) are shown (Table 3.2).

CTR0 is fed into OSC2, where OSC2 is a linear ramp generator (schematic in Fig. 3.5). The Labview VI forces CTR0 low, thereby causing a linear ramp to be generated. The duration over which CTR0 is low is the up-ramp pulse time (t_{sample}) of the radar. The ramp output from OSC2 is fed into the V_{tune} input of OSC1. OSC1 is a voltage tuned YiG oscillator.

OSC1 chirps from 7.835 GHz to 12.817 GHz over the duration of the up-ramp. This chirp is fed into the radar front end (Fig. 2.10) where it is passed through the directional coupler CLPR1, through the circulator CIRC1, and into ANT1 where the chirp is radiated toward the target scene. CIRC1 is configured as an isolator, where a 50 ohm load is placed on the third port. Placing a circulator between the transmit signal and transmit antenna reduces reflections from antenna mismatch at the edges of the antenna's frequency range.

Scattered energy from the target scene is collected by ANT2, amplified by the low noise amplifier LNA1, and fed through ATT1 into the RF port of frequency mixer MXR1. The output of LNA1 is attenuated by 5 dB through ATT1 because the gain of LNA1 would cause MXR1 to saturate due to the direct coupling from ANT1 to ANT2.

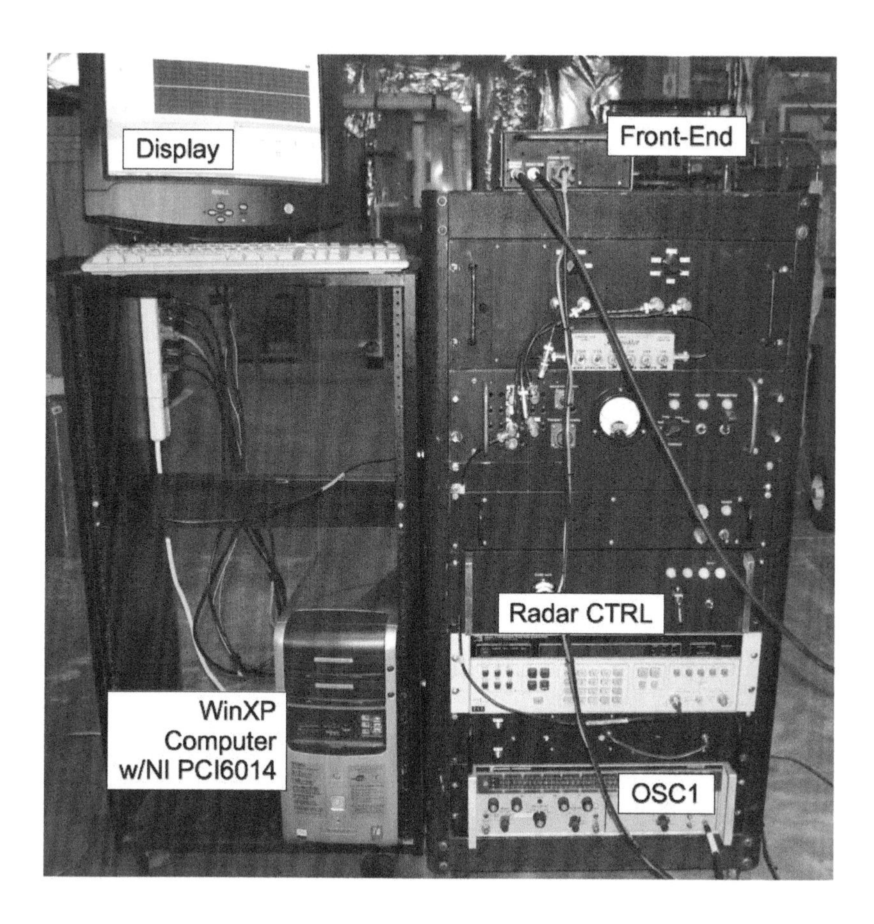

FIGURE 3.2
Photo of the radar testbed configured to run the X-band UWB FMCW radar front-end.

Some of the transmit chirp is coupled off CLPR1 through CIRC2 and into the LO port of MXR1 where it is multiplied by scattered signal collected by the receive antenna. CIRC2 is configured as an isolator to absorb reflected power from the mismatch from the LO port of MXR1. Double balanced mixers at microwave frequencies typically are not matched to 50 ohms at the LO port. The IF product is amplified, and low-pass filtered through the video amplifier. The −3 dB cut-off point of this filter is 80 KHz, thereby providing an anti-alias filter for the digitizer in the PCI-6014 data acquisition card.

The output of the video amplifier is fed into ACH0 port of the I/O breakout which is then fed to the PIC-6014 digitizer running at a sample rate of 200

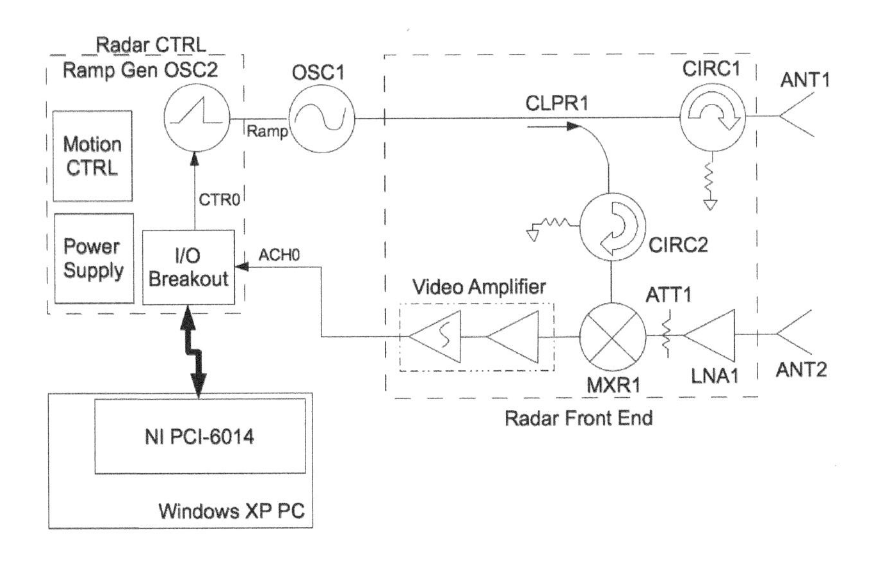

FIGURE 3.3
Block diagram of the X-band UWB FMCW radar system.

KSPS. The VI acquires 2000 samples at the same time as it causes CTR0 to go low, thereby chirping the radar at the same time as it acquires data. This data is recorded for analysis.

3.3.1.1 Expected Performance of the X-Band UWB FMCW Radar System

Substituting specifications for the X-band UWB FMCW radar (Table 3.3) into the radar range equation (3.5) provides the maximum range estimate of 1190 m for a 10 dBsm automobile target using the MATLAB script [20].

Antenna gains G_{rx} and G_{tx} are from the data sheets for ANT1 and ANT2. Noise figure is estimated based on the apparent age of the surplus microwave amplifier LNA1. With these parameters and the fact that the IDFT is applied to 10 ms intervals of data; an effective noise bandwidth of 100 Hz is doubled to 200 Hz because of the direct conversion architecture of this UWB X-band FMCW radar.

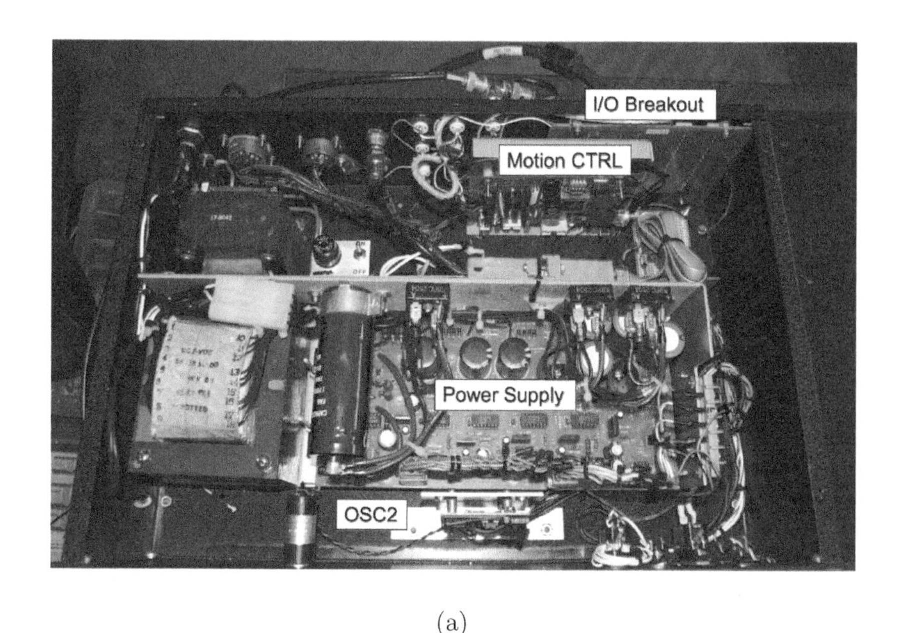

(a)

(b) (c)

FIGURE 3.4
Radar controller inside callouts (a), front panel (b), rear panel (c).

3.3.1.2 Working Example of the X-Band UWB FMCW Radar System

The radar sensor was directed down range and the author and his colleague as radar targets walked down range and back. The demo is shown [21]. The video output was recorded for every up-chirp. Data and MATLAB script for processing are provided [22].

TABLE 3.2
X-band CW radar system RF modular component list.

Component	Description
OSC1	Weinschel Engineering 430A Sweep Oscillator with 434A RF Unit
OSC2	Figure 3.5
CLPR1	Narda Model 4015C-10, -10 dB, 7-12.4 GHz
CIRC1	Raytron Model 300049 Circulator
CIRC2	Raytron Model 300049 Circulator
ANT1	M/A-COM MA86551, 17 dBi gain 25 deg E and H plane beamwidths
ANT2	M/A-COM MA86551, 17 dBi gain 25 deg E and H plane beamwidths
LNA1	Dexcel, Inc, 58985, NF = 4 dB (educated guess), gain = 30 dB
ATT1	Midwest Microwave 5 dB attenuator
MXR1	X-band mixer of unknown manufacture
Video amplifier	Figure 2.13

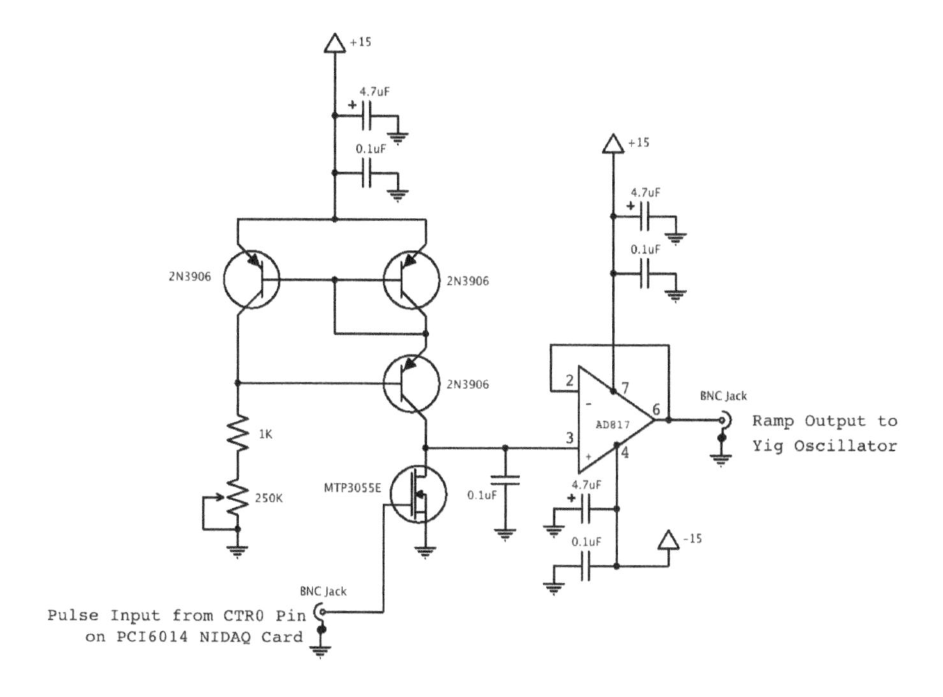

FIGURE 3.5
Schematic of the ramp generator, OSC2.

TABLE 3.3
X-band FMCW radar system radar range equation parameters.

$$
\begin{aligned}
P_{ave} &= 31 \cdot 10^{-3} \text{ (watts)} \\
G_{tx} &= 17 \text{ dBi antenna gain} \\
G_{rx} &= 17 \text{ dBi antenna gain} \\
A_{rx} &= G_{rx}\lambda_c^2/(4\pi) \text{ (m}^2\text{)} \\
\lambda_c &= c/f_c \text{ (m) wavelength of carrier frequency} \\
f_c &= 10 \text{ GHz center frequency of radar} \\
\rho_{rx} &= 1 \text{ because antenna efficiency is accounted for in antenna gain} \\
\sigma &= 10 \text{ (m}^2\text{) for automobile at 10 GHz} \\
L_s &= 6 \text{ dB miscellaneous system losses} \\
\alpha &= 0 \text{ attenuation constant of propagation medium} \\
F_n &= 4 \text{ dB receiver noise figure} \\
B_n &= 2/t_{sample} \text{ system noise bandwidth (Hz) where } t_{sample} = 10 \text{ ms} \\
(SNR)_1 &= 13.4 \text{ dB}
\end{aligned}
$$

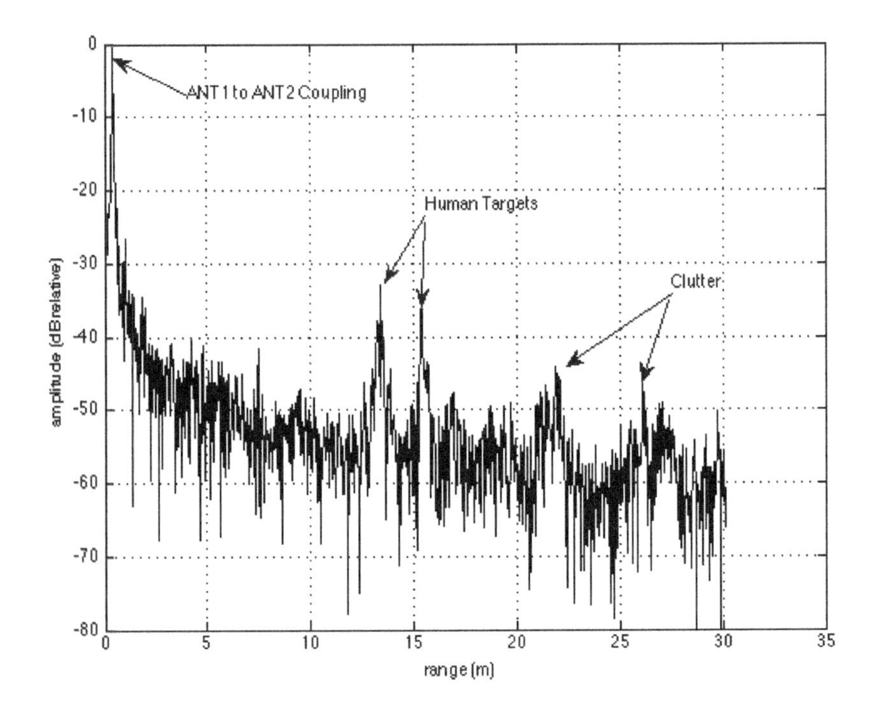

FIGURE 3.6
Range profile of two human targets using the X-band UWB FMCW radar system; each target position is clearly shown but unfortunately the range profile contains a good deal of clutter and the largest signal present is the transmit-to-receive (ANT1 to ANT2) antenna coupling.

The IDFT was applied to each range profile, providing a time domain representation of the scattered radar signal. Speed of light was multiplied by the round-trip time to calculate range to target (Fig. 3.6). Direct coupling from the transmit antenna to the receive antenna (ANT1 to ANT2) was observed as a strong signal near zero range. The location of both human targets is clearly shown at approximately 12.5 m and 15 m nearly 20 dB above the clutter floor. Clutter is evident throughout, especially at 22 m and 26 m.

Each range profile was stacked so as to make a 2D array of range profiles, range versus time with amplitude in the z-axis. This is known as a range time intensity (RTI) plot (Fig. 3.7). This is a common data product provided by many instrumentation radar systems. It clearly shows the range and amplitude of each target in time moving away from and back toward the radar sensor. Clutter that does not move (stationary clutter) appears as vertical streaks throughout the plots. The range to each target is clearly shown until the

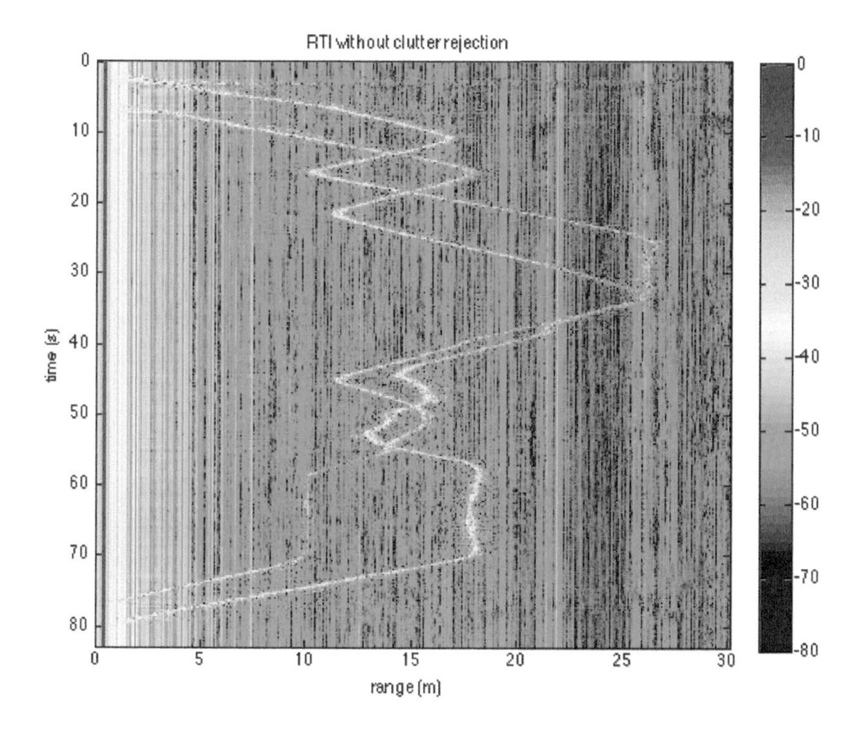

FIGURE 3.7
RTI plot of two human targets walking down range and back using the UWB X-band FMCW radar system.

targets reach 25 m where the magnitude of the clutter is greater than that of targets. The strongest signal is the transmit-to-receive antenna coupling, shown as a bright read vertical streak near zero range.

One effective method of reducing clutter is to apply CCD by subtracting the current range profile from the previous one before the IDFT is applied (Sec. 3.1.3.3). A range profile is shown after applying CCD (Fig. 3.8). The location of both human targets is clearly shown and the amplitude of each target is approximately 30 dB above the noise floor. The transmit-to-receive coupling is no longer present and the clutter is greatly reduced.

Clutter is not completely eliminated using CCD. The degree to which clutter is eliminated is a function of the stationary clutter movement (e.g., trees move in the wind), the modulation (or equivalent pulse rate) frequency of the radar, and the phase noise of the radar (e.g., how stable the master oscillator is from pulse-to-pulse).

CCD was applied to all range profiles in the resulting RTI (Fig. 3.9). The location of both human targets is clear and clutter is significantly reduced. The transmit-to-receive coupling at zero range is all but eliminated. However, when the targets are nearly stationary their magnitude is greatly reduced. When

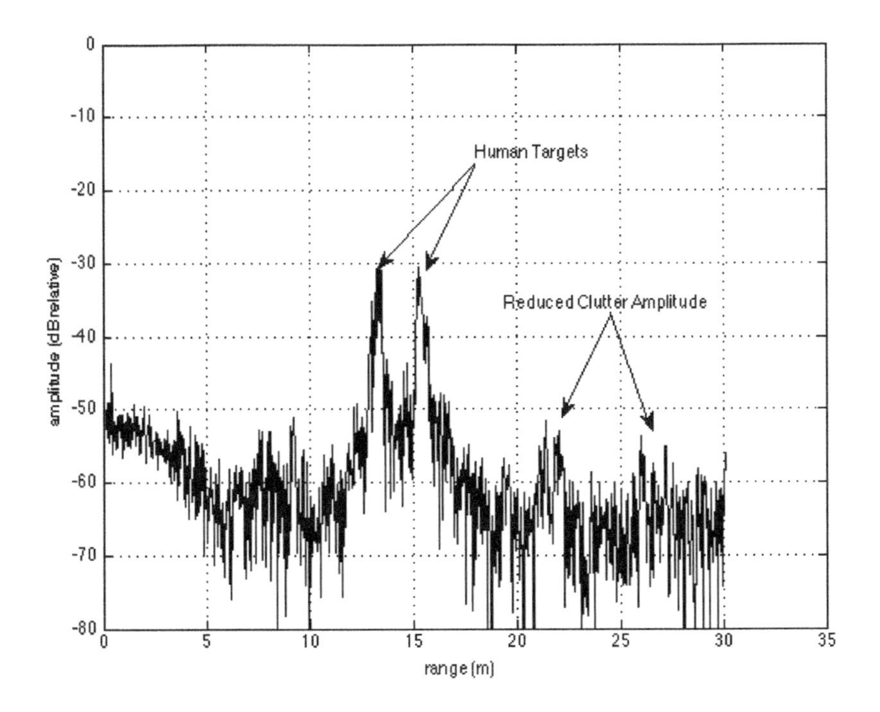

FIGURE 3.8
Range profile of two human targets using the X-band UWB FMCW radar system with CCD processing; each target position is clearly shown and clutter is significantly reduced; no transmit-to-receive coupling is evident.

using CCD, target magnitude is also a function of its movement from pulse-to-pulse but the overall result provides a significant improvement in detectability of moving targets. Often times moving targets are the only relevant targets of interest.

3.3.2 The MIT Coffee Can Radar System in Ranging Mode

The coffee can radar (Fig. 2.4) was developed as part of a short radar course at MIT, where the cost of parts is low enough for college students to afford. It operates in three modes, Doppler, ranging, and SAR imaging. In the course, students learn about radar, build their own, then perform field tests. In this section the radar will be used in FMCW ranging mode.

Modulator1 is a free-running saw-tooth generator which feeds the voltage tune input of OSC1 while also sending synchronization pulses out to the left channel of an audio output cable back to a laptop computer's audio input port (Fig. 3.10, call-out diagram shown in Fig. 2.5b and BOM Table 3.4). Modulator1 causes OSC1 to be FM up-chirped from 2.260 GHz (f_{start}) to 2.590 GHz (f_{stop}) in 20 ms.

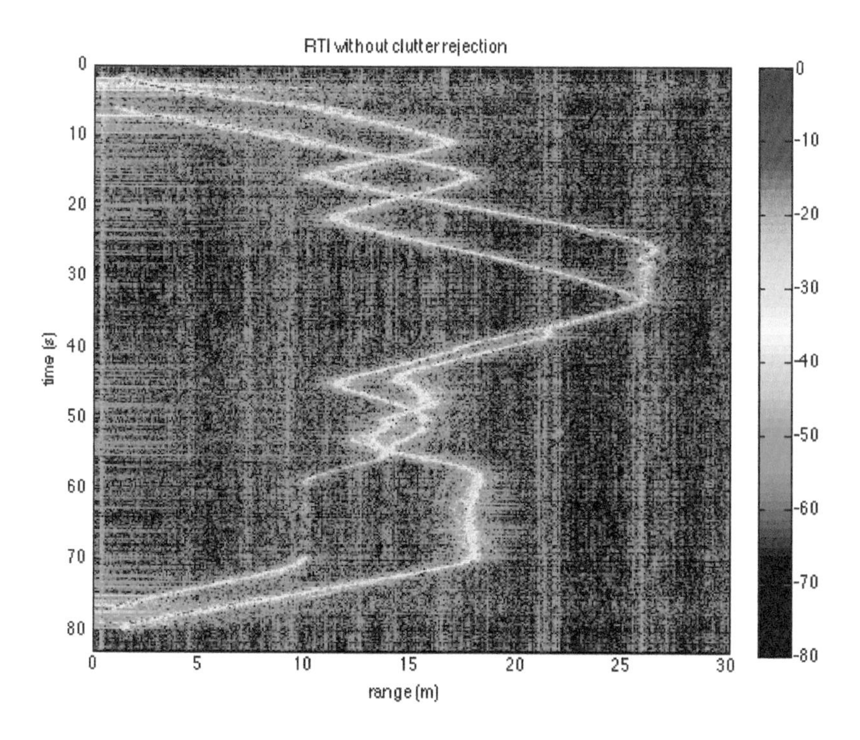

FIGURE 3.9
RTI plot using CCD of two human targets walking down range and back using the UWB X-band FMCW radar system.

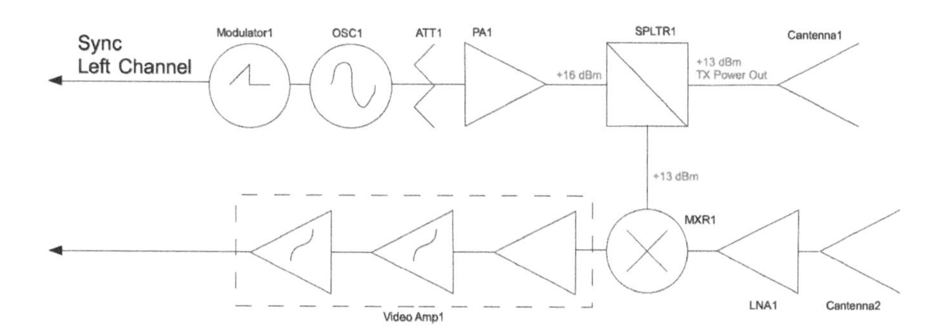

FIGURE 3.10
Block diagram of the coffee can radar in FMCW ranging mode.

TABLE 3.4
MIT coffee can radar RF modular component list.

Component	Description
Modulator1	Figure 3.11
OSC1	Mini-Circuits ZX95-2536C+, 2315-2536 MHz VCO, +6 dBm out
ATT1	Mini-Circuits VAT-3+, 3 dB attenuator
PA1	Mini-Circuits ZX60-272LN-S+ LNA, gain = 14 dB, NF = 1.2 dB, IP1 = 18.5 dBm
ANT1	Figure 2.6
ANT2	Figure 2.6
LNA1	Mini-Circuits ZX60-272LN-S+ LNA, gain = 14 dB, NF = 1.2 dB, IP1 = 18.5 dBm
MXR1	Mini-Circuits ZX05-43MH-S+, 13 dBm LO, RF-LO loss = 6.1 dB, IP1 = 9.1 dBm
Video amplifier	Figure 3.11

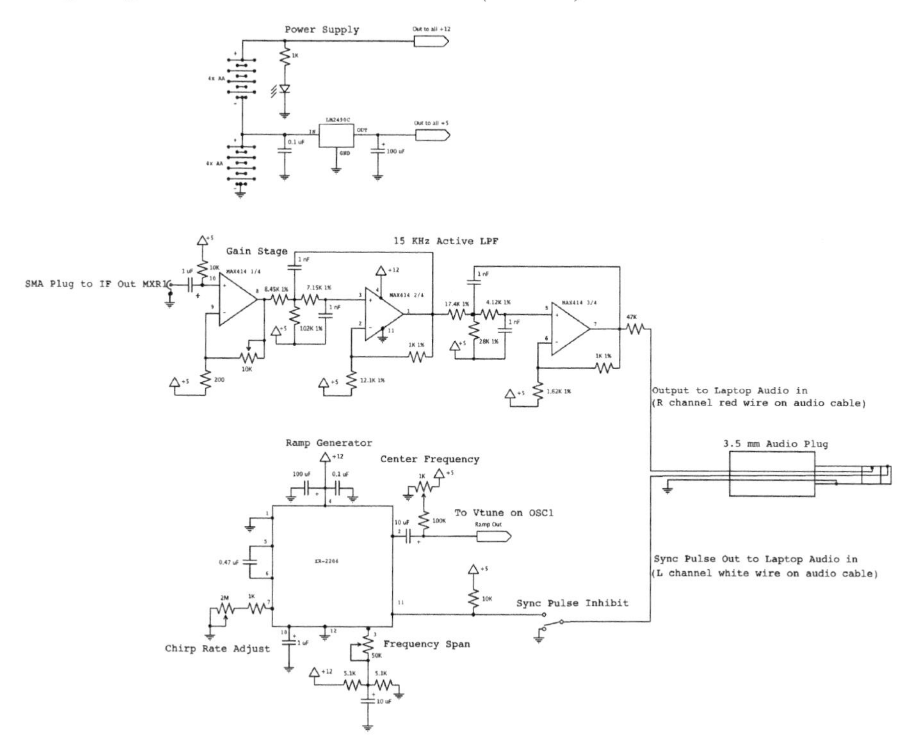

FIGURE 3.11
Analog signal chain of the coffee can radar in FMCW ranging mode showing the video output and trigger lines fed to the right and left audio input channels of a laptop computer.

The chirp output of OSC1 is attenuated by 3 dB through ATT1 and amplified by PA1. ATT1 serves to reduce the drive amplitude to PA1 so that PA1 is not saturated. The output of PA1 is fed through power splitter SPLTR1 which divides the transmit power between the transmit antenna ANT1 and the LO port of the mixer MXR1.

Transmit chirp is radiated out ANT1 toward the target scene. Scattered returns from the target scene are collected by the receive antenna ANT2. LNA1 amplifies the received signal and feeds the RF port of MXR1. The transmit chirp is multiplied by the receive chirp in MXR1, de-chirping the received signal. The IF output of MXR1 is amplified and low-pass filtered (-3 dB cut-off at 15 KHz) then fed to the right channel of the audio input to a laptop computer.

Using the audio input to a laptop for digitization is a unique innovation that increases accessibility to radar technology by eliminating costs of data acquisition and real-time processing. Schematic of this analog circuitry is shown (Fig. 3.11). For the start of every up-ramp of the sawtooth output from Modulator1, a synchronization pulse is fed out to the left audio input to a laptop computer.

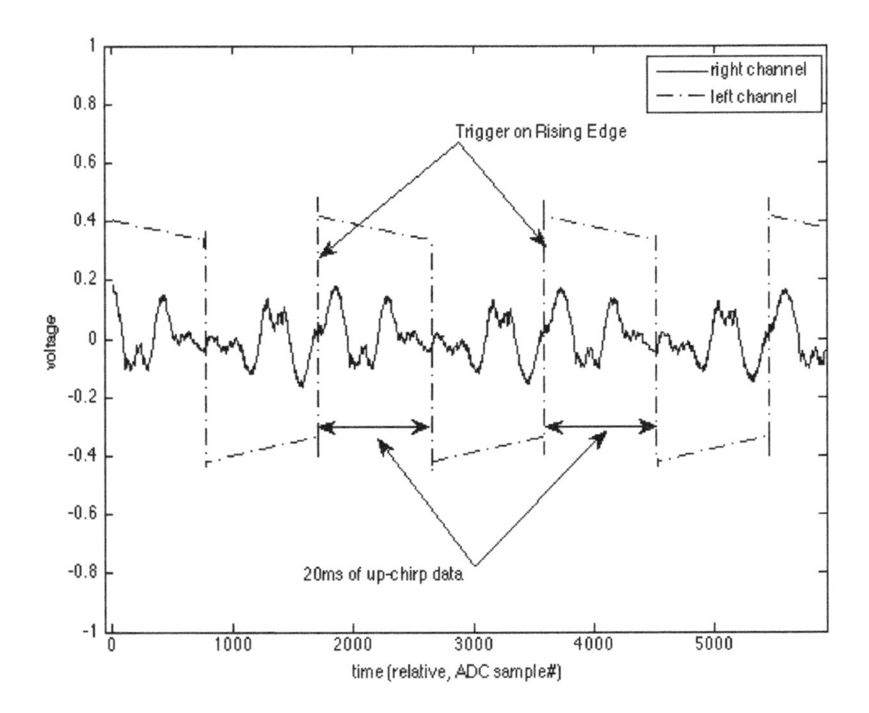

FIGURE 3.12
The output of the coffee can radar is digitized by both the right and left audio input channels of a laptop computer, where the right channel contains the video output and the left channel is a synchronization pulse where the rising edges indicate the start of an up-chirp.

The video output is fed to the right audio input channel of a laptop. A wave recorder is used to acquire uncompressed right and left channels. A MATLAB script then searches for rising edges on the left channel then creates an array of 20 ms groups of samples from the right channel that coincide with the rising edge (Fig. 3.12). This MATLAB script applies the IDFT to each of the 20 ms range profiles and displays a DTI plot with and without CCD [23].

3.3.2.1 Expected Performance of the MIT Coffee Can Radar System in Ranging Mode

The antenna gains G_{rx} and G_{tx} were measured [25] and the IDFT was applied to 20 ms intervals of data. Substituting specifications for coffee can FMCW radar (Table 3.5) into the radar range equation (3.5), the maximum range is

TABLE 3.5
System specifications for the MIT coffee can radar in FMCW ranging mode.

$$
\begin{aligned}
P_{ave} &= 10^{-3} \text{ (watts)} \\
G_{tx} &= 7.2 \text{ dBi measured antenna gain} \\
G_{rx} &= 7.2 \text{ dBi measured antenna gain} \\
A_{rx} &= G_{rx}\lambda_c^2/(4\pi) \text{ (m}^2\text{)} \\
\lambda_c &= c/f_c \text{ (m) wavelength of carrier frequency} \\
f_c &= 2.4 \text{ GHz center frequency of radar} \\
\rho_{rx} &= 1 \text{ because antenna efficiency is accounted for in measured antenna gain} \\
\sigma &= 10 \text{ (m}^2\text{) for target of interest} \\
L_s &= 6 \text{ dB miscellaneous system losses} \\
\alpha &= 0 \text{ attenuation constant of propagation medium} \\
F_n &= 1.2 \text{ dB receiver noise figure} \\
B_n &= 2/t_{sample} \text{ system noise bandwidth (Hz) where } t_{sample} = 20 \text{ ms} \\
(SNR)_1 &= 13.4 \text{ dB}
\end{aligned}
$$

estimated to be 829 m for a 10 dBsm automobile target using the MATLAB script [20].

3.3.2.2 Working Example of the MIT Coffee Can Radar System in Ranging Mode

In this experiment the coffee can radar is configured for FMCW ranging and the author serves as a moving target, walking down range and back, while data is recorded (video of this demo can be viewed [26]). A range profile is computed for each pulse and plotted with and without two-pulse CCD. Data from this experiment and a MATLAB script for processing can be found [27].

The location of the author is clearly shown (Fig. 3.13). Stationary (non-moving) clutter is also present in these results causing the return of the author to be 20 dB above the noise and clutter floor.

An RTI plot is made by displaying all range profiles (Fig. 3.13). The range to target is clearly shown until the target is at 30 m where the clutter begins to dominate all scattered returns. The clutter is stationary because it is persistent from pulse to pulse as indicated by the vertical streaks.

CCD is applied to this data, where the current pulse is subtracted from the previous pulse to reduce stationary clutter (Fig. 3.15). This has a significant effect on the results, where the clutter is greatly attenuated and the target return is now approximately 30 dB above the noise and clutter floor.

An RTI plot is shown using CCD range profiles (Fig. 3.16) and the human target is clearly shown walking out to 50 m because of significant reduction of stationary clutter amplitude.

A more interesting example is shown when two human targets walk then run down range and back (Fig. 3.17). CCD processing is also applied to this

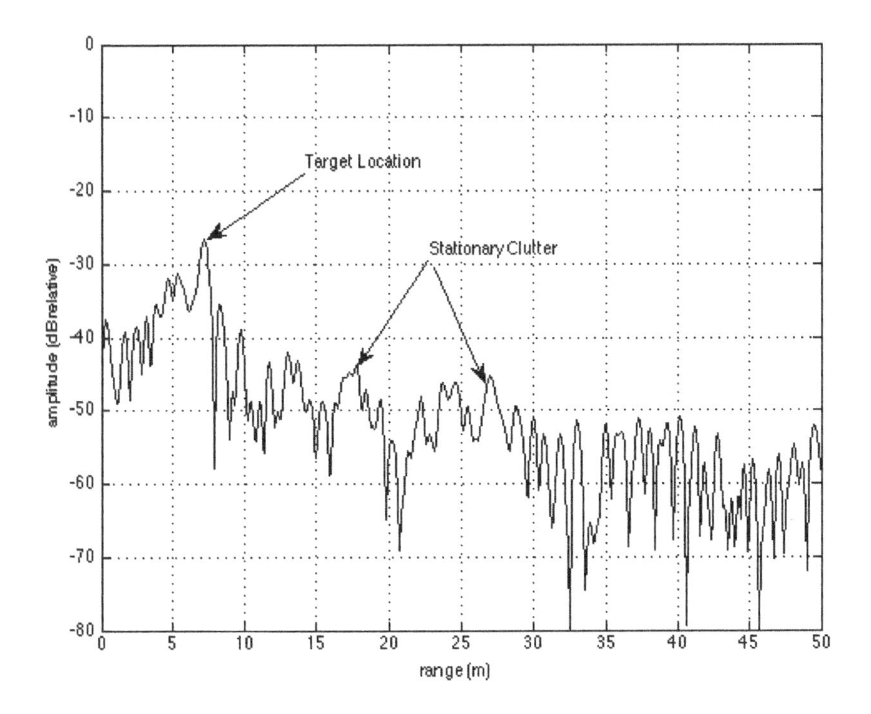

FIGURE 3.13
Range profile of one human target walking down range using the coffee can radar in FMCW ranging mode; the position of this target is shown along with stationary clutter.

data, thereby reducing the clutter and showing targets past 70 m in range. This data is provided [27].

3.3.3 Range-Gated UWB FMCW Radar System

Range gating is a method by which a radar system ignores (gates out) scattered returns before and after a specified time duration. For example, if the radar designer wants to know the location of a targets that are 20 to 30 ft down range but does not want to know if anything is present from 0 to 20 ft or beyond 30 ft. It could be that intense clutter exists from 0 to 20 ft and anything beyond 30 ft would not provide actionable information.

Range gating for FMCW radar systems is difficult. A stepped CW pulsed radar is often used which includes a frequency synthesizer that steps through known frequencies, thereby making a discrete up-chirp. For each frequency step the radar is pulsed. Scattered pulses are collected for each step where the

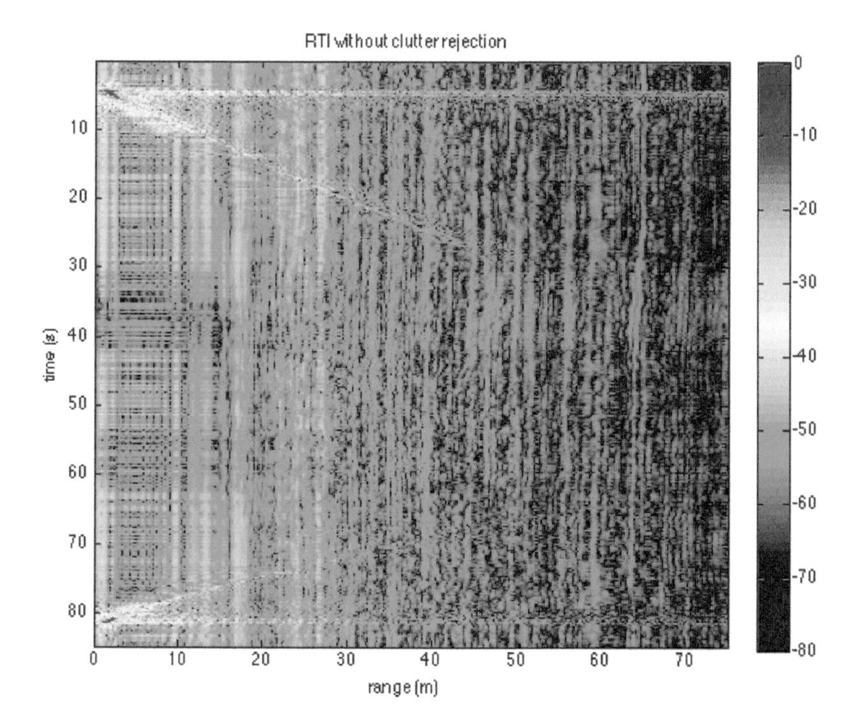

FIGURE 3.14
RTI plot of one human target walking down range and back where stationary clutter dominates returns past 30 m.

up-chirp can be reconstructed on the receive side and processed in a similar fashion to FMCW by the use of Fourier analysis. These systems are effective but expensive.

An additional method includes pulsing the input and output of an FMCW radar system. This requires the use of precision digital control and a reduction of average transmit power proportional to both the duty cycle of the transmitter and receiver gate.

In this section a low-cost range gate for UWB FMCW radar is shown. This is implemented by using high-Q bandpass filters passing only spatial frequencies corresponding to a desired range swath, providing a range gate without pulsing the radar. This range gate is valuable for through-wall radar imaging [28]-[35], where it facilitates the use of long duration FMCW chirps so that a low peak transmit power can be used while gating out the air-wall response. This technology could also be useful in radar imaging small target scenes in highly cluttered environments.

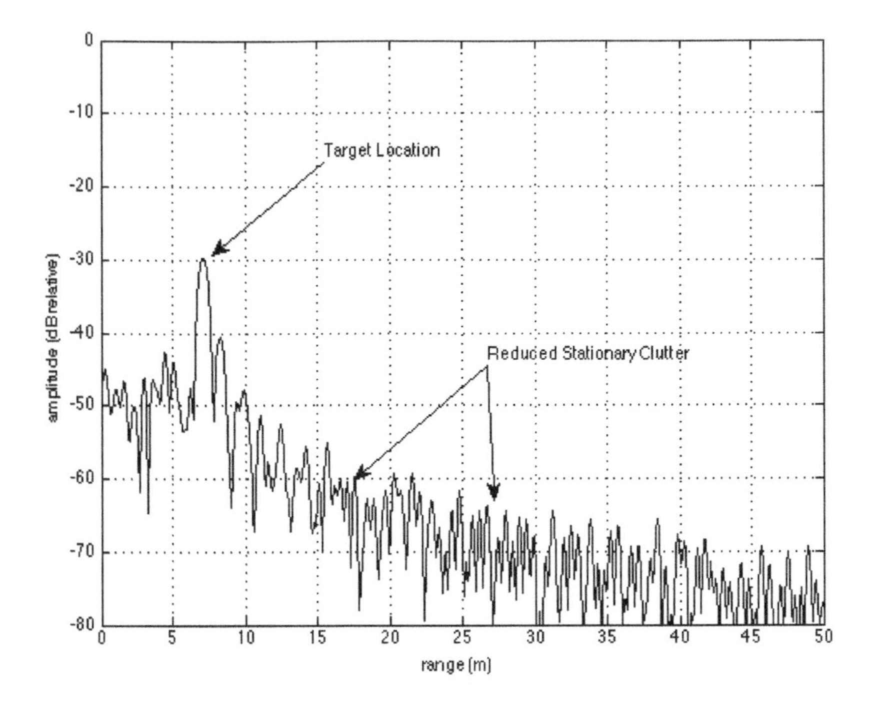

FIGURE 3.15
Range profile using CCD of one human target walking down range using the coffee can radar in FMCW ranging mode; the position of this target is much more easily shown and the clutter return has been significantly reduced.

3.3.3.1 Analog Range Gate

A range gate can be implemented by placing a bandpass filter (BPF) on the output of a video amplifier (Fig. 3.1), thereby passing only those beat frequencies representing a range swath corresponding to the pass band of filter. Unfortunately it is difficult to design high-Q bandpass filters at base band. Higher performing BPF's are available in the form of IF communications filters that operate at high frequencies.

These filters are found in two way radios, TV sets, and radio receivers. Examples include crystal, ceramic, surface acoustic wave, and mechanical filters operating at frequencies of 10.7 MHz, 21.4 MHz, 455 KHz, 49 MHz etc. These communications filters are high Q, where $Q = f_c/B$, f_c = center frequency of the filter, and $B = -3$ dB bandwidth of the filter [36]. A common operating frequency of a crystal filter is $f_c = 10.7$ MHz with a bandwidth of $B = 7.5$ KHz, where the result would be $Q = 1427$.

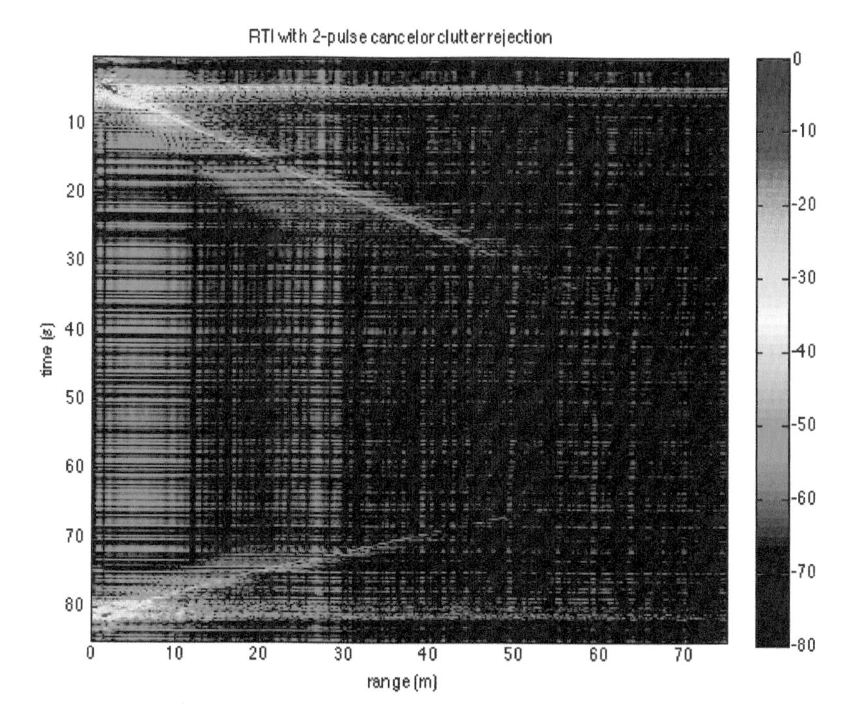

FIGURE 3.16
RTI plot using CCD of one human target walking down range and back where stationary clutter is significantly reduced and the target is now detectable out to 60 m.

A simplified block diagram is shown (Fig. 3.18). OSC1 is a high frequency tunable oscillator set to just above or below the IF filter's center frequency. The frequency output of OSC1 is f_{BFO} represented by (amplitude coefficients will be ignored in the following derivation)

$$BFO(t) = \cos\big(2\pi f_{BFO} t\big). \tag{3.7}$$

The output of OSC1 is fed into the IF port of MXR1. The LO port of MXR1 is driven by OSC2. OSC2 is a wide-band voltage tuned YiG oscillator tunable across a wide range of microwave frequencies. OSC2 is FM modulated by a linear ramp input, where the output of OSC2 can be represented by

$$LO(t) = \cos\big(2\pi(\text{fosc} + c_r t)t\big). \tag{3.8}$$

OSC1 and OSC2 are mixed together in MXR1 to produce the transmit signal which is then amplified by power amplifier PA1. The output of PA1 is fed into the transmit antenna ANT1 and radiated toward the target scene. The

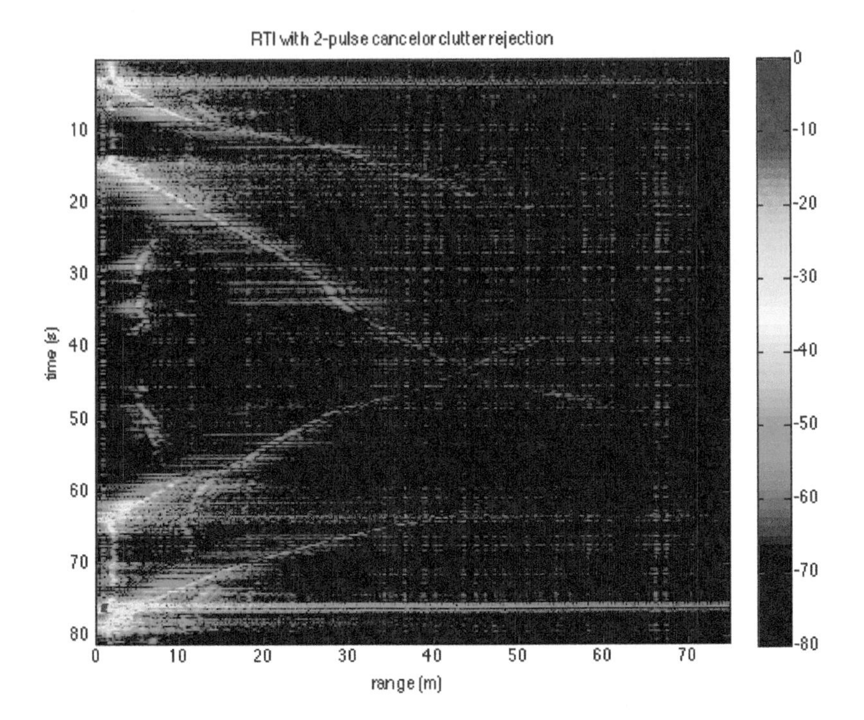

FIGURE 3.17
RTI plot using CCD of two human targets shown clearly out to 70 m.

transmitted signal out of ANT1 is $TX(t)$, where

$$TX(t) = LO(t) \cdot BFO(t),$$

$$TX(t) = \cos\big(2\pi(\text{fosc} + c_r t)t\big) \cdot \cos\big(2\pi f_{BFO} t\big).$$

After some simplification this becomes

$$TX(t) = \cos\big(2\pi(\text{fosc} + c_r t)t + 2\pi f_{BFO} t\big) + \cos\big(2\pi(\text{fosc} + c_r t)t - 2\pi f_{BFO} t\big). \tag{3.9}$$

The transmitted waveform $TX(t)$ consists of two tones spaced in frequency by $2f_{BFO}$ that are linear FM modulated. This is radiated out to the target scene, reflected off of a target, delayed by some round trip time t_{delay} and propagated back to the receiver antenna ANT2. The received signal at ANT2 is represented by

$$\begin{aligned} RX(t) = {} & \cos\big(2\pi(\text{fosc} + c_r t)(t - t_{delay}) + 2\pi f_{BFO}(t - t_{delay})\big) \\ & + \cos\big(2\pi(\text{fosc} + c_r t)(t - t_{delay}) - 2\pi f_{BFO}(t - t_{delay})\big). \end{aligned} \tag{3.10}$$

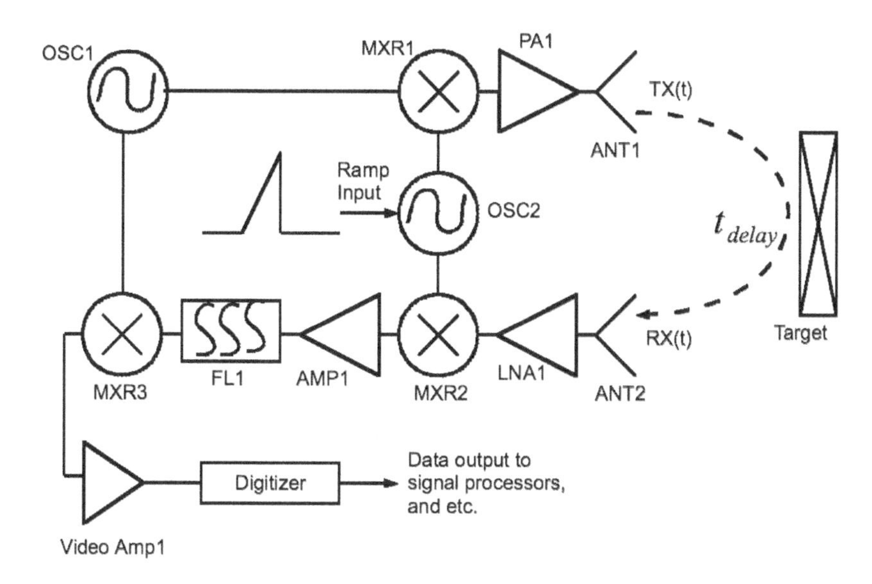

FIGURE 3.18
Block diagram of the FMCW radar architecture that supports an analog range gate.

The output of ANT2 is amplified by LNA1 and fed into MXR2. The LO port of MXR2 is fed by OSC2. The IF output of MXR2 is the product

$$IF(t) = LO(t) \cdot RX(t).$$

The resulting product

$$\begin{aligned} IF(t) &= \\ &\cos\left(2\pi(\text{fosc} + c_r t)t\right) \cdot \cos\left(2\pi(\text{fosc} + c_r t)(t - t_{delay}) + 2\pi f_{BFO}(t - t_{delay})\right) \\ &+\cos\left(2\pi(\text{fosc} + c_r t)t\right) \cdot \cos\left(2\pi(\text{fosc} + c_r t)(t - t_{delay}) - 2\pi f_{BFO}(t - t_{delay})\right). \end{aligned} \tag{3.11}$$

Multiplying out the terms in the above equation expands into

$$\begin{aligned} IF(t) &= \\ &\cos\left(2\pi(\text{fosc} + c_r t)(t - t_{delay}) + 2\pi f_{BFO}(t - t_{delay}) + 2\pi(\text{fosc} + c_r t)t\right) \\ &+ \cos\left(2\pi(\text{fosc} + c_r t)(t - t_{delay}) + 2\pi f_{BFO}(t - t_{delay}) - 2\pi(\text{fosc} + c_r t)t\right) \\ &+ \cos\left(2\pi(\text{fosc} + c_r t)(t - t_{delay}) - 2\pi f_{BFO}(t - t_{delay}) + 2\pi(\text{fosc} + c_r t)t\right) \\ &+ \cos\left(2\pi(\text{fosc} + c_r t)(t - t_{delay}) - 2\pi f_{BFO}(t - t_{delay}) - 2\pi(\text{fosc} + c_r t)t\right). \end{aligned} \tag{3.12}$$

As a practical consideration the IF port of MXR2 cannot output microwave frequencies so the high frequency terms can be dropped,

$$\begin{aligned}IF(t) &= \\ &\cos\big(2\pi(\text{fosc}+c_r t)(t-t_{delay})+2\pi f_{BFO}(t-t_{delay})-2\pi(\text{fosc}+c_r t)t\big) \\ &+\cos\big(2\pi(\text{fosc}+c_r t)(t-t_{delay})-2\pi f_{BFO}(t-t_{delay})-2\pi(\text{fosc}+c_r t)t\big).\end{aligned} \tag{3.13}$$

Expanding out the terms inside of the cosine argument,

$$\begin{aligned}IF(t) = \cos\Big[&2\pi(\text{fosc}+c_r t)t-2\pi(\text{fosc}+c_r t)t_{delay} \\ &+2\pi f_{BFO}(t-t_{delay})-2\pi(\text{fosc}+c_r t)t\Big] \\ +\cos\Big[&(2\pi(\text{fosc}+c_r t)t-2\pi(\text{fosc}+c_r t)t_{delay} \\ &-2\pi f_{BFO}(t-t_{delay})-2\pi(\text{fosc}+c_r t)t\Big].\end{aligned} \tag{3.14}$$

Letting the high frequency terms cancel out,

$$\begin{aligned}IF(t) = \cos\Big[&-2\pi(\text{fosc}+c_r t)t_{delay}+2\pi f_{BFO}(t-t_{delay})\Big] \\ +\cos\Big[&-2\pi(\text{fosc}+c_r t)t_{delay}-2\pi f_{BFO}(t-t_{delay})\Big].\end{aligned} \tag{3.15}$$

As another practical consideration, the DC blocking capacitors in the IF amplifier AMP1 will reject the DC phase terms,

$$IF(t) = \cos\big(-2\pi c_r t t_{delay}+2\pi f_{BFO}t\big)+\cos\big(-2\pi c_r t t_{delay}-2\pi f_{BFO}t\big).$$

Simplifying arguments in the cosine terms,

$$IF(t) = \cos\big(2\pi(f_{BFO}-c_r t_{delay})t\big)+\cos\big(2\pi(f_{BFO}+c_r t_{delay})t\big). \tag{3.16}$$

$IF(t)$ is fed into the high Q IF filter FL1. FL1 has a center frequency of f_c and a bandwidth of BW. OSC1 is set to a frequency such that $f_{BFO} \geq \frac{BW}{2}+f_c$ causing FL1 to pass only the lower sideband of $IF(t)$. The output of FL1 becomes

$$FIL(t) = \begin{cases}\cos\big(2\pi(f_{BFO}-c_r t_{delay})t\big) & \text{if } \frac{-BW}{2}+f_c < f_{BFO}-c_r t_{delay} < \frac{BW}{2}+f_c \\ 0 & \text{for all other values}\end{cases}. \tag{3.17}$$

Only beat frequencies in the range of $\frac{-BW}{2}+f_c < f_{BFO}-c_r t_{delay} < \frac{BW}{2}+$

f_c are passed through IF filter FL1; therefore the band limited IF signal is effectively a range-gate, passing only those spatial frequencies corresponding to round-trip time delays within a specific range set by FL1.

Increasing the bandwidth of FL1 increases the range-gate duration. Decreasing the bandwidth of FL1 decreases the range-gate duration. As f_{BFO} is increased the range gate is moved further down range because FL1 passes only signals that fit the equality, making the gate adjustable in range.

The output of FL1 is downconverted to base band through MXR3 where the LO port of MXR3 is driven by OSC1. The IF output of MXR3 is fed through Video Amp1 and can be represented by the equation:

$$Video(t) = BFO(t) \cdot FIL(t).$$

Video Amp1 is an active low pass filter, rejecting the higher frequency component of the cosine multiplication,

$$Video(t) = \begin{cases} \cos\left(2\pi c_r t_{delay} t\right) & \text{if } \frac{-BW}{2} + f_c - f_{BFO} < c_r t_{delay} < \frac{BW}{2} + f_c - f_{BFO} \\ 0 & \text{for all other values} \end{cases}. \tag{3.18}$$

The result is a base-band FMCW video signal similar to (except without the DC phase term) Equation (3.2).

For example, when applied to through-wall imaging, this range-gate circuit is used to significantly attenuate the spatial frequency response that corresponds to the wall, thereby easing the dynamic range requirements on the digitizer allowing for the use of low-cost commercial digitizers. The measured losses through solid concrete alone typically exceed the maximum dynamic range available from most digitizers of reasonable cost. For example, a 10 cm thick solid concrete slab provides 45 dB and a 20 cm thick solid concrete slab provides 90 dB two-way path loss at 3 GHz [30]; additional materials and free space path loss further increase attenuation.

One of the range-gate filters used in this book was measured (Fig. 3.19), this one having a bandwidth of 7.5 KHz. Using the through-wall model to be discussed later in Sec. 10.2.2, a range profile of a 10 cm thick solid concrete wall 6 m and a 7.6 cm radius perfect electric conductor cylinder at a range of 6 m and 9.1 m was simulated at S-band (1.926 to 4.069 GHz chirp). The frequency response of the IF filter in time domain was shifted thereby placing the wall in the stop band (Fig. 3.20). The product of the filter and time domain response show that the magnitude of the cylinder is significantly increased compared to the wall, thus requiring less dynamic range for a through-wall measurement (Fig. 3.21). Multi-path response between the back-side of the wall and cylinder is also increased relative to the wall. The IF filter plays the major role in reducing the wall reflection, in this case, reducing the wall reflection by approximately 55 dB.

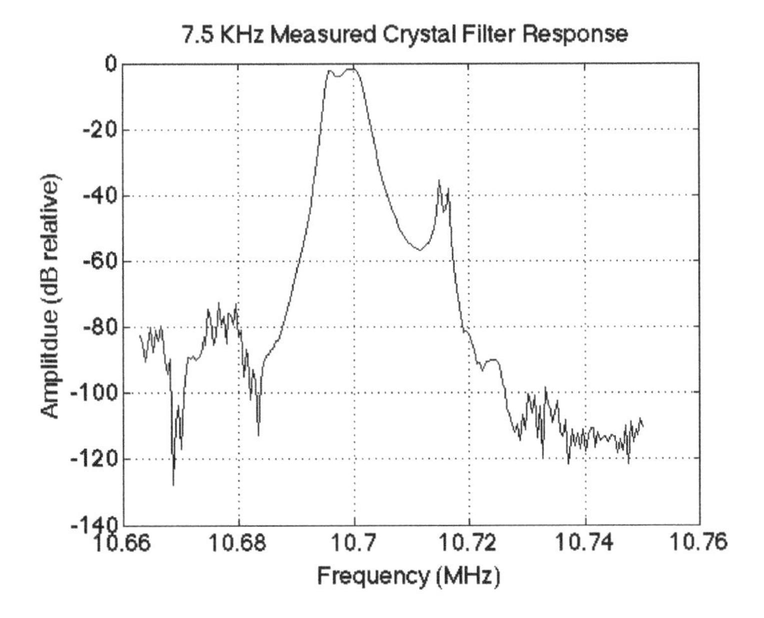

FIGURE 3.19
Measured S21 magnitude response of the crystal filter used for the spatial frequency range gate, having a center frequency of 10.7 MHz and a bandwidth of 7.5 KHz.

In summary, this range-gating technique facilitates the use of long duration LFM waveforms, thereby providing increasing average transmit power to be radiated while using a low peak-power (and therefore lower cost) transmitter. In addition to this, the long duration LFM waveform eases data acquisition requirements so that inexpensive digitizers can be used. Specific examples of three different through-wall imaging systems using this architecture will be shown in Chapter 10.

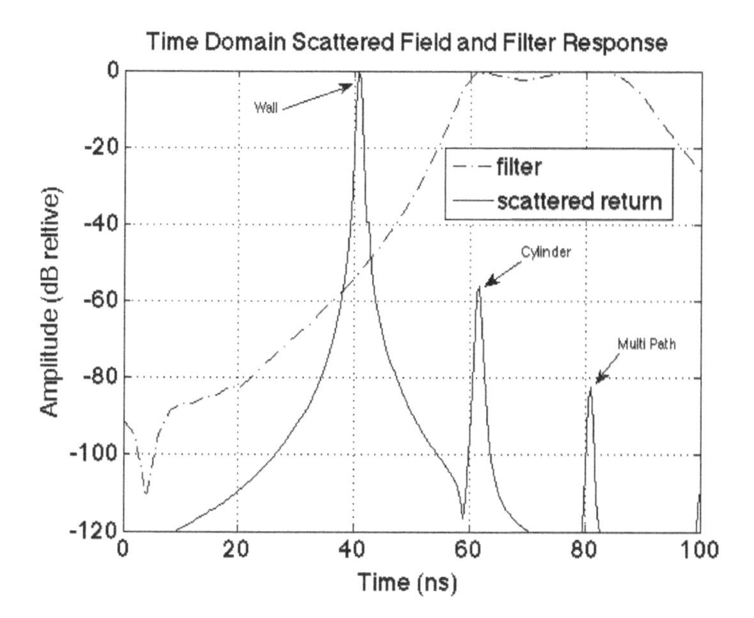

FIGURE 3.20
Time domain simulated scattered field and spatial frequency filter response.

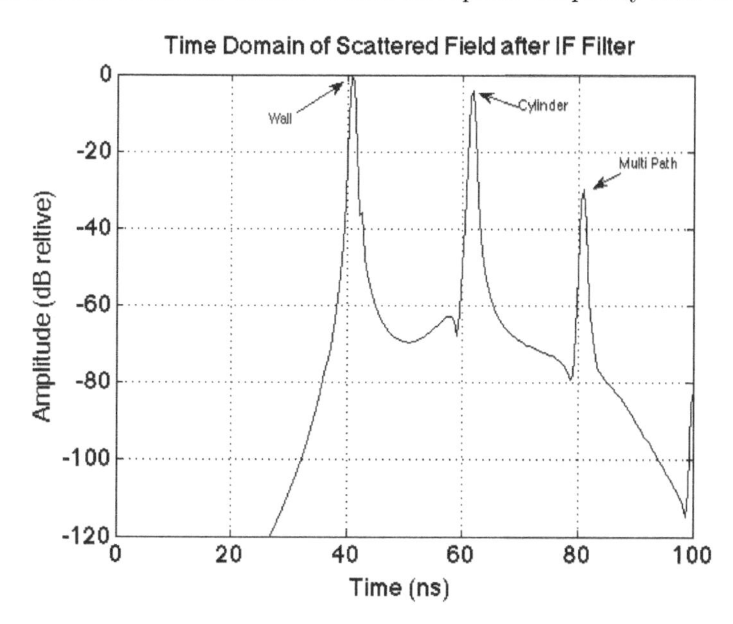

FIGURE 3.21
Simulated time domain response after spatial frequency crystal filter response is applied.

3.3.3.2 S-Band Implementation

Range-gated FMCW architecture is ideal for through-wall imaging, where the large reflection from the wall can be greatly reduced and only spatial frequencies corresponding to targets behind the wall pass onto the digitizer. For this reason an S-band system was developed for use in a through-wall radar application [31].

A complete block diagram of this system (Fig. 3.22) and callout diagram of all major subsystems (Fig. 3.23) are shown. A Windows XP computer controls the radar system through a National Instruments PCI6014 data acquisition card, where a VI is programmed to modulate the YiG oscillator while digitizing (200 KSPS) simultaneously. Unconnected to this, a frequency synthesizer serves as the BFO.

The CTRO0 pin from the PCI6014 causes the ramp generator (Fig. 3.5) to produce an up-ramp from 1.926 to 4.069 GHz which is fed into the voltage tune input of the YiG oscillator. The output of the YiG oscillator is split between the transmitter and receiver front ends.

The transmitter front end multiplies the output from the YiG oscillator by the BFO to produce the two-tone transmit waveform described previously. Half of the output from the splitter is fed into the LO port of MXR1 in the transmit front end (photos and callouts in Fig. 3.24) where it is multiplied by the BFO within MXR1. This product is amplified by AMP1 and low-pass filtered by FL2 to be radiated out toward the target scene by antenna ANT1. Some power is coupled off through CLPR2 for diagnostic purposes.

A simple directional UWB antenna was developed which was easy to fabricate and provided sufficient gain and bandwidth for UWB operation. The transmit and receive antennas (ANT1 and ANT2) are identical linear tapered slot antennas [38]–[44] following the layout on FR4 substrate shown (Fig. 3.25). This antenna has better than -10 dB S11 from 2 to 4 GHz and an assumed gain of 12 dBi (the gain and antenna patterns were never measured for this antenna).

The receive front end multiplies the scattered signal by the YiG oscillator's output then feeds this product to the IF. The chirped waveform is radiated out of ANT1 toward the target scene. Scattered returns are collected by ANT2 which is co-located with the receiver front end (photos and call-outs shown in Fig. 3.26), filtered by low-pass filter FL3, amplified by low noise amplifier LNA1, and band-pass filtered by FL4 and 5. The result is multiplied by chirp from the YiG OSC within MXR2. The product of MXR2 is filtered by low-pass filter FL1 and amplified by AMP2 then fed down to the IF chassis.

The IF chassis is where the output from the receiver front end is band limited (e.g., range gated) and shifted down to base band. The BFO is fed into the IF chassis, where some of it is coupled out through CLPR1, through a transmit IF attenuator ATTN3, located externally to the IF chassis, and out to the IF input of the transmitter front end.

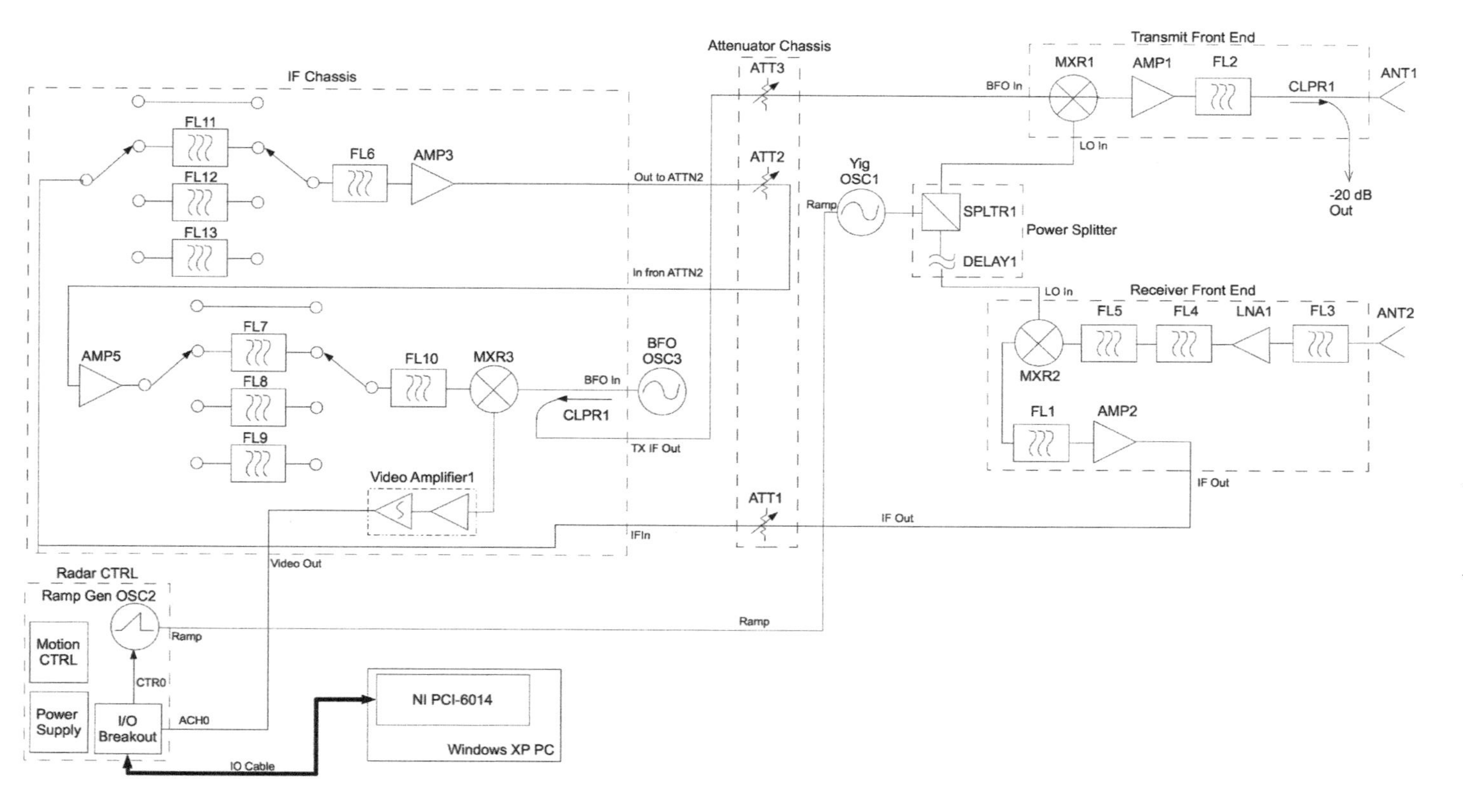

FIGURE 3.22
Block diagram of the S-band analog range-gated FMCW radar system.

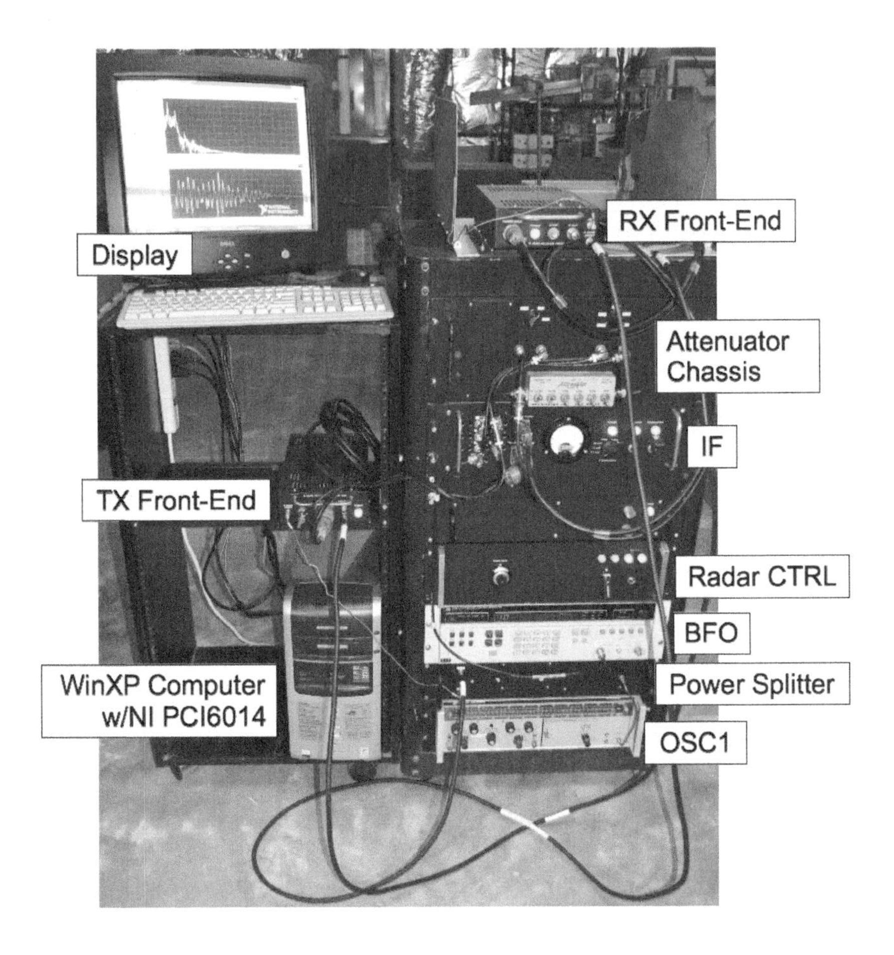

FIGURE 3.23
Call-out diagram of the range-gated FMCW radar system configured for S-band operation.

The IF output of the receive front end is fed into the IF chassis where it is filtered by one of the crystal filters (FL11 through 13) at a center frequency of 10.7 MHz having bandwidths of 7.5 KHz, 15 KHz, and 30 KHz respectively. Filters are switched in and out of the filter bank by the use of PiN diode switches. The output of this filter bank is fed through FL6 which is an 11 MHz low pass filter followed by amplifier AMP3. The output of AMP3 is fed out of the IF chassis to the step attenuator ATTN2, where the IF gain is manually controlled.

(a)

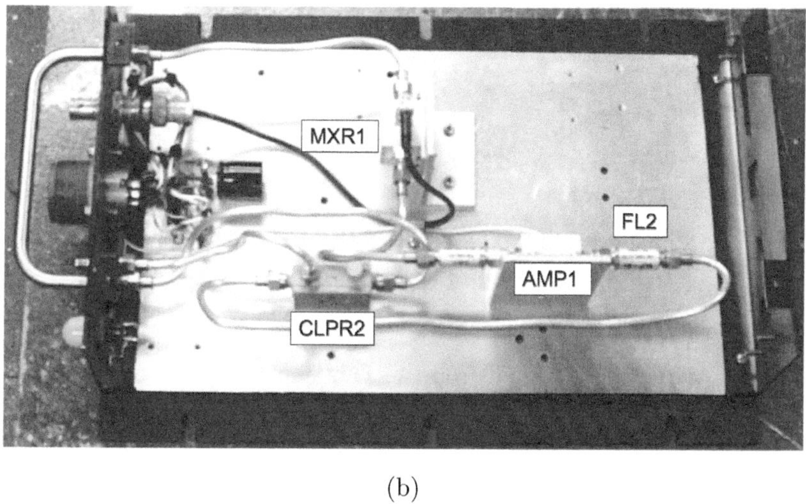

(b)

FIGURE 3.24
S-band transmitter front end (a) and inside view with call-outs (b).

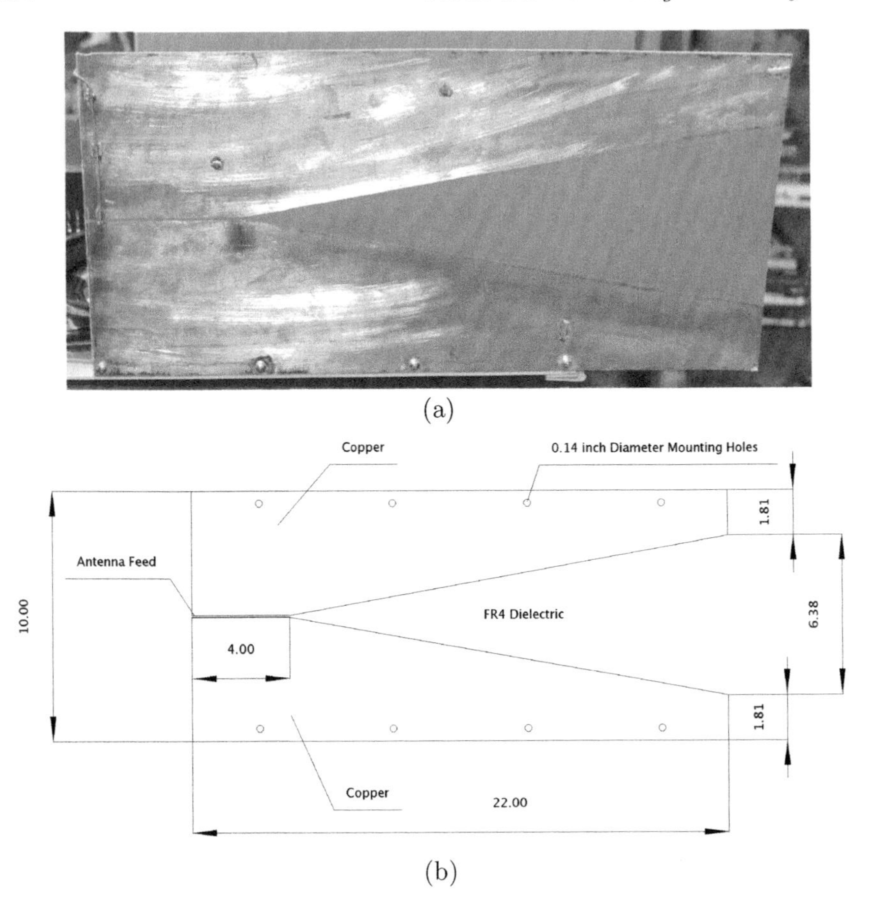

FIGURE 3.25
S-band antenna (a) and layout (b) where two are used, one for transmit and one for receive. Gap between upper and lower half of antenna is 0.125 inches.

AMP5 amplifies the signal from ATTN2 and feeds this to another filter bank containing crystal filters FL7 through 9 which are identical to FL11 through 13 and are also switched by PiN diode switches. The output of this filter bank is fed through FL10, which is another 11 MHz low pass filter. FL10 feeds the RF input of MXR3, which is a high dynamic range double balanced mixer.

Most of the BFO signal is fed to the LO port of MXR3. The product of the BFO and the radar IF result in a base-band video signal. This signal is amplified by VideoAmp1 and fed out to the PCI6014.

A complete bill of material for this radar system is provided (Tables 3.6 and 3.7) and additional schematics are shown (Figs. 3.28 and 3.29).

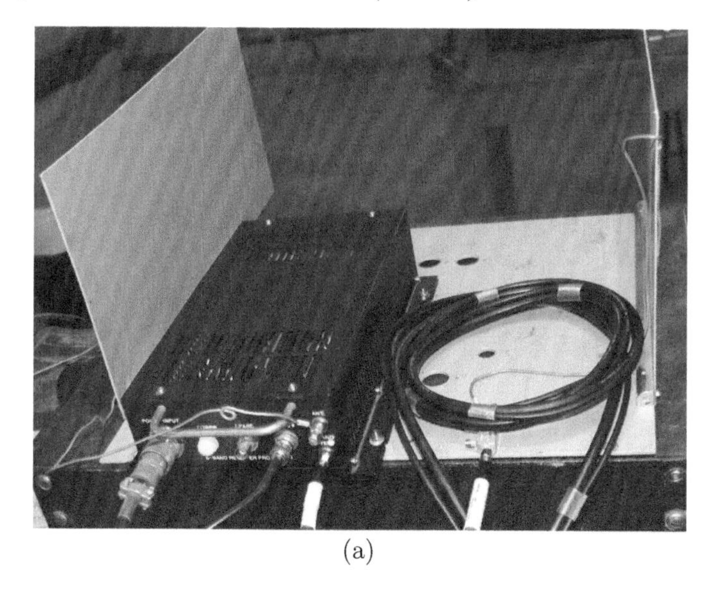

(a)

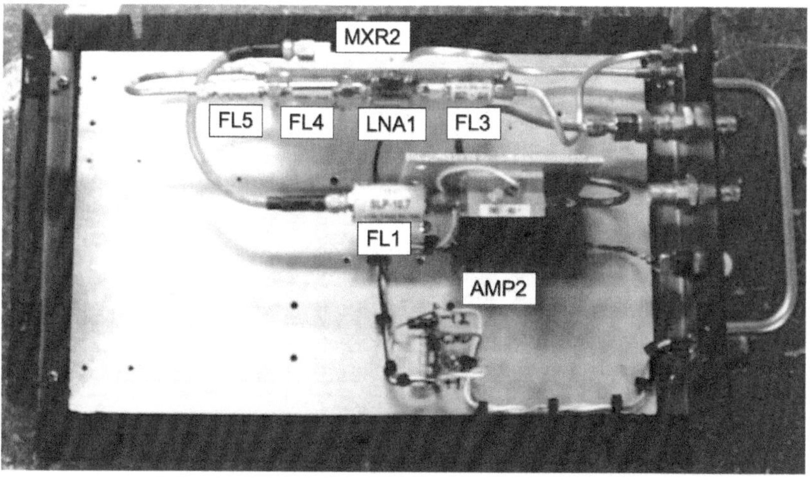

(b)

FIGURE 3.26
S-band receiver front end (a) and inside view with call-outs (b).

3.3.3.3 Expected Performance of the Range-Gated S-Band FMCW Radar System

Substituting specifications for the range-gated S-band FMCW radar (Table 3.8) into the radar range equation (3.5), the maximum range is estimated to be 1083 m for a 10 dBsm automobile target using the MATLAB script [20].

TABLE 3.6
S-band bill of material.

Component	Description
AMP1	Mini-Circuits ZJL-4G, gain = 11 dB, IP1 = 12 dBm, NF = 5.5 dB
AMP2	Mini-Circuits ZFL-1000VH, gain = 20 dB, IP1 = 25 dBm, NF = 4.5 dB
AMP3	Mini-Circuits ZKL-2R5, gain = 30 dB, IP1 = 15 dBm, NF = 5 dB
AMP5	Mini-Circuits ZHL-6A, gain = 25 dB, IP1 = 22 dBm, NF = 9.5 dB
ANT1	Linear tapered slot antenna, vertically polarized [38]-[44]
ANT2	Linear tapered slot antenna, vertically polarized [38]-[44]
ATTN1	Surplus 3-position attenuator: thru/-30 dB/load
ATTN2	Kay Model 20 adjustable attenuator
ATTN3	Surplus rotary attenuator
Beat frequency oscillator	Hewlett Packard HP3325A Synthesizer/Function Generator
CLPR1	Mini-Circuits ZX30-12-4, 5 MHz to 1 GHz, -12 dB directional coupler
CLPR2	Midwest Microwave, 2 GHz to 4 GHz, -20 dB directional coupler
Delay line	Miscellaneous microwave coaxial cables
FL1	Mini-Circuits SLP-10.7, 11 MHz LPF
FL2	Mini-Circuits VLP-41, 4.1 GHz LPF
FL3	Mini-Circuits VHF-1200, 1200 MHz HPF
FL4	Mini-Circuits VHF-1200, 1200 MHz HPF
FL5	Mini-Circuits VLP-41, 4.1 GHz LPF
FL6	Mini-Circuits PLP-10.7, 11 MHz LPF (Fig. 3.28)
FL7	ECS-10.7-7.5B, 4 pole crystal filter, $f_c = 10.7$ MHz, $B = 7.5$ KHz (Fig. 3.28)
FL8	ECS-10.7-15B, 4 pole crystal filter, $f_c = 10.7$ MHz, $B = 15$ KHz (Fig. 3.28)
FL9	ECS-10.7-30B, 4 pole crystal filter, $f_c = 10.7$ MHz, $B = 30$ KHz (Fig. 3.28)

TABLE 3.7
S-band bill of material (continued).

Component	Description
FL10	Mini-Circuits PLP-10.7, 11 MHz LPF (Fig. 3.28)
FL11	ECS-10.7-7.5B, 4 pole crystal filter, $f_c = 10.7$ MHz, $B = 7.5$ KHz (Fig. 3.28)
FL12	ECS-10.7-15B, 4 pole crystal filter, $f_c = 10.7$ MHz, $B = 15$ KHz (Fig. 3.28)
FL13	ECS-10.7-30B, 4 pole crystal filter, $f_c = 10.7$ MHz, $B = 30$ KHz (Fig. 3.28)
LNA1	Mini-Circuits ZX60-6013E, 20 MHz to 6 GHz, gain=14 dB, NF=3.3 dB
Motor controller	RMV SPRT232-ST stepper motor controller
MXR1	Mini-Circuits ZEM-4300MH, 300 MHz to 4300 MHz, Level 13 (+13 dBm LO)
MXR2	Mini-Circuits ZEM-4300MH, 300 MHz to 4300 MHz, Level 13 (+13 dBm LO)
MXR3	Mini-Circuits RAY-6U, Level 23 (+23 dBm LO)
OSC1	Weinschel Engineering 430A Sweep Oscillator and 432A RF Unit
PC	PC running Labview, all system control software was written in Labview
PCI-6014	National Instruments PCI-6014 data acquisition and IO card
Ramp generator	Wilson current mirror based linear ramp generator (Fig. 3.5)
SPLTR1	Mini-Circuits ZN2PD2-50-S, 500 MHz to 5 GHz, 2-way −3 dB splitter
VideoAmp1	Active LPF, based using the MAX414 op-amp (Fig. 3.29)
XTAL Filter Mux1	Pin diode selectable crystal filter mux (Fig. 3.28)
XTAL Filter Mux2	Pin diode selectable crystal filter mux (Fig. 3.28)

(a)

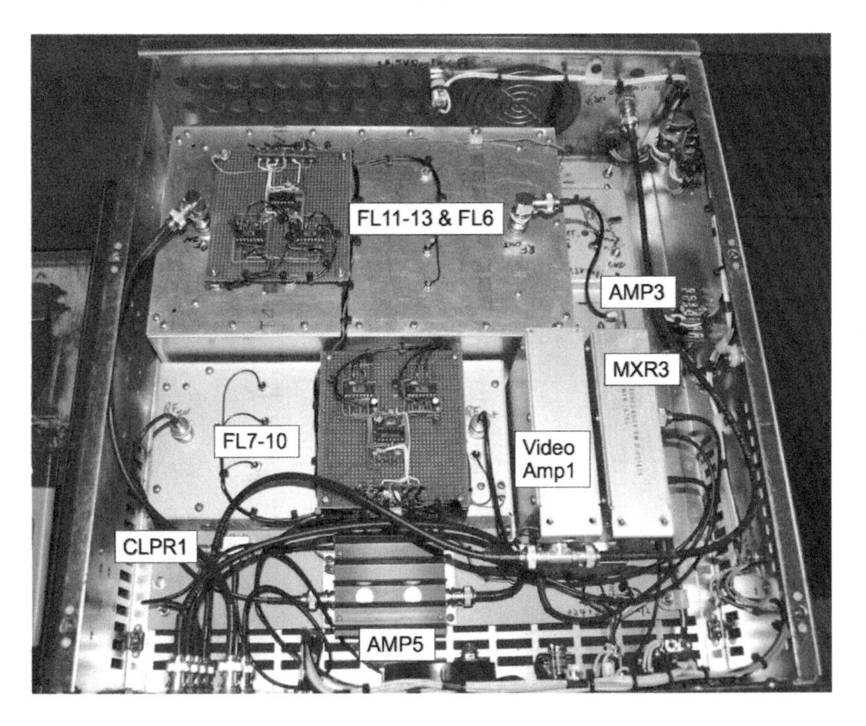

(b)

FIGURE 3.27
IF chassis front panel (a) and inside view with call-outs (b).

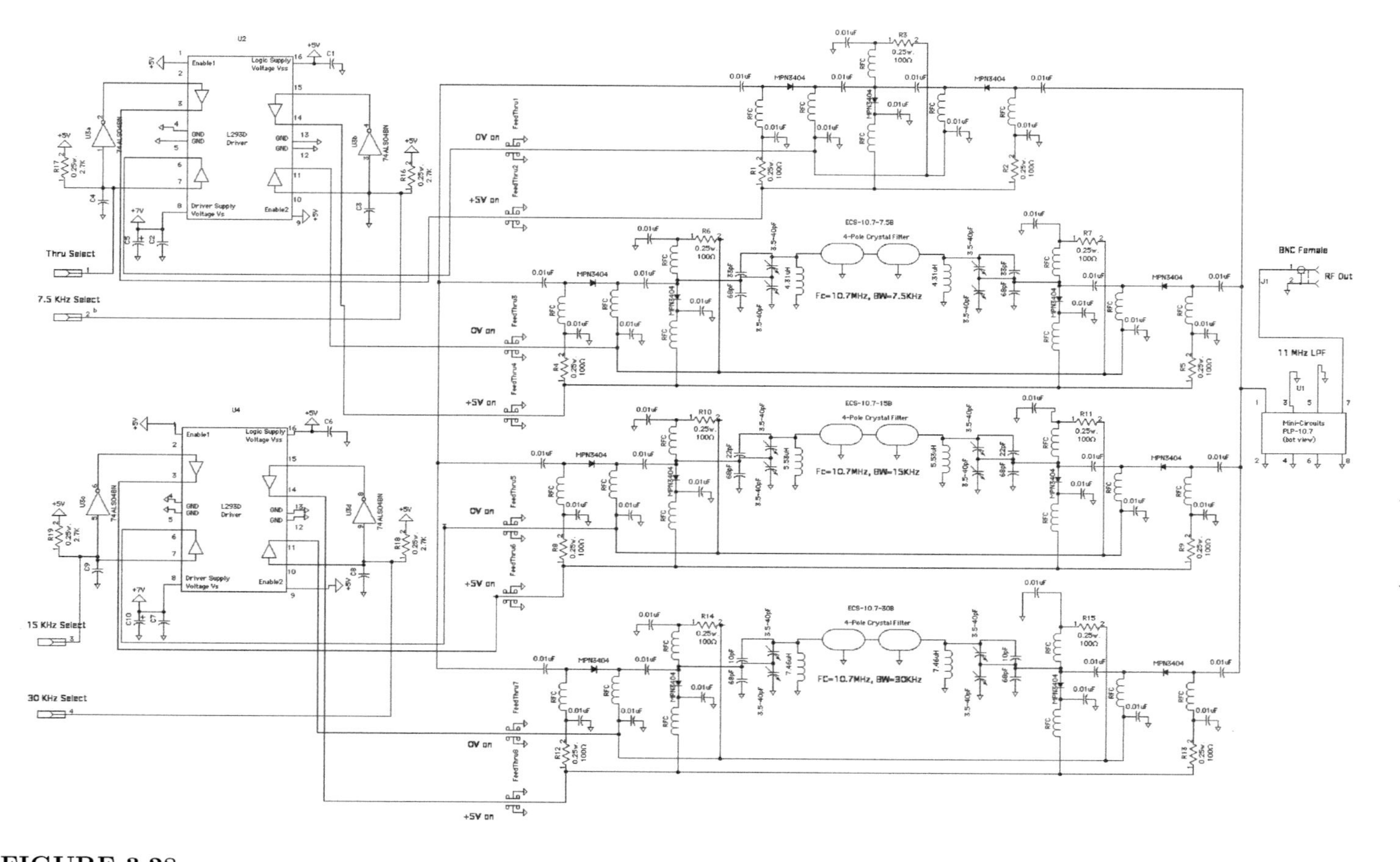

FIGURE 3.28
Schematic of IF filter switch matrix XTAL Filter Mux1 and XTAL Filter Mux2.

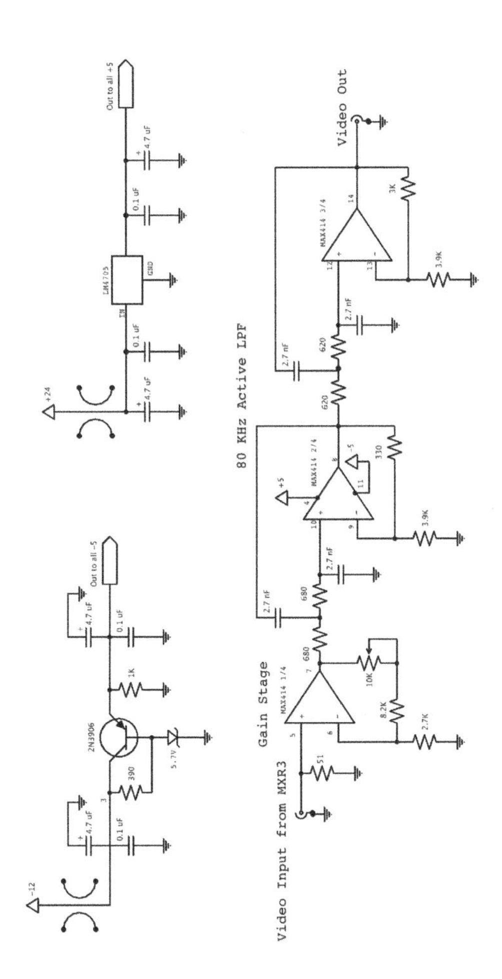

FIGURE 3.29
Schematic of the video amplifier VideoAmpl.

TABLE 3.8
X-band CW radar system radar range equation parameters.

$$
\begin{aligned}
P_{ave} &= 10 \cdot 10^{-3} \text{ (watts)} \\
G_{tx} &= 12 \text{ dBi antenna gain (estimated)} \\
G_{rx} &= 12 \text{ dBi antenna gain (estimated)} \\
A_{rx} &= G_{rx}\lambda_c^2/(4\pi) \text{ (m}^2\text{)} \\
\lambda_c &= c/f_c \text{ (m) wavelength of carrier frequency} \\
f_c &= 3 \text{ GHz center frequency of radar} \\
\rho_{rx} &= 1 \text{ because antenna efficiency is accounted for in antenna gain} \\
\sigma &= 10 \text{ (m}^2\text{) for automobile at 10 GHz} \\
L_s &= 6 \text{ dB miscellaneous system losses} \\
\alpha &= 0 \text{ attenuation constant of propagation medium} \\
F_n &= 3.5 \text{ dB receiver noise figure} \\
B_n &= 2/t_{sample} \text{ system noise bandwidth (Hz) where } t_{sample} = 10 \text{ ms} \\
(SNR)_1 &= 13.4 \text{ dB}
\end{aligned}
$$

The antenna gains G_{rx} and G_{tx} were an educated guess based on performance of similar antennas in the literature [38]–[44]. The IDFT is applied to 10 ms up-chirps but the front end is a single-conversion receiver providing an effective noise bandwidth to 200 Hz.

3.3.3.4 Working Example of the Range-Gated S-Band FMCW Radar System

In this experiment two human targets walk down range and back while the S-band radar acquires range profile data. Two different range gate filters were used, a 7.5 KHz and a 15 KHz filter. A video demo of these experiments is shown [45] and data is available with a MATLAB processing script [46].

We first consider the results when using the 7.5 KHz filter. The range profile of one human target is shown at time t = 20 s (Fig. 3.30). The range gate filter transfer function passes one portion of the range profile, greatly attenuating targets outside of its passband. The human target position is clearly shown 15 dB above clutter at 12.5 m. Clutter is relatively high compared to the target.

A RTI plot of the range profiles is shown (Fig. 3.31), where the down range location of the range gate filter is clearly shown by prominent red streaks between 10 and 15m. The returns of both human targets are strongest when they fall within the range gate filter. Stationary clutter within the range gate is high compared to target returns.

CCD processing is applied, where the current pulse is subtracted from the last (Sec. 3.1.3.3). The range profile at t = 20 s is plotted (Fig. 3.32). The transfer function of the range gate filter continues to be evident on this plot but stationary clutter is significantly attenuated and the target return is 27 dB above clutter.

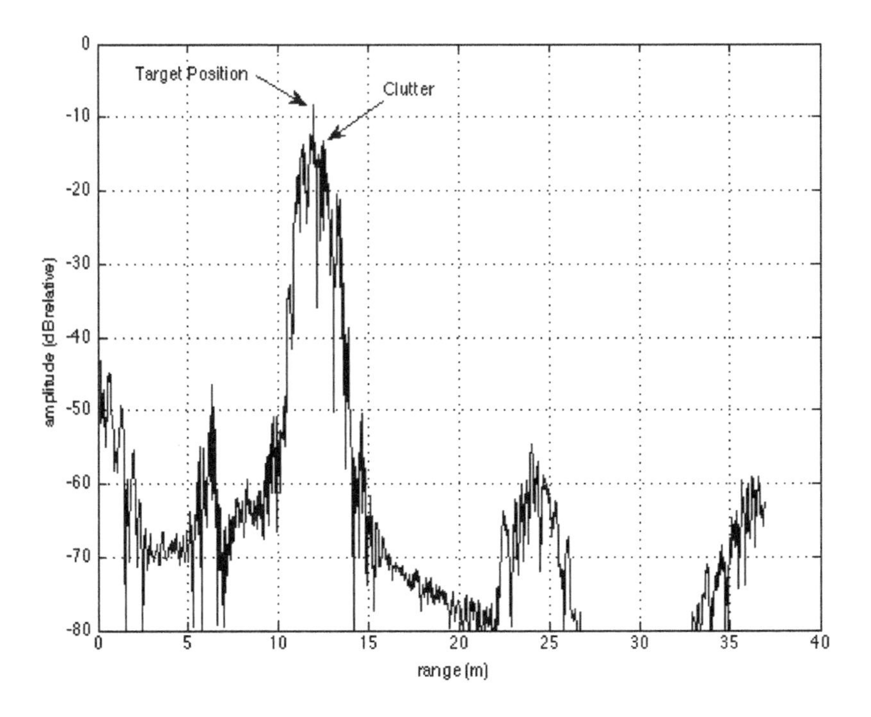

FIGURE 3.30
Range profile plot of one target walking down range using the range-gated S-band FMCW radar with a 7.5 KHz range gate filter.

An RTI plot of the CCD range profiles is shown (Fig. 3.33). The yellow streak between 10 and 15 m is the range gate filter transfer function passing what is left of stationary clutter. The amplitude of the human target returns relative to clutter passed through by the range gate filter is significantly higher, although stationary clutter is still shown passing through the filter. The positions of both targets are clearly shown within the range gate filter and significantly attenuated tracks of each target are shown before the filter and returns after the filter are all but removed from the data, showing the utility of this analog range-gating technique.

We now consider the results when using the 15 KHz range gate filter. Two human targets walked down range and back while range profiles were recorded. A single range profile is plotted at t = 33.3 s, where the transfer function of the range gate filter is clearly shown, passing and attenuating portions of the range profile. The position of both human targets is difficult to notice among the clutter (Fig. 3.34).

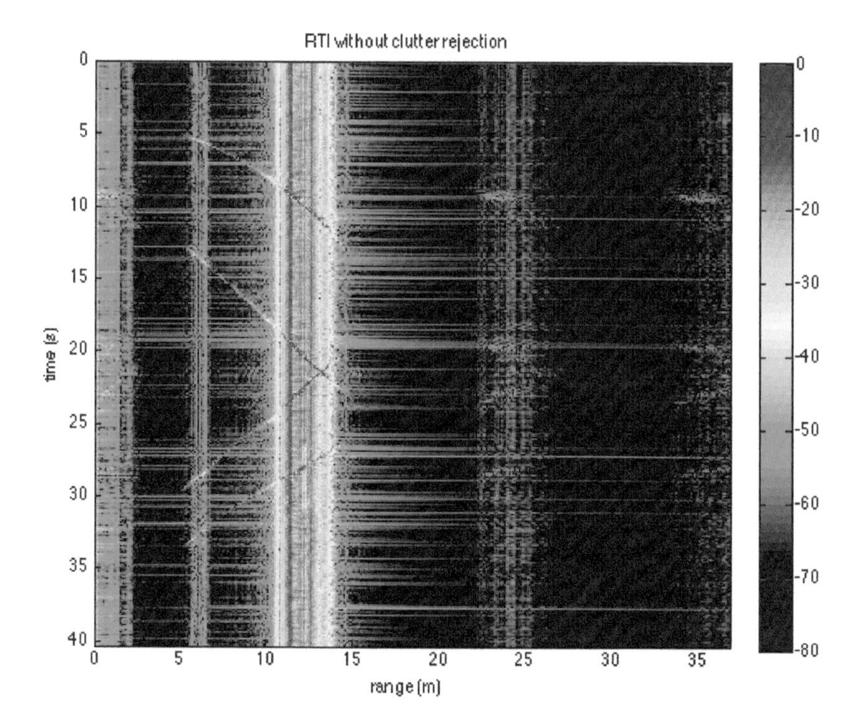

FIGURE 3.31
RTI plot of two targets walking down range using the range-gated S-band FMCW radar with a 7.5 KHz range gate filter.

An RTI plot of these range profiles shows the target positions most of the time; however the targets compete with clutter returns (Fig. 3.35). Red streaks through this RTI plot are clutter returns passing through the range gate filter.

CCD processing is applied to the single range profile at t = 33.3 s, where the previous pulse is subtracted from the current one. Stationary clutter is greatly reduced and the location of both moving human targets is shown (Fig. 3.36).

An RTI plot was made using CCD range profiles (Fig. 3.37). The location of both moving targets is clearly shown above the clutter for all range profiles. The transfer function of the range gate filter is still evident, where attenuated clutter returns are passed through the filter. Target returns before or after the filter are greatly attenuated, thereby showing the utility of the range gate.

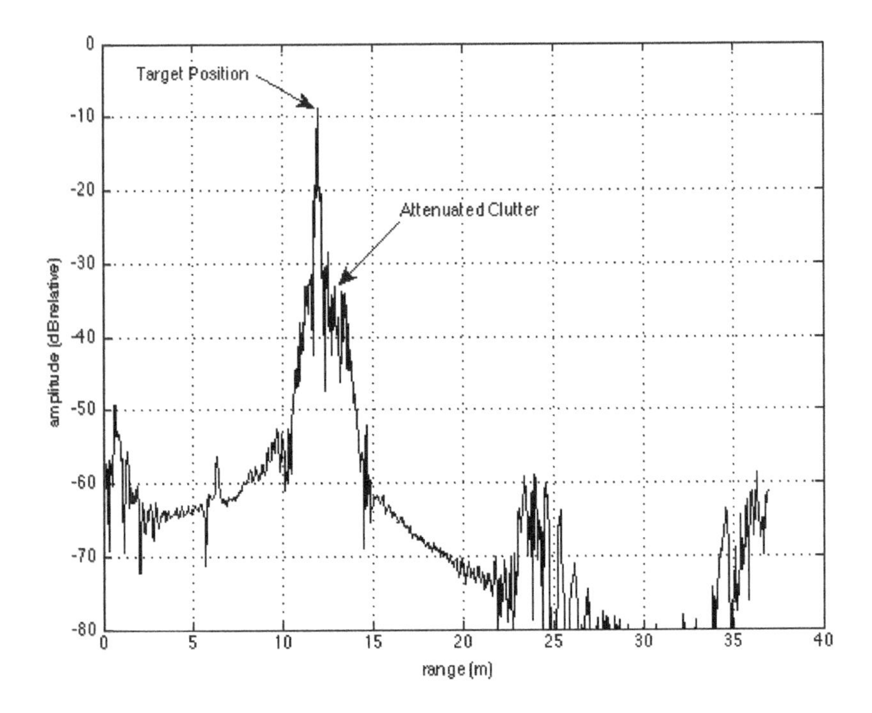

FIGURE 3.32
Range profile plot of one target walking down range with CCD clutter rejection using the range-gated S-band FMCW radar with a 7.5 KHz range gate filter.

3.3.3.5 X-Band Implementation of a Range-Gated FMCW Radar System

It is useful to be able to acquire high resolution radar imagery or data indoors or within a high clutter environment, where a limited target area can be range gated. For this reason an X-band UWB system was developed using the range-gated architecture [37].

This is a front-end substitute to the S-band system described previously (Sec. 3.3.3.2). When the X-band front end is installed the system is configured as shown (Fig. 3.38).

The CTRO0 pin from the PCI6014 causes the ramp generator (Fig. 3.5) to produce an up-ramp from 7.835 to 12.817 GHz which is fed into the voltage tune input of the YiG oscillator. The output of the YiG oscillator is fed directly to the X-band front end (Fig. 3.39, photo and call-outs in Fig. 3.40).

The swept output from the YiG oscillator is fed through ATTN4, 6, CIRC1, and into the LO port of MXR4. The attenuated BFO output from the IF

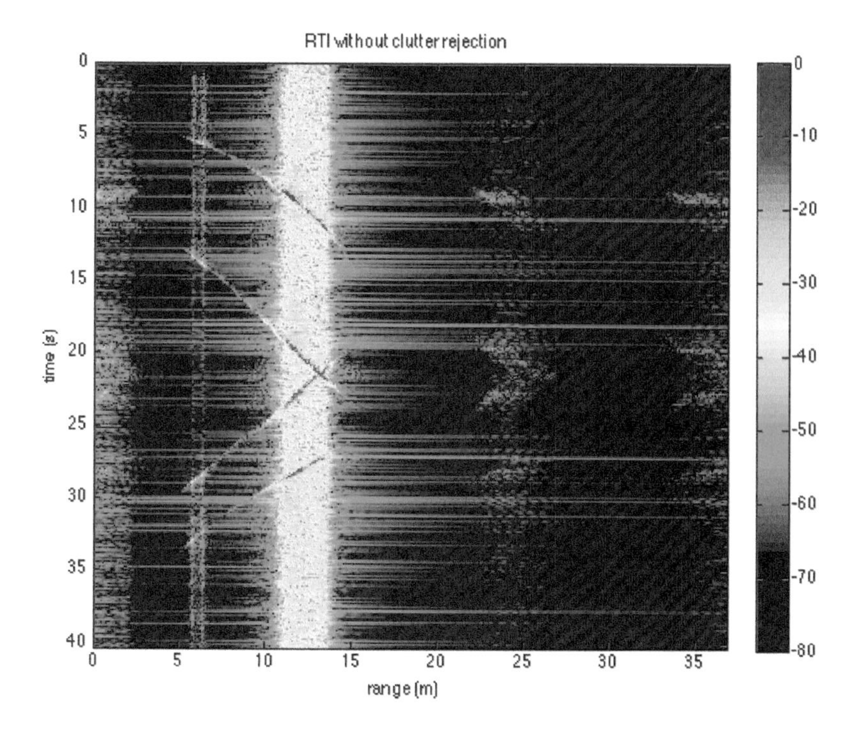

FIGURE 3.33
RTI plot of two targets walking down range with CCD clutter rejection using the range-gated S-band FMCW radar with a 7.5 KHz range gate filter.

chassis is fed into the IF port of MXR4. The RF product output of MXR4 is amplified by AMP6 and fed through CIRC3 and CLPR3 to ANT3, where it is radiated out toward the target scene. Some power is coupled off of CLPR3 for testing purposes. Circulators CIRC1-3 are configured as isolators, where a 50 ohm load is placed in the third port. The purpose of CIRC1 and 2 is to provide a good impedance match to MXR4 and 5 respectively because the LO ports of double balanced mixers at microwave frequencies typically do not provide an ideal impedance match to 50 ohms. Similarly, CIRC3 provides an ideal match to the transmit antenna ANT3 over the wide chirp bandwidth, reducing reflections within the radar that might cause false target returns. An external coaxial jumper is present between CIRC3 and CLPR3 so that an external attenuator can be added to further reduce transmit power.

The scattered field is collected by ANT4, amplified by LNA2, then fed to the RF port of MXR5. Some power is coupled off of the YiG oscillator via CLPR4. This is fed through CIRC2 and delayed through DELAY2 to phase match with the LO port of MXR4. This sweep is then fed into the LO port of MXR5 where it is multiplied by the RF input.

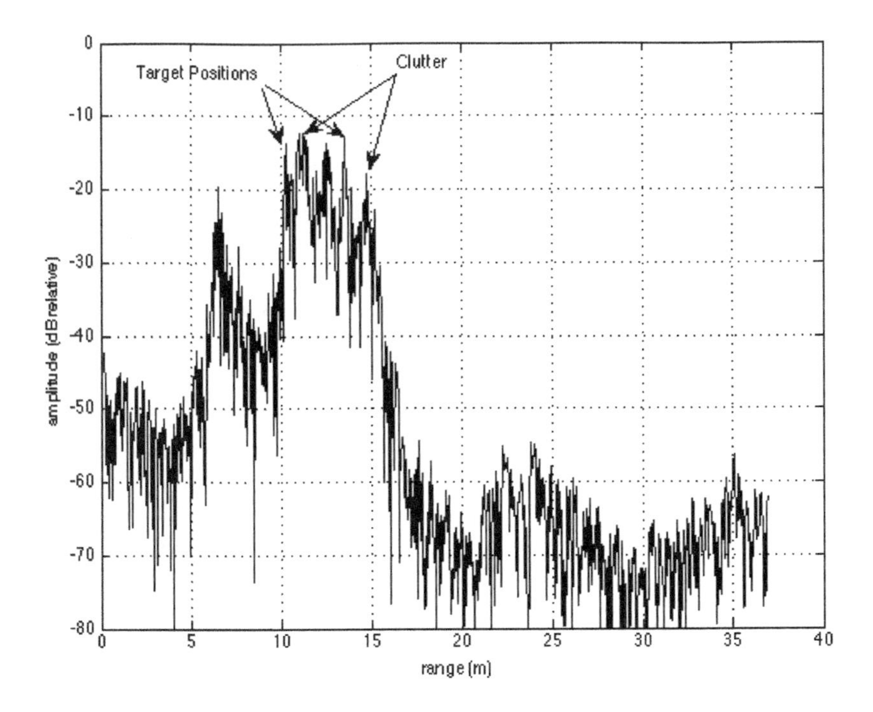

FIGURE 3.34
Range profile plot of one target walking down range using the range-gated S-band FMCW radar with a 15 KHz range gate filter.

The product of MXR5 is amplified by AMP7 and low-pass filtered by FL14, an 11 MHz low pass filter. The output of FL14 is fed to the IF chassis where it is amplified and range gate filters are applied, converted down to baseband and digitized by the PCI6014.

Complete bill of material for this front end is provided (Figs. 3.9).

3.3.3.6 Expected Performance of the Range-Gated X-Band FMCW Radar System

Substituting the specifications for the range-gated X-band FMCW radar (Table 3.10) into the radar range equation (3.5), the maximum range is estimated to be 1055 m for a 10 dBsm automobile target using the MATLAB script [20].

The antenna gains G_{rx} and G_{tx} were estimated based on aperture size, the noise figure is an educated guess based on the apparent age of the surplus microwave amplifier LNA2, and the IDFT is applied over 10 ms up-chirps with

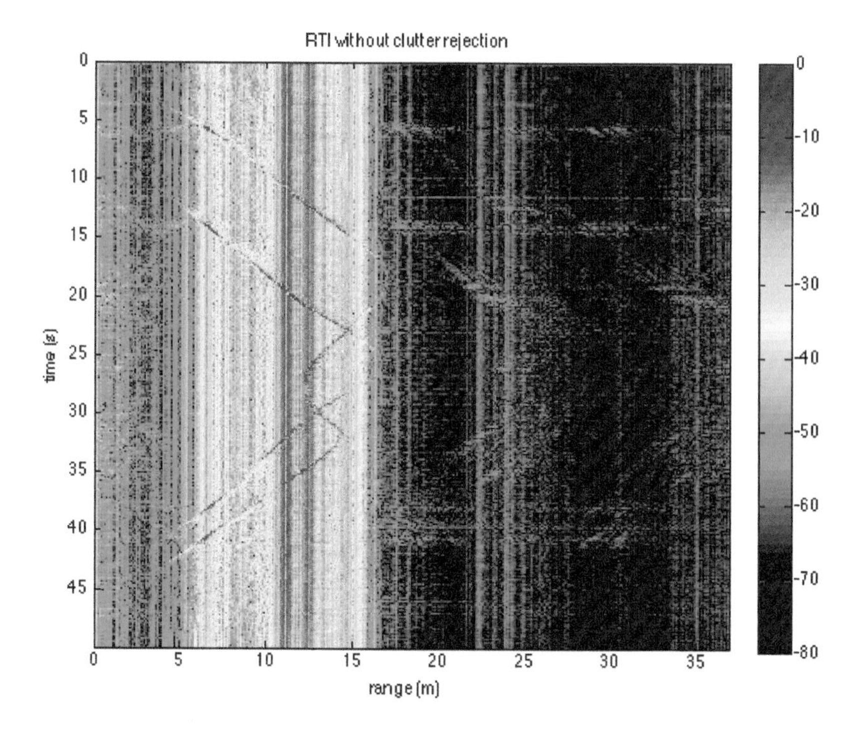

FIGURE 3.35
RTI plot of two targets walking down range using the range-gated S-band FMCW radar with a 15 KHz range gate filter.

a direct conversion receive architecture resulting in a 200 Hz effective noise bandwidth.

3.3.3.7 Working Example of the Range-Gated X-Band FMCW Radar System

Two human targets walk down range then back while recording range profiles. A video of this experiment is shown [47] and data can be downloaded with a MATLAB processing script [48].

We first consider the 7.5 KHz range gate filter. A single range profile at t = 11 s is plotted (Fig. 3.41), where the target's location is shown at 11 m down range. The transfer function of the range gate filter is attenuating clutter before and after the filter's pass band. Some clutter is shown in the pass band.

An RTI plot is made from all range profiles (Fig. 3.42), where the target returns are most intense when they are passing through the range gate filter pass band between 8 and 12 meters. Targets are shown slightly before starting at 5 m and are all but eliminated after 12 m.

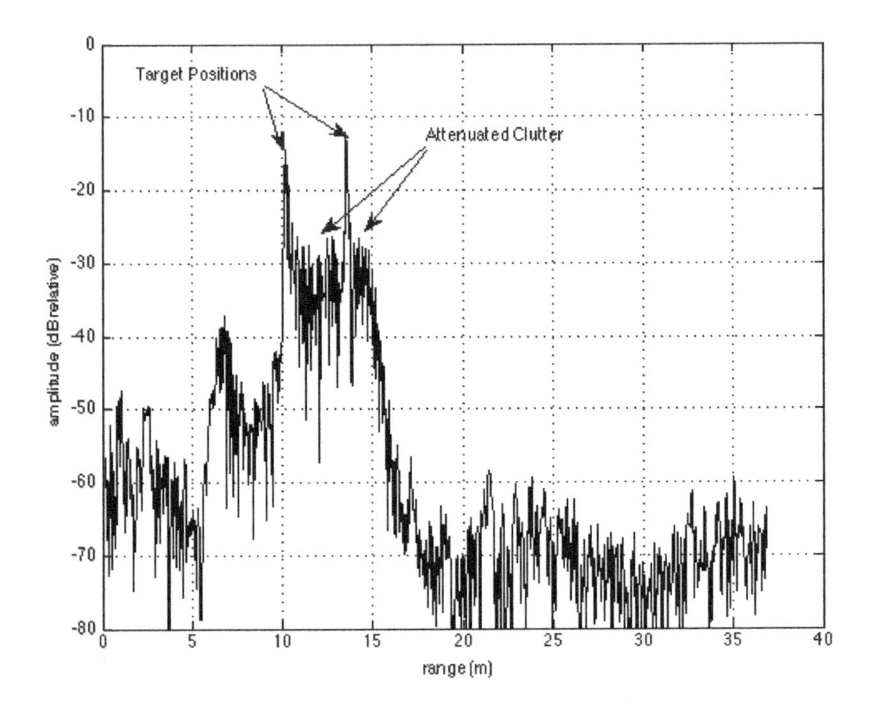

FIGURE 3.36
Range profile plot of one target walking down range with CCD clutter rejection using the range-gated S-band FMCW radar with a 15 KHz range gate filter.

Coherent change detection is applied by subtracting the current pulse from the previous one, resulting in the range profile shown (Fig. 3.43). In this, the clutter is significantly attenuated below the target compared to the range profile without CCD (Fig. 3.41).

An RTI plot of the CCD results is plotted (Fig. 3.44), where the range returns are well attenuated before and after the range gate. Clutter is attenuated compared to results without the range gate. Additional 'ghosting' can be see down range, likely due to in-band non-linearities in MXR3 in the IF chassis.

We now consider the 15 KHz range gate filter, where two human targets walk down range then back while range profiles are recorded. A range profile at t = 26 s is shown (Fig. 3.45), where the target return is 20 dB above the clutter floor. In this plot, clutter is present and the transfer function of the range gate filter attenuates target returns before and after its passband.

An RTI plot is made from all range profiles (Fig. 3.46). The location of both targets is shown. The range gate filter manifests itself as vertical streaks in

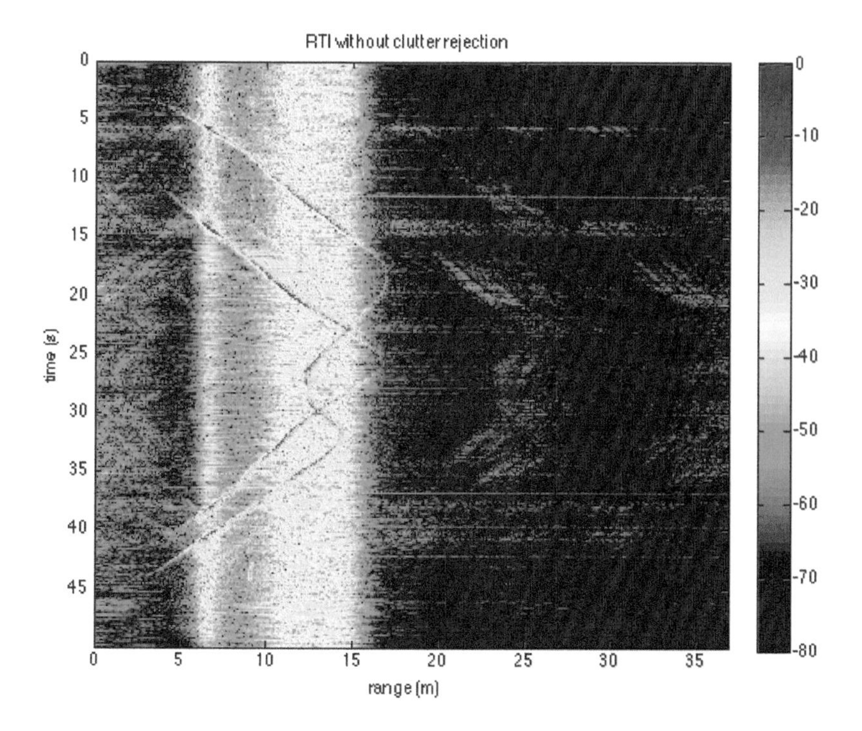

FIGURE 3.37
RTI plot of two targets walking down range with CCD clutter rejection using the range-gated S-band FMCW radar with a 15 KHz range gate filter.

the plot passing stationary clutter returns. Both targets are greatly attenuated before and after the range gate, showing the effectiveness of the range gate.

CCD processing is applied, subtracting the current pulse from the last for the range profile at t = 26 s (Fig. 3.47). After this process the target return is significantly greater than the clutter and the transfer function of the range gate filter is also there but less prevalent.

An RTI plot is made from the CCD range profiles (Fig. 3.48). Clutter in the pass band of the range gate filter is greatly attenuated. Target returns before and after the range gate are significantly attenuated. Some 'ghosting' is evident down range, likely due to nonlinearities in the IF frequency mixer.

These results show the efficacy of the analog range gate for FMCW radar systems. Such a range gate allows FMCW radar operation while also gating out unwanted target returns. This architecture is valuable for applications where radar data must be acquired in a high-clutter indoor environment.

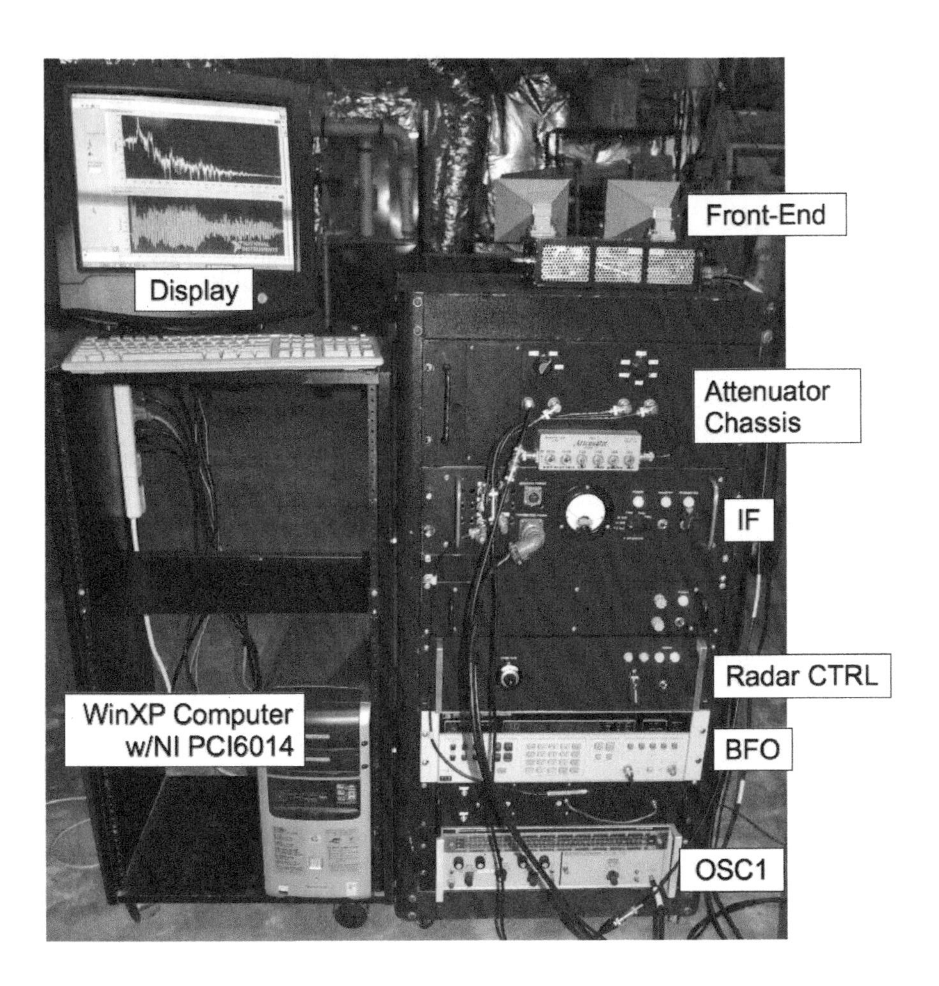

FIGURE 3.38
X-band range gated FMCW radar system call-outs.

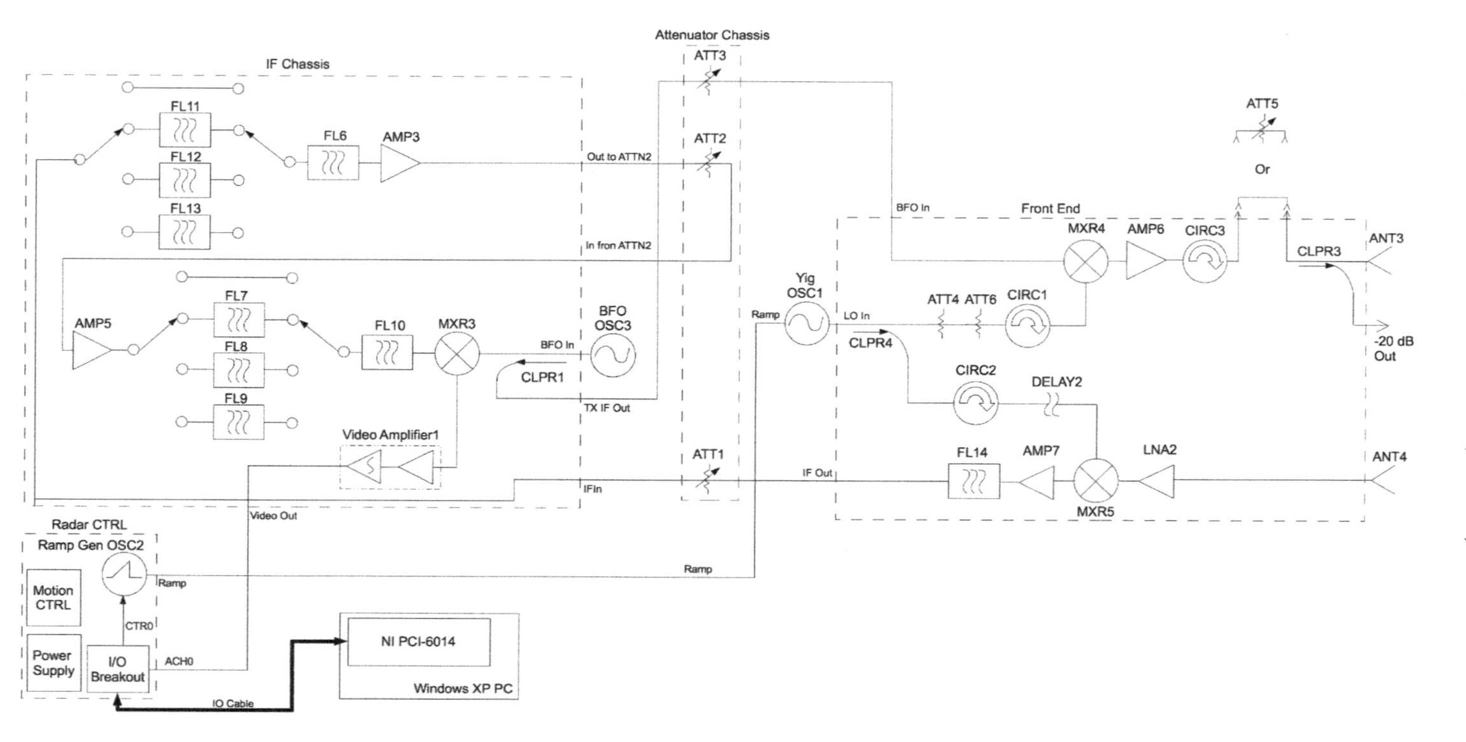

FIGURE 3.39
Block diagram of the X-band range-gated FMCW radar system.

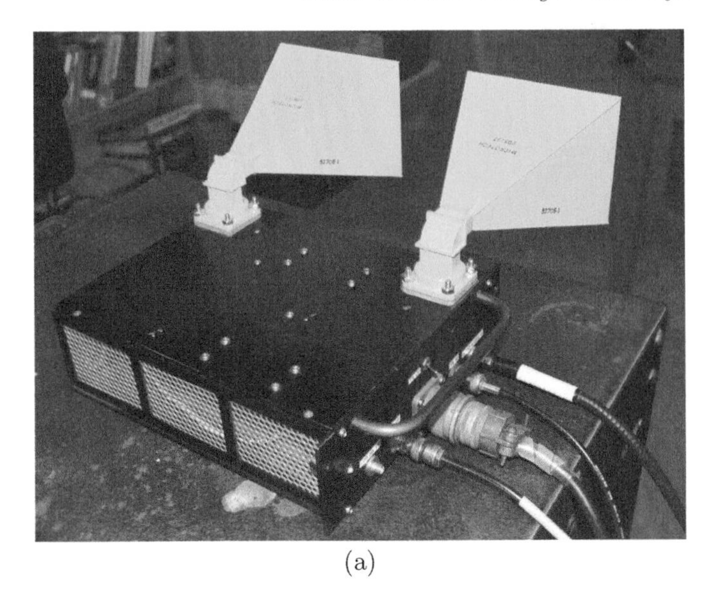

(a)

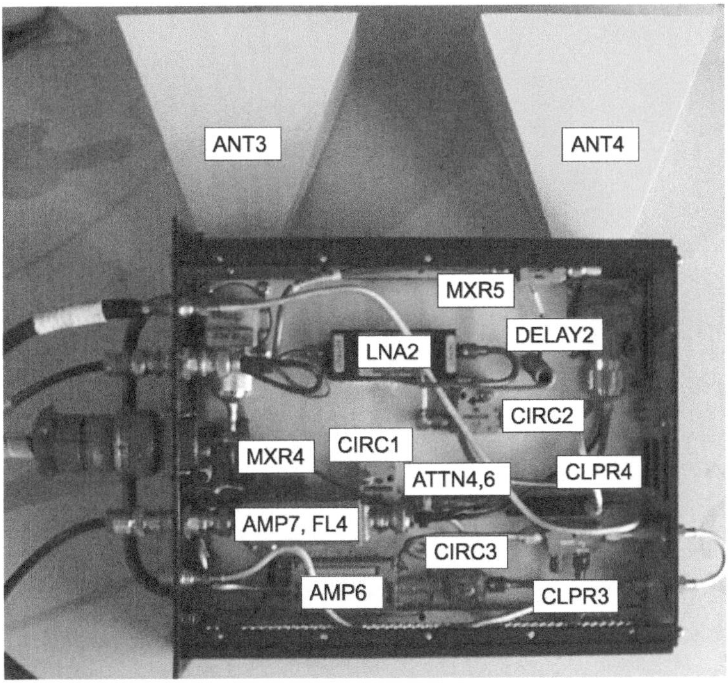

(b)

FIGURE 3.40
X-band front end (a) and inside view with call-outs (b).

TABLE 3.9
X-band front end modular components list.

Component	Description
AMP6	Microwave Components Corporation MH858231, gain = 25 dB, IP1 = 23 dBm
AMP7	Mini-Circuits MAR-4, gain = 8, NF = 6 dB, IP1 = 12.5 dBm
ANT3	Microtech 205297, X-band horn with WR90 waveguide flange
ANT4	Microtech 205297, X-band horn with WR90 waveguide flange
ATTN4	3 dB in line attenuator
ATTN5	Narda Microline Step Attenuator, Model 705-69
ATTN6	Midwest Microwave 6 dB in line attenuator
CIRC1	UTE Microwave X-band isolator
CIRC2	UTE Microwave X-band isolator
CIRC3	Unknown surplus X-band isolator
CLPR3	Unknown surplus X-band -20 dB directional coupler
CLPR4	Omni Spectra X-band -10 dB directional coupler
DELAY2	Coaxial delay line
FL14	Mini-Circuits PBP-10.7, 10.7 MHz bandpass filter
LNA1	Amplica, Inc. XM553403, Gain = 20dB, IP1 = 25 dBm
MXR4	Watkins Johnson M31A
MXR5	TRW Microwave MX18533

TABLE 3.10
Range-gated X-band FMCW radar system parameters.

$$
\begin{aligned}
P_{ave} &= 10 \cdot 10^{-3} \text{ (watts)} \\
G_{tx} &= 17 \text{ dBi antenna gain (estimated)} \\
G_{rx} &= 17 \text{ dBi antenna gain (estimated)} \\
A_{rx} &= G_{rx}\lambda_c^2/(4\pi) \text{ (m}^2\text{)} \\
\lambda_c &= c/f_c \text{ (m) wavelength of carrier frequency} \\
f_c &= 10 \text{ GHz center frequency of radar} \\
\rho_{rx} &= 1 \text{ because antenna efficiency is accounted for in antenna gain} \\
\sigma &= 10 \text{ (m}^2\text{) for automobile at 10 GHz} \\
L_s &= 6 \text{ dB miscellaneous system losses} \\
\alpha &= 0 \text{ attenuation constant of propagation medium} \\
F_n &= 4 \text{ dB receiver noise figure} \\
B_n &= 2/t_{sample} \text{ system noise bandwidth (Hz) where } t_{sample} = 10 \text{ ms} \\
(SNR)_1 &= 13.4 \text{ dB}
\end{aligned}
$$

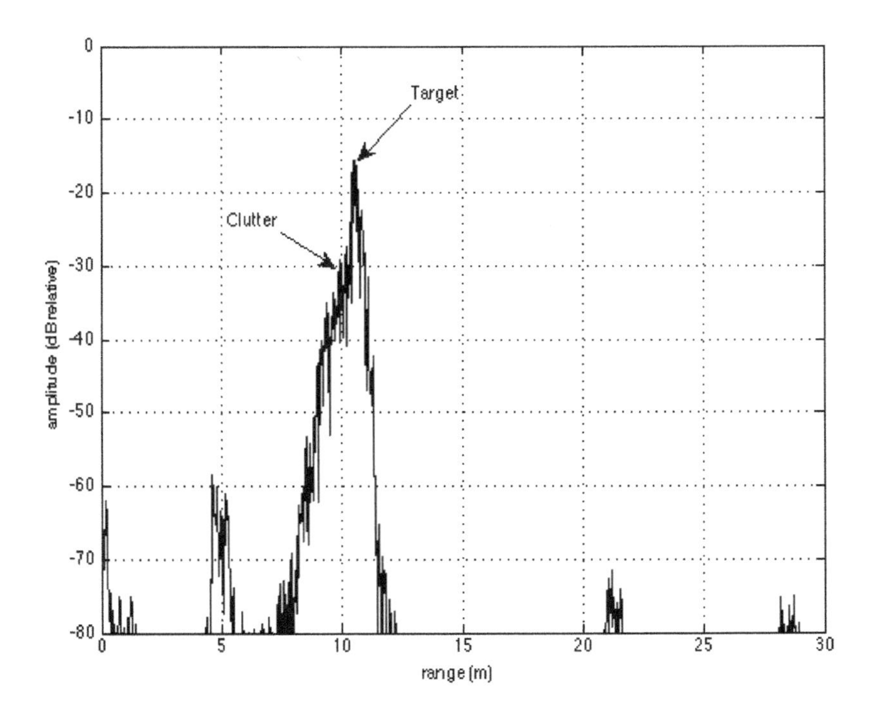

FIGURE 3.41
Range profile plot of one target walking down range using the range-gated X-band FMCW radar with a 7.5 KHz range gate filter.

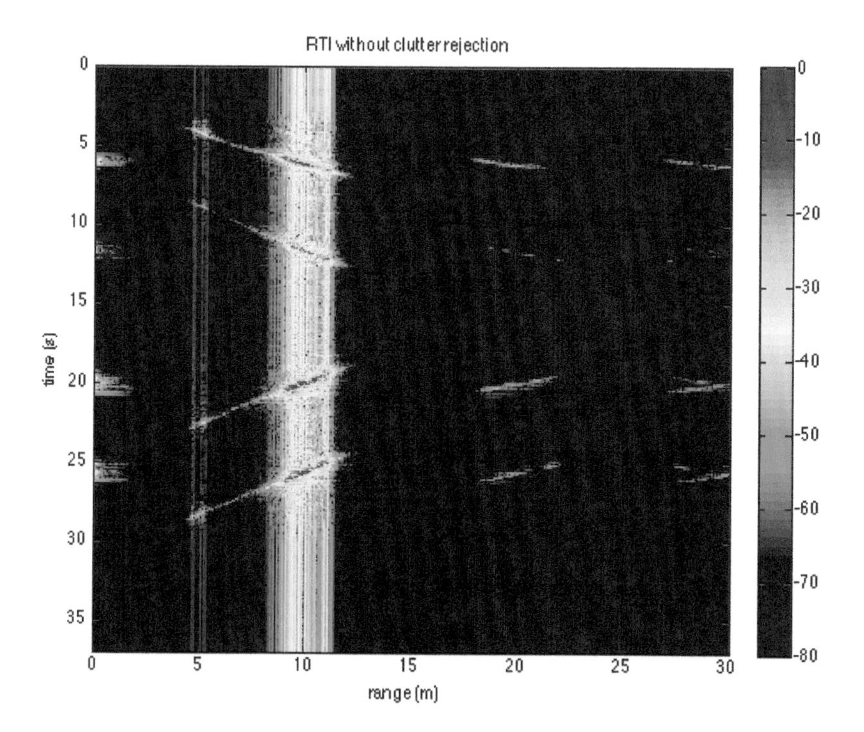

FIGURE 3.42
RTI plot of two targets walking down range using the range-gated X-band FMCW radar with a 7.5 KHz range gate filter.

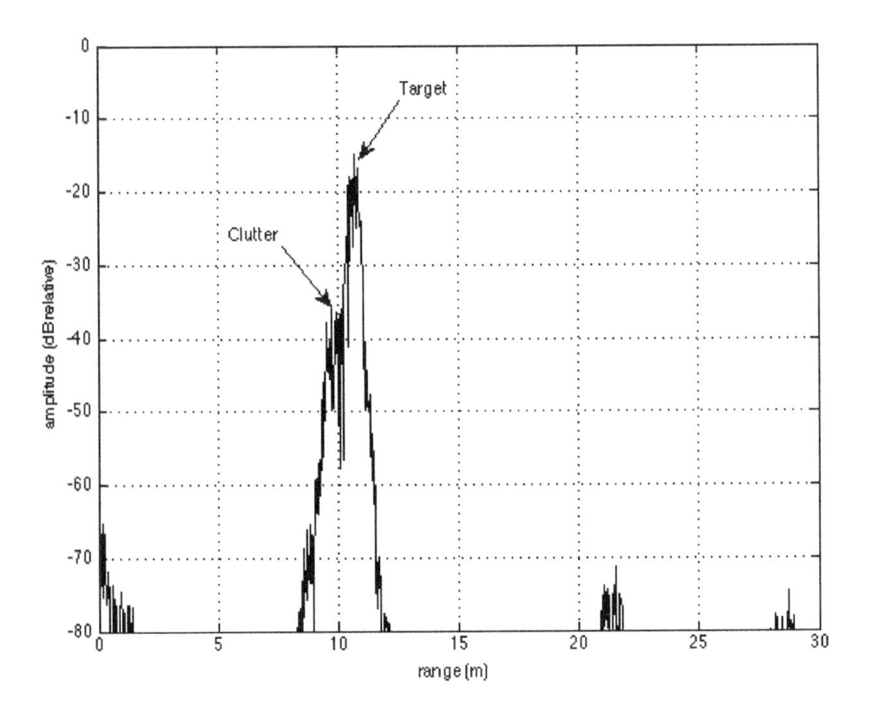

FIGURE 3.43
Range profile plot of one target walking down range with CCD clutter rejection using the range-gated X-band FMCW radar with a 7.5 KHz range gate filter.

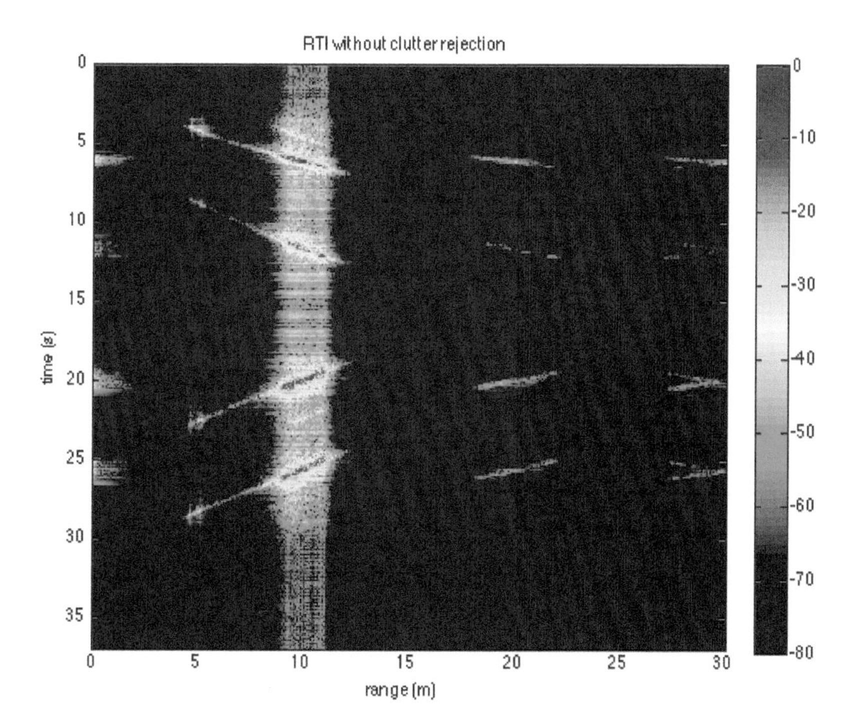

FIGURE 3.44
RTI plot of two targets walking down range with CCD clutter rejection using the range-gated X-band FMCW radar with a 7.5 KHz range gate filter.

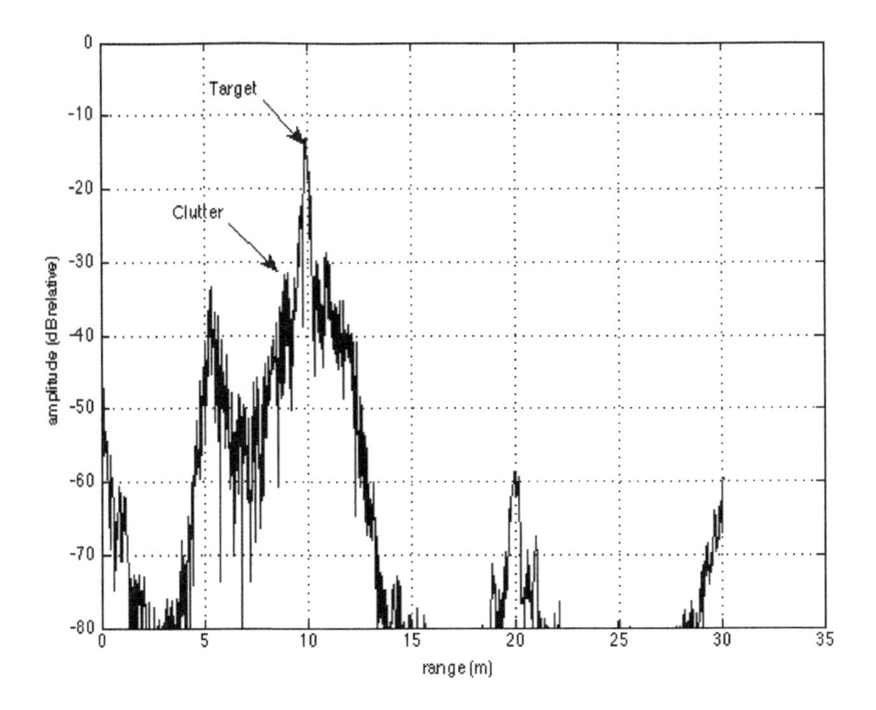

FIGURE 3.45
Range profile plot of one target walking down range using the range-gated X-band FMCW radar with a 15 KHz range gate filter.

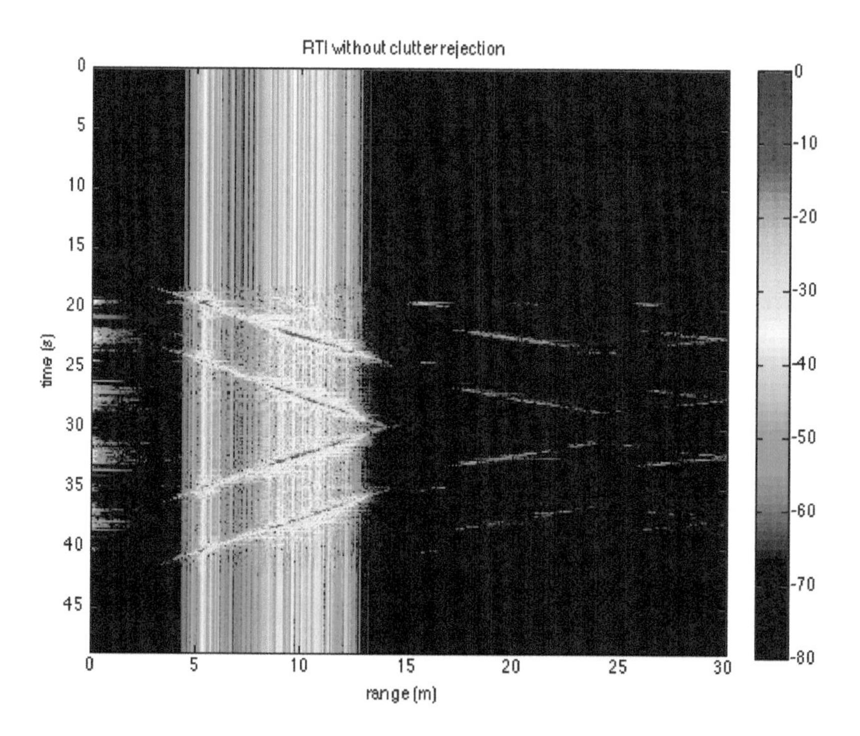

FIGURE 3.46
RTI plot of two targets walking down range using the range-gated X-band FMCW radar with a 15 KHz range gate filter.

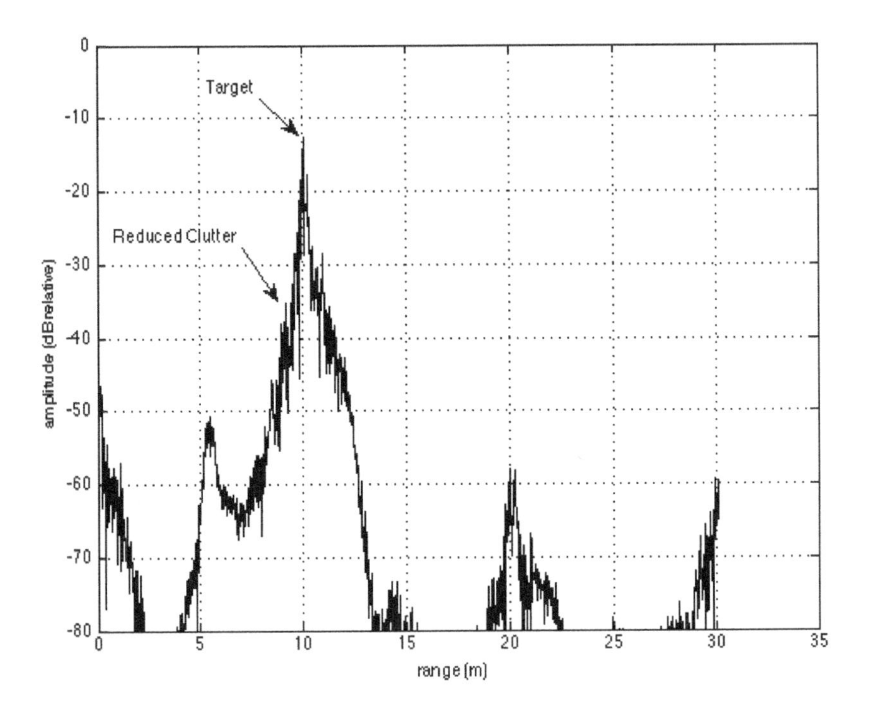

FIGURE 3.47
Range profile plot of one target walking down range with CCD clutter rejection using the range-gated X-band FMCW radar with a 15 KHz range gate filter.

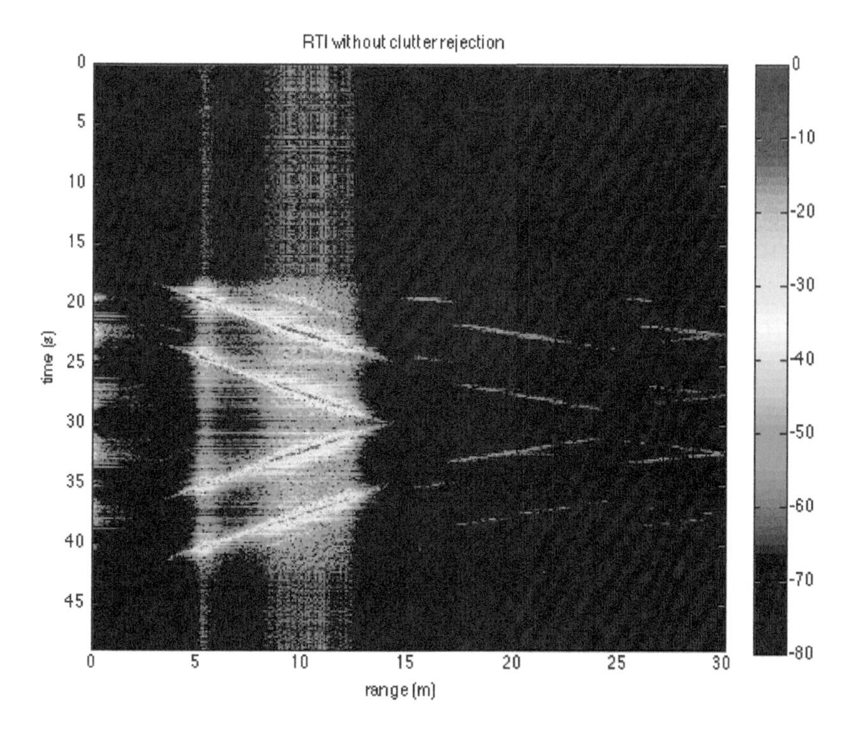

FIGURE 3.48
RTI plot of two targets walking down range with CCD clutter rejection using the range-gated X-band FMCW radar with a 15 KHz range gate filter.

3.4 Summary

When the oscillator of a CW coherent radar system is frequency modulated range data can be computed. FMCW radar architectures were shown, performance was estimated, and numerous examples of radar systems were demonstrated. It was also shown that range gating in the frequency domain with a simple crystal filter and a well thought out architecture is possible. Additional complexity will be added in the next chapter, where FMCW radar systems will be used to image a target scene, providing both down range and cross range target information.

Bibliography

[1] W. K. Saunders, "CW and FM Radar," M. I. Skolnik, editor, *Radar Handbook, Second Edition,* McGraw Hill, New York, NY, 1990 chap. 14.

[2] I. V. Komarov, S. M. Smolskiy, *Fundamentals of Short-Range FM Radar,* Artech House, Boston MA, 2003.

[3] M. Jamkiraman, *Design of Multi-Frequency CW Radars,* SciTech Publishing, 2007.

[4] G. M. Brooker, *Introduction to Sensors for Ranging and Imaging,* SciTech Publishing, 2009.

[5] P. E. Pace, *Detecting and Classifying Low Probability of Intercept Radar,* Artech House, Boston MA, 2009.

[6] M. P. G. Capelli, "Radio Altimeter," *IRE Transactions on Aeronautical and Navigational Electronics,* Vol. 1, June 1954, pp. 3–7.

[7] F. T. Wimberly, J. F. Lane, "The AN/APN-22 Radio Altimeter," *IRE Transactions on Aeronautical and Navigational Engineering,* Vol. 1, June 1954, pp. 8–14.

[8] A. Black, K. E. Buecks, A. H. Heaton, "Improved Radio Altimeter," *Wireless World,* Vol. 60, March 1954, pp. 138–140.

[9] A. G. Stove, "Linear FMCW radar techniques," *IEE Proceedings of Radar and Signal Processing,* Vol. 139, October 1992, pp. 343–350.

[10] W. L. Briggs, V. E. Henson, *The DFT, an Owner's Manual for the Discrete Fourier Transform,* Society for Industrial and Applied Mathematics, Philadelphia, PA, 1995, p. 28, eq. 2.9.

[11] W. L. Briggs, V. E. Henson, *The DFT, an Owner's Manual for the Discrete Fourier Transform,* Society for Industrial and Applied Mathematics, Philadelphia, PA, 1995, p. 22, eq. 2.5.

[12] S. O. Piper, "FMCW Systems," J. A. Scheer, J. L. Kurtz, editors, *Coherent Performance Estimation,* Artech House, Boston MA, 1993 chap. 14.

[13] S. O. Piper, J. Wiltse, "Continuous Wave Radar," M. Golio, Janet Golio, editors, *The RF and Microwave Handbook, Second Edition,* CRC Press, 2008 chap. 14.

[14] W. G. Carrara, R.S. Goodman, and R.M. Majewski, *Spotlight Synthetic Aperture Radar Signal Processing Algorithms,* Artech House, Boston MA, 1995, p. 28, eq. 2.2.

[15] W. G. Carrara, R.S. Goodman, and R.M. Majewski, *Spotlight Synthetic Aperture Radar Signal Processing Algorithms,* Artech House, Boston MA, 1995, Appendix D, pp. 507–529.

[16] G. L. Charvat, L. C. Kempel. "Low-Cost, High Resolution X-Band Laboratory Radar System for Synthetic Aperture Radar Applications," East Lansing, MI, IEEE Electro/Information Technology Conference, May 2006.

[17] G. L. Charvat, "Low-Cost, High Resolution X-Band Laboratory Radar System for Synthetic Aperture Radar Applications," Antenna Measurement Techniques Association Conference, Austin, Texas, October 2006.

[18] G. L. Charvat. "Build a high resolution synthetic aperture radar imaging system in your backyard," MIT Haystack Observatory, May 12, 2010.

[19] T. S. Ralston, G. L. Charvat, S. G. Adie, B. J. Davis, S. Carney, S. A. Boppart. "Interferometric synthetic aperture microscopy, microscopic laser radar," *Optics and Photonics News*, June 2010, Vol. 21, No. 6, pp. 32–38.

[20] radar_range_eq_FMCW.m, `http://glcharvat.com/shortrange/frequency-modulated-continuous-wave-fmcw-radar/`

[21] X-Band FMCW Radar Experiment, `http://glcharvat.com/shortrange/frequency-modulated-continuous-wave-fmcw-radar/`

[22] xband_FMCW.zip, `http://glcharvat.com/shortrange/frequency-modulated-continuous-wave-fmcw-radar/`

[23] G. L. Charvat, J. H. Williams, A. J. Fenn, S. M. Kogon, J. S. Herd, "RES.LL-003 Build a Small Radar System Capable of Sensing Range, Doppler, and Synthetic Aperture Radar Imaging, January IAP

2011," (Massachusetts Institute of Technology: MIT OpenCourseWare), http://ocw.mit.edu (Accessed 02 Sep, 2012). License: Creative Commons BY-NC-SA.

[24] G. L. Charvat, J. H. Williams, A. J. Fenn, S. M. Kogon, J. S. Herd. "The MIT IAP 2011 radar course: build a small radar system capable of sensing range, Doppler, and SAR," The Boston Chapters of the IEEE Life Members, AES, and GRSS, May 24, 2011.

[25] G. L. Charvat, A. J. Fenn, B. T. Perry, "The MIT IAP radar course: build a small radar system capable of sensing range, Doppler and synthetic aperture radar (SAR) imaging," Atlanta, GA: IEEE Radar Conference, May 2012.

[26] Coffee Can FMCW Ranging Demo, `http://glcharvat.com/shortrange/frequency-modulated-continuous-wave-fmcw-radar/`

[27] coffee_can_ranging_demo.zip, `http://glcharvat.com/shortrange/frequency-modulated-continuous-wave-fmcw-radar/`

[28] G. L. Charvat, "A Low-Power Radar Imaging System," Ph.D. dissertation, Department of Electrical and Computer Engineering, Michigan State University, East Lansing, MI, 2007.

[29] G. L. Charvat, T. S. Ralston, J. E. Peabody, "A Through-Wall Real-Time MIMO Radar Sensor for Use at Stand-off Ranges," Orlando FL: MSS Tri-Services Radar Symposium, June 2010.

[30] P. R. Hirschler-Marchand, "Penetration losses in construction materials and buildings," MIT Lincoln Laboratory Project Report TrACC-1 Rev. 1, July 2006.

[31] G. L. Charvat, L. C. Kempel, E. J. Rothwell, C. Coleman, and E. L. Mokole, "A through-dielectric radar imaging system," *IEEE Transactions on Antennas and Propagation*, Vol. 58, No. 8, pp. 2594–2603, 2010.

[32] G. L. Charvat, L. C. Kempel, E. J. Rothwell, C. Coleman, and E. L. Mokole, "An ultrawideband (UWB) switched-antenna-array radar imaging system," 2010 International Symposium on Phased Array Systems and Technology, October 12–15, Waltham, MA.

[33] T. S. Ralston, G. L. Charvat, J. E. Peabody, "Real-time through-wall imaging using an ultrawideband multiple- input multiple-output (MIMO) phased array radar system," Waltham, MA: IEEE International Symposium on Phased Array Systems & Technology, October 2010.

[34] G. L. Charvat, J. E. Peabody, J. Goodwin, M. Tobias, "A real-time through-wall imaging system," *MIT Lincoln Laboratory Journal*, June 2012.

[35] G. L. Charvat, L. C. Kempel, E. J. Rothwell, C. Coleman, and E. L. Mokole, "A through-dielectric ultrawideband (UWB) switched-antenna-array radar imaging system," *IEEE Transactions on Antennas and Propagation*, Vol. 60, No. 11, pp. 5495–5500, 2012.

[36] *The ARRL Handbook*, 71st *Edition*, The American Radio Relay League, Inc., Newington, CT, 1994.

[37] G. L. Charvat, L. C. Kempel, and C. Coleman, "A low-power high-sensitivity X-band rail SAR imaging system." *IEEE Antennas and Propagation Magazine*, June 2008, pp. 108–115.

[38] R. Janaswamy, D. H. Schaubert, and D. M. Pozar, "Analysis of the transverse electromagnetic mode linearly tapered slot antenna," *Radio Science*, Vol. 21, No. 5, pp. 797–804, 1986.

[39] R. Janaswamy, D. H. Schaubert, "Analysis of the Tapered Slot Antenna," *IEEE Transactions on Antennas and Propagation*, Vol. AP-35, No. 9, 1987, pp. 1058–1065.

[40] R. Janaswamy, "An Accurate Moment Method Model for the Tapered Slot Antenna," *IEEE Transactions on Antennas and Propagation*, Vol. 37, No. 12, 1989.

[41] Y. S. Kim, S. Yngvesson, "Characterization of Tapered Slot Antenna Feeds and Arrays," *IEEE Transactions on Antennas and Propagation*, Vol. 38, No. 10, 1990.

[42] P. S. Kooi, T. S. Yeo, M. S. Leong, "Parametric Studies of the Linearly Tapered Slot Antenna (LTSA)," *IEEE Microwave and Optical Technology Letters*, Vol. 4, No. 5, 1991.

[43] X. D. Wu, and K. Chang, "Compact Wideband Integrated Active Slot Antenna Amplifier," *IEEE Electronics Letters*, Vol. 29, No. 5, 1993.

[44] E. Thiele, A. Taflove, "FD-TD Analysis of Vivaldi Flared Horn Antennas and Arrays," *IEEE Transactions on Antennas and Propagation*, Vol. 42, No. 5, 1994.

[45] S-Band FMCW Radar with Range Gate, `http://glcharvat.com/shortrange/frequency-modulated-continuous-wave-fmcw-radar/`

[46] sband_FMCW_rangegated.zip, `http://glcharvat.com/shortrange/frequency-modulated-continuous-wave-fmcw-radar/`

[47] X-Band FMCW Radar with Range Gate, `http://glcharvat.com/shortrange/frequency-modulated-continuous-wave-fmcw-radar/`

[48] xband_FMCW_rangegated.zip, `http://glcharvat.com/shortrange/frequency-modulated-continuous-wave-fmcw-radar/`

4

Synthetic Aperture Radar

Synthetic aperture radar (SAR) was developed in the late 1950's for the purpose of airborne ground mapping at a stand-off distance. This is valuable for military applications where an aircraft can be flying at a stand-off range and acquire a high precision ground map where each pixel corresponds to the relative location of the aircraft.

The radar is mounted onto an aircraft and acquires range profiles of a ground target scene at known increments (x_n) along its flight path (Fig. 4.1) [1]. The radar's field of view (FOV), determined by its antenna pattern, illuminates the target scene for each pulse. Each range profile is recorded and groups of range profiles are processed into an image. Within the imaging algorithm, the SAR radar synthesizes a linear array across the aircraft's flight path that is focused on every point, providing the best possible cross range resolution.

SAR Imaging can also be implemented on a small laboratory scale. In the

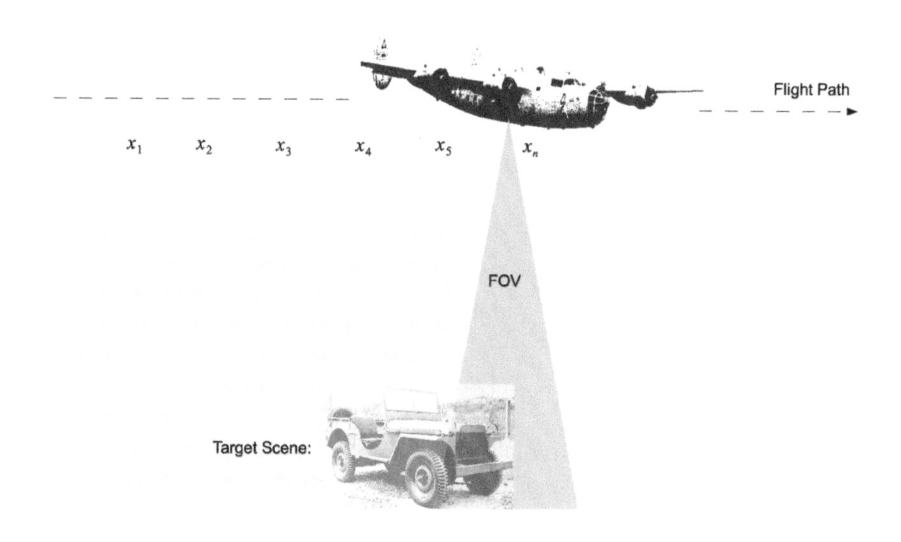

FIGURE 4.1
Airborne SAR.

laboratory environment, high resolution is required to image smaller target scenes. Small laboratory SARs do not fly through the lab on known flight paths, instead they are mounted to linear or circular rails that are moved by precision stepper or servo motors so that the radar's location is well known.

Rail SAR imaging systems are useful for RCS measurements, studying EM propagation phenomenology, through-wall radar imaging, and many other short-range imaging applications. In this section a SAR imaging algorithm will be described that can be used with small-scale laboratory SAR imaging systems mounted on linear rails and near-field phased array systems.

Description of the rail SAR measurement geometry is discussed (Sec. 4.1). The range migration algorithm (RMA) is the SAR imaging algorithm that will be described (Sec. 4.2). The RMA will be demonstrated on a simulated point target data (Sec. 4.3). Performance will be estimated (Sec. 4.4), and additional processing will discussed (Sec. 4.5).

4.1 Measurement Geometry

This book will focus on linear rail SAR and near-field phased array imaging systems where the position of the radar is well known and assumed to be correct. This rail SAR measurement geometry is shown (Fig. 4.2). A small radar device is mounted on a straight rail of length L. The radar uses a wide beamwidth antenna that is directed toward an unknown target scene parallel to the rail. The radar begins its journey at the end of the rail near the drive motor. The drive motor moves the radar to a location, stops, and the radar acquires a range profile. This process is repeated at known intervals along the rail until the radar device reaches the end. The range profiles over each rail position produce a 2D data matrix of position versus frequency in time

$$s\big(x_n, \omega(t)\big),$$

where $s\big(x_n, \omega(t)\big)$ is the frequency domain range profile data matrix, x_n is the nth cross range position of the radar on the automated rail, and

$$\omega(t) = 2\pi\left(c_r t + f_c - \frac{BW}{2}\right) \tag{4.1}$$

is the instantaneous radial frequency of the scattered chirp where f_c is the radar center frequency, BW is the chirp bandwidth (Hz), and T_p is the chirp duration (s). This signal represents the IF output on a direct conversion FMCW radar (or the DFT of the received waveform from an impulse radar).

For linear FM radar systems (e.g., FMCW radar systems) c_r (Hz/s) is the chirp rate of the linear frequency modulated (FM) chirp

$$c_r = \frac{BW}{T_p}, \tag{4.2}$$

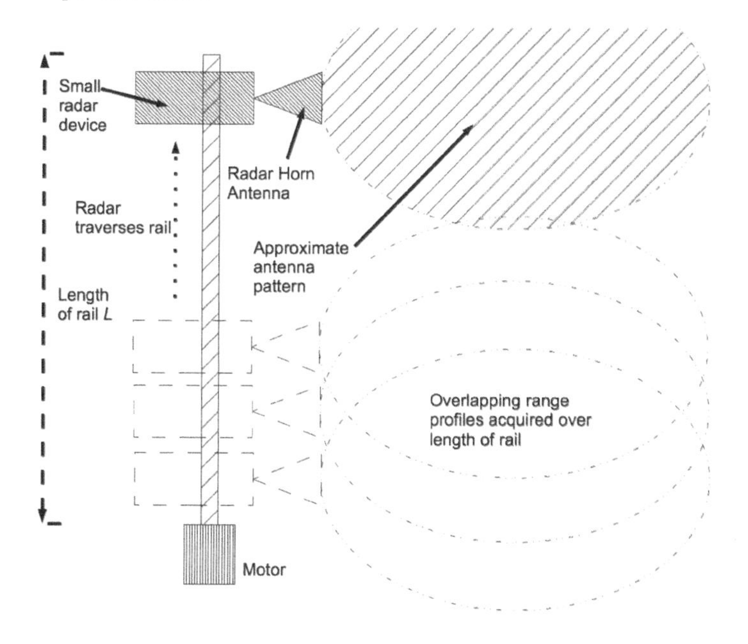

FIGURE 4.2
Rail SAR data collection geometry.

TABLE 4.1
Imaging parameters for an example S-band synthetic aperture radar imaging system.

$$
\begin{aligned}
BW &= f_{stop} - f_{start} \text{ transmit bandwidth} \\
f_{start} &= 1.926 \text{ GHz start frequency of chirp} \\
f_{stop} &= 4.069 \text{ GHz stop frequency of chirp} \\
T_p &= 10 \text{ ms transmit pulse length} \\
c_r &= BW/T_p = 214.3 \text{ GHz/second chirp rate} \\
L &= 8 \text{ ft length of linear rail} \\
\Delta x &= 2 \text{ inch spacing between range profiles}
\end{aligned}
$$

and $BW = f_{stop} - f_{start}$ is the transmit bandwidth from the transmit start frequency to the transmit stop frequency.

To demonstrate how the RMA algorithm works, an example will be shown for an S-band UWB FMCW rail SAR imaging system using the parameters outlined (Table 4.1).

4.2 The Range Migration Algorithm (RMA)

As its input, the RMA accepts the measured data matrix $s(x_n, \omega(t))$ and processes this into an image matrix $S(X, Y)$, providing a SAR image of what is within the radar's field of view. Additional information on the RMA, other SAR processing techniques, and further in-depth analysis can be found [2].

To process an image using the RMA, these steps must be followed:

1. Cross range DFT
2. Matched filter
3. Stolt interpolation
4. 2D IDFT into image domain

The details of each step will be discussed in this section by examples using a single point scatterer.

4.2.1 Simulation of a Point Scatterer

Assuming measurement geometry (Sec. 4.1), a single point scatterer located cross range and down range at (x_t, y_t) with a scattered magnitude of a_t is represented by the data matrix

$$s\big(x_n, \omega(t)\big) = a_t e^{-j2\frac{\omega(t)}{c}\sqrt{(x_n - x_t)^2 + y_t^2}}, \tag{4.3}$$

where $c = 3 \cdot 10^8$ m/s is the speed of light.

Without loss of generality let $a_t = 1$. Locating the point scatter at $x_t = 0$ in cross range from rail center (where $x = 0$ is the center of the rail) and $y_t = -10$ feet down range produces a simulated data matrix where the real values of the SAR data matrix points are shown (Fig. 4.3) and the phase of the SAR data matrix points is shown (Fig. 4.4).

The down range IDFT of the data matrix (Fig. 4.5) shows the wave-front curvature of the single point scatterer in an arc-like contour. This plot shows some intuition behind SAR imaging, where we can expect the range to a point target centered with respect to the rail to be further away while the radar is at $-L/2$, shortest distance when the radar is at 0, and further away when at $L/2$, with a parabolic range curvature between them. This arc shape is what we would expect from a point target located at the middle and some distance away from the rail.

4.2.2 Cross Range Fourier Transform

The first step in the RMA is to calculate the DFT of the spatial component—the SAR data matrix $s\big(x_n, \omega(t)\big)$ (e.g., apply the DFT to all values x_n for each $\omega(t)$) resulting in the spatial frequency domain data matrix $s\big(k_x, \omega(t)\big)$, where t_x varies linearly from $\frac{-\pi}{\Delta x}$ to $\frac{\pi}{\Delta x}$.

In addition to the DFT, the substitution is made: $k_r = \omega(t)/c$. This results in the SAR data matrix $s(k_x, k_r)$.

With these operations complete, the magnitude after the cross range DFT is shown (Fig. 4.6) and the phase after the cross range DFT is shown (Fig. 4.7).

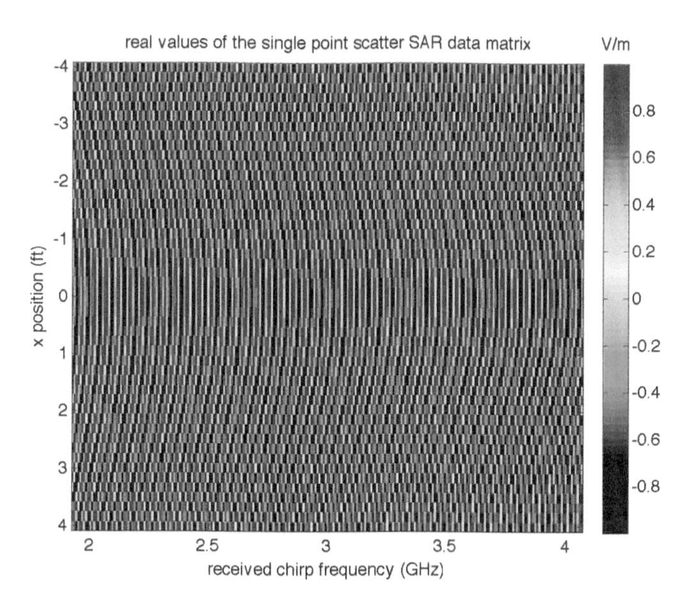

FIGURE 4.3
Real values of the SAR data matrix for a single point scatterer.

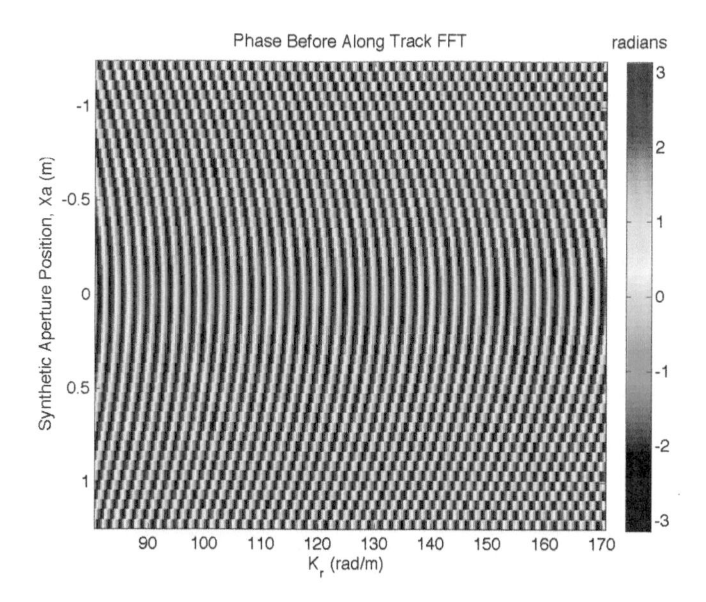

FIGURE 4.4
Phase of the SAR data matrix for a single point scatterer.

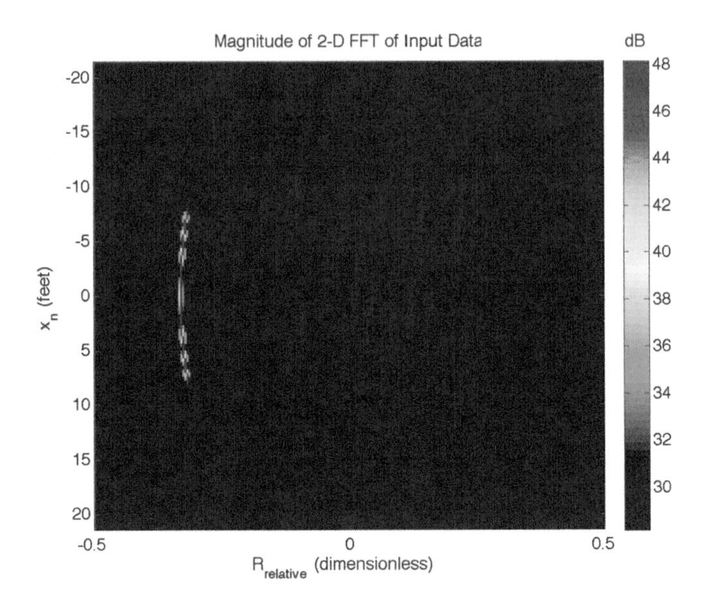

FIGURE 4.5
Magnitude of simulated point scatterer after down range DFT, showing the wave-front curvature of the point scatterer.

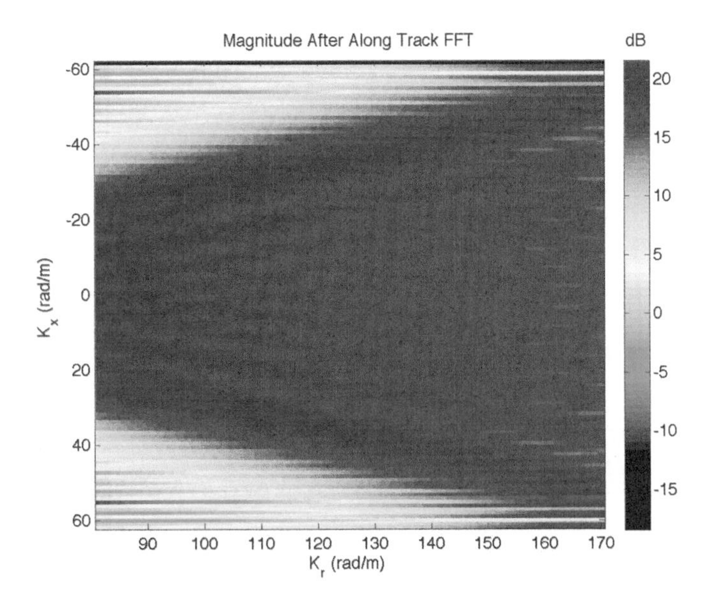

FIGURE 4.6
Magnitude after the cross range DFT.

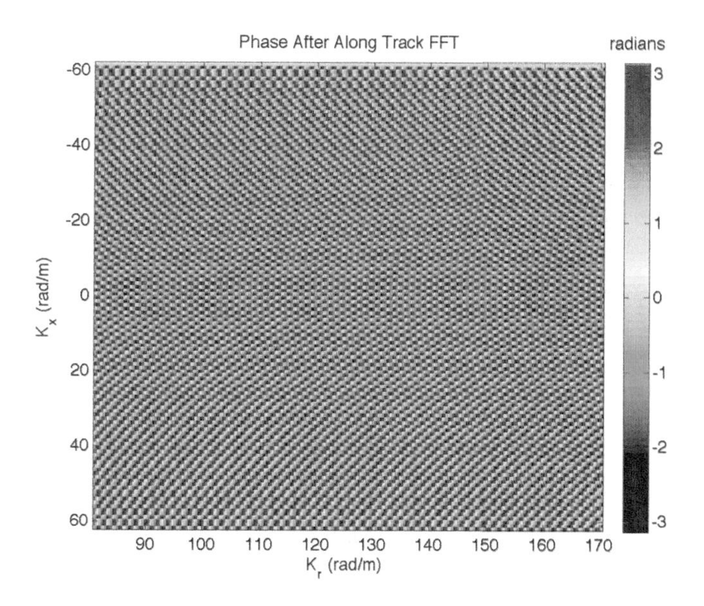

FIGURE 4.7
Phase after the cross range DFT.

4.2.3 Matched Filter

The next step is to apply a matched filter to the $s(k_x, k_r)$ matrix. This matched filter is represented by

$$s_{mf}(k_x, k_r) = e^{jR_s\sqrt{k_r^2 - k_x^2}}, \tag{4.4}$$

where R_s is the down range distance to scene center. In this example and for most radar examples in this book $R_s = 0$, but for long range SAR imaging devices, R_s is a key parameter for accurately processing de-chirp data. Multiplying Equation (4.4) by $s\big(k_x, k_r\big)$ results in

$$s_{matched}(k_x, k_r) = s_{mf}(k_x, k_r) \cdot s\big(k_x, k_r\big). \tag{4.5}$$

But when $R_s = 0$, the matched filter $s_{matched}(k_x, k_r) = 1$ and is not relevant to the SAR algorithm.

The resulting phase of $s_{matched}(k_x, k_r)$ is plotted (Fig. 4.8, where the phase remains unchanged after the matched filter because $R_s = 0$). The down range DFT magnitude is shown (Fig. 4.9).

4.2.4 Stolt Interpolation

Stolt interpolation is the critical step in the RMA. Stolt transforms the 2D SAR data after the matched filter ($s_{matched}(k_x, k_r)$) from the spatial frequency

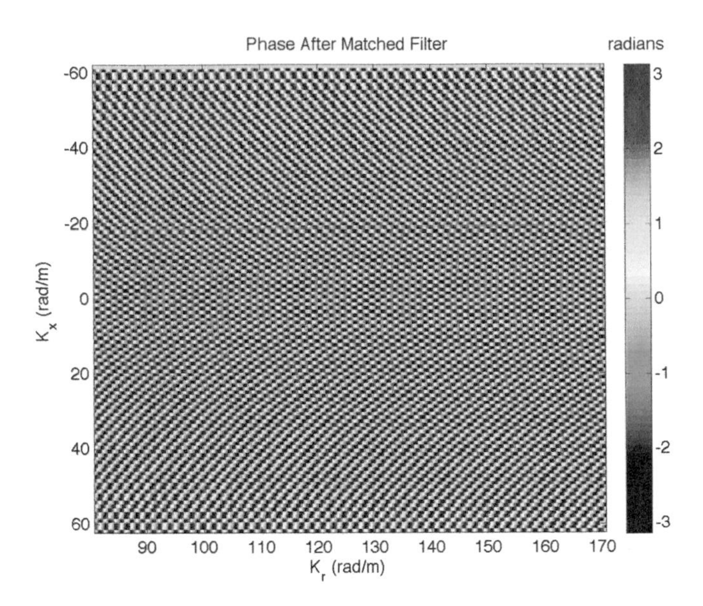

FIGURE 4.8
Phase after the matched filter.

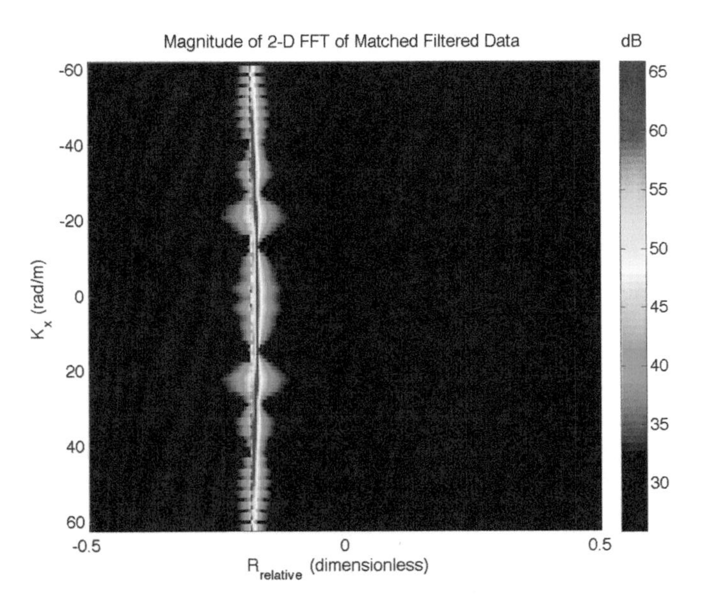

FIGURE 4.9
Magnitude after 2D DFT of matched filtered data.

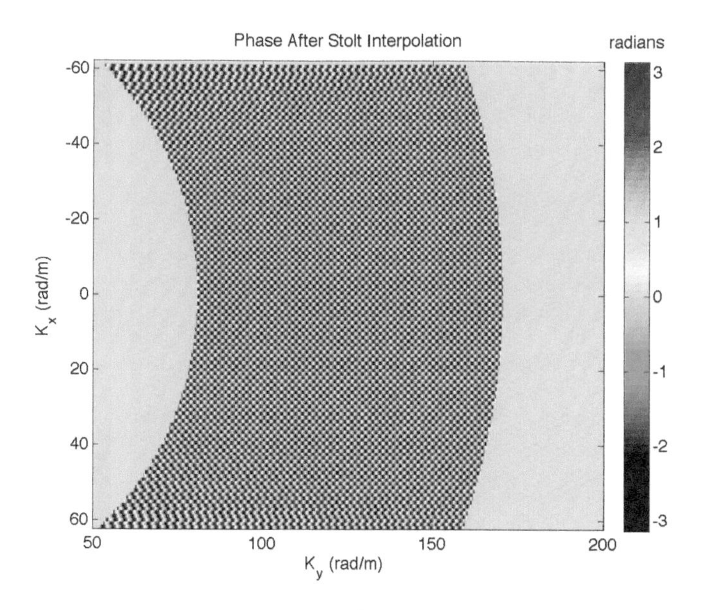

FIGURE 4.10
Phase after Stolt interpolation.

and wave number domain (k_x, k_r) to the the spatial wave number domain (k_x, k_y).

The Stolt relationship between k_y, k_r, and k_x is given by

$$k_y = \sqrt{k_r^2 - k_x^2}. \tag{4.6}$$

With this, a one dimensional interpolation is applied across all the wave numbers (k_r) to map them onto k_y resulting in the Stolt interpolated matrix $s_{st}(k_x, k_y)$. The resulting phase after the Stolt interpolation of the point target is shown (Fig. 4.10).

Stolt interpolated data forms an annulus of data where the lower k_y wave numbers have smaller radii of curvature than the larger wave numbers. Stolt provides a correction for scattered wavefront curvature, meaning that as a wave scatters it has curvature as any other transmitted electromagnetic wave. The larger the radar aperture, the more the curvature is noticeable in the radar data and the larger the wavelength, the more the curvature is noticeable. The Stolt corrects for wavefront curvature by curving the data, in software, back in the opposite direction for each frequency component of the measured scattered field so that Fourier analysis can be used to compute an image.

4.2.5 Inverse Fourier Transform to Image Domain

To convert the Stolt matrix $s_{st}(k_x, k_y)$ into image domain $S(X, Y)$ a subsection of the curved Stolt interpolated data (Fig. 4.10) must be taken. This

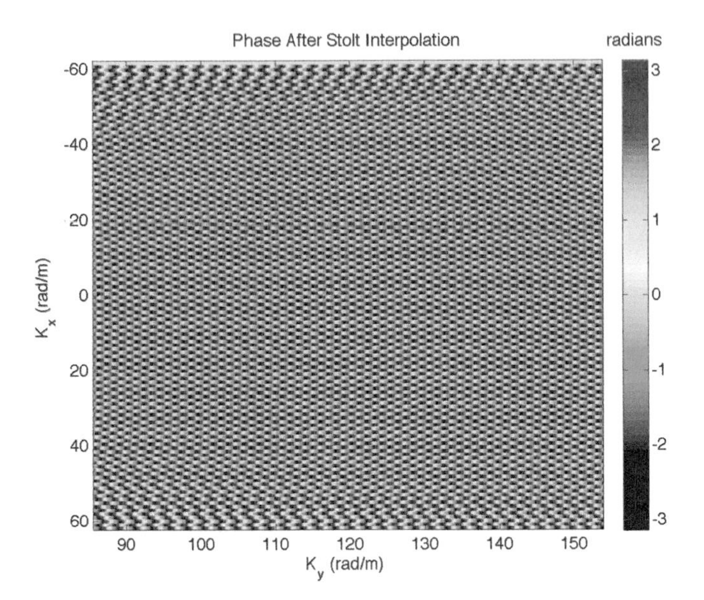

FIGURE 4.11
A subsection of the phase after Stolt interpolation.

subsection is a rectangle. The subsection generally must be taken so that the rectangle is completely filled with data. Sometimes this is not possible when working with narrow band SAR imaging systems, in which case a subsection would not be used and all areas where there are no values in the annulus would be set to 0.

The resulting subsection of the data for the point target is shown (Fig. 4.11). The phase pattern shown is representative of the single point target.

The 2D IDFT is applied (to the rows then columns separately) and the resulting SAR image matrix $S(X, Y)$ is computed. The dB relative magnitude of this SAR image is shown (Fig. 4.12), where the location of the point target is clearly shown at its expected location.

4.3 Simulation of Multiple Point Targets

It is useful to simulate multiple point scatterers to test a SAR imaging algorithm. For this reason Equation (4.3) was modified to represent N number of scatterers

$$s\big(x_n, \omega(t)\big) = \sum_{i=1}^{N} a_{ti} e^{-j2\omega(t)\sqrt{(x_n - x_{ti})^2 + y_{ti}^2}}. \tag{4.7}$$

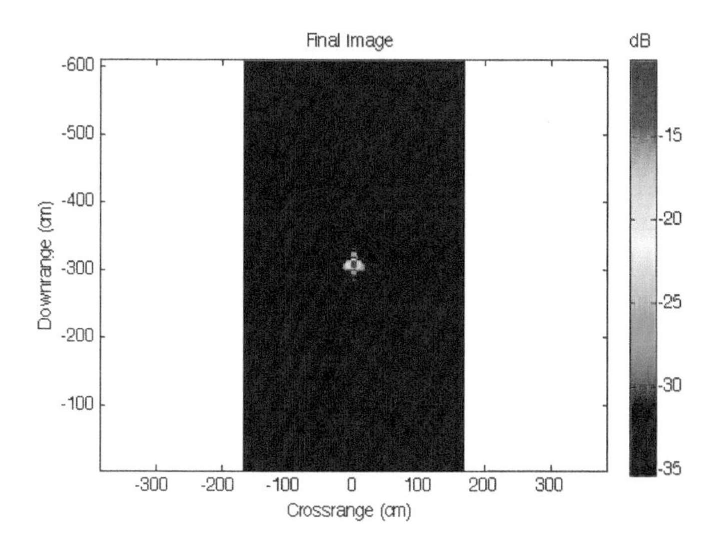

FIGURE 4.12
SAR image of a simulated point scatterer.

Specifically three point scatterers are simulated in this example with scattered amplitudes of $a_{t1} = a_{t2} = a_{t3} = 1$. The location of each scatterer with respect to the rail is

$$(x_{t1}, y_{t1}) = (3, -10) \text{ feet},$$
$$(x_{t2}, y_{t2}) = (-3, -15) \text{ feet},$$
$$(x_{t3}, y_{t3}) = (-2, -10) \text{ feet}.$$

This simulated SAR data matrix was fed into the RMA imaging algorithm described above resulting in the dB relative image shown (Fig. 4.13). The location of each point target is clearly shown demonstrating this algorithm's effectiveness to image multiple point targets at various ranges.

4.4 Estimating Performance

Two important performance parameters for a SAR imaging system are the maximum range to detect a given target RCS (Sec. 4.4.1) and its resolution (Sec. 4.4.2). There are many additional metrics that can also be considered but these are beyond the scope of this book.

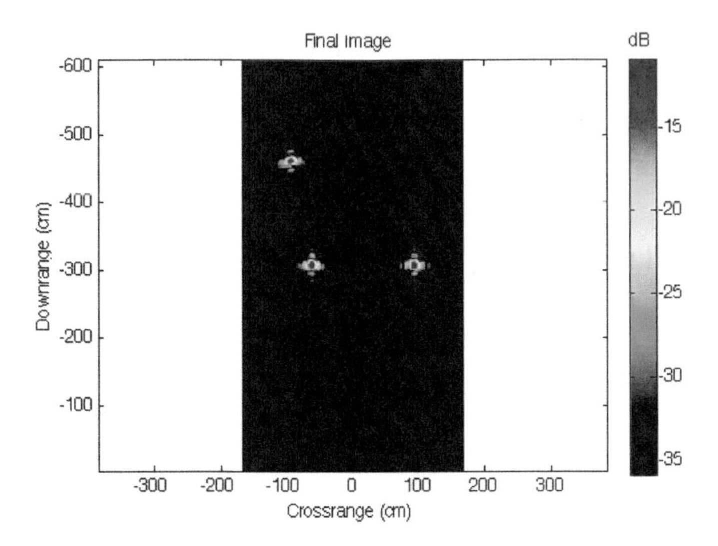

FIGURE 4.13
SAR image of three simulated point scatterers.

4.4.1 The Radar Range Equation Applied to SAR

The maximum range for a small SAR imaging system, like those discussed in this book, where the radar antennas illuminate the target scene across the entire length of aperture, can be estimated by multiplying the performance of an FMCW radar by N, the number of range profiles acquired across the aperture.

The maximum range to target for a small SAR is given by

$$R_{max}^4 = \frac{N P_{ave} G_{tx} A_{rx} \rho_{rx} \sigma e^{(2\alpha R_{max})}}{(4\pi)^2 k T_o F_n B_n \tau F_r (SNR)_1 L_s}, \tag{4.8}$$

where:

$R_{max} =$ maximum range of radar system (m)
$P_{ave} =$ average transmit power (watts)
$G_{tx} =$ transmit antenna gain
$A_{rx} =$ receive antenna effective aperture (m^2)
$\rho_{rx} =$ receiver antenna efficiency
$\sigma =$ radar cross section (m^2) for target of interest
$L_s =$ miscellaneous system losses
$\alpha =$ attenuation constant of propagation medium
$F_n =$ receiver noise figure (derived from procedure outlined in Sec 1.1.5.4)

$k =$ $1.38 \cdot 10^{-23}$ (joul/deg) Boltzmann's constant
$T_o =$ 290°K standard temperature
$B_n =$ system noise bandwidth (Hz)
$\tau F_r =$ 1, radar duty cycle which is assumed to be CW radar for most examples in this book
$(SNR)_1 =$ single-pulse signal-to-noise ratio requirement
$N =$ number of range profiles used in synthesizing the aperture

For an FMCW radar, the noise bandwidth is inversely proportional to the discrete sample length $B_n = 1/t_{sample}$. For an impulse radar, the noise bandwidth is simply the -3 dB roll-off frequency (f_{-3db}) of the anti-aliasing filter.

For a direct conversion radar (this applies to all FMCW radar examples in this book) the noise bandwidth is twice as wide as a radar following an image rejection architecture where $B_n = 2/t_{sample}$ for an FMCW radar or simply f_{-3db} for an impulse radar.

4.4.2 Resolution of SAR Imagery

Range resolution for SAR imaging systems using linear FM waveforms is the same as previously described (Sec. 3.2.2).

Cross range resolution is the minimum cross range distance over which a SAR can differentiate between two targets. Cross range resolution depends on the length of the array (or aperture) L and the location of the point target being measured relative to the front of the array in both down range and cross range. Cross range resolution is calculated by [4]

$$\rho_{cr} = \frac{\lambda_c K_a R_t}{2L \sin \phi_{dc} \cos(\Delta\theta/2)}, \tag{4.9}$$

where R_t is the range to the point target, ϕ_{dc} is the angle from the center of the aperture to the point target. For linear rail SAR and near-field phased array imaging systems in this book $\phi_{dc} = \pi/2$. $\Delta\theta$ is the change in target aspect angle from 0 to L across the aperture, and K_a is the cross range excess bandwidth factor which depends on the weighting function applied before the IDFT (Table 3.1).

For linear rail SAR imaging sensors, a useful equation for computing $\Delta\theta$ given a target's range R_t and cross range x_t location, is

$$\Delta\theta = \left[\frac{\pi}{2} - \arctan\left(\frac{R_t}{(L/2 + x_t)}\right)\right] + \left[\frac{\pi}{2} - \arctan\left(\frac{R_t}{(L/2 - x_t)}\right)\right]. \tag{4.10}$$

4.5 Additional Processing

In this book, examples of imaging sensors are shown where the position of the radar sensor is inherently well known because each radar traverses a linear rail or is electronically scanned across an antenna array. Processing discussed in this section will include a method of calibration (Sec. 4.5.1) and coherent background subtraction to reduce clutter (Sec. 4.5.2). In addition to these methods, there is a great deal of processing beyond the scope of this book to improve SAR images including, auto focus, multi-spectral imaging, cross-polarized imaging, and others.

4.5.1 Calibration

Practical radar imaging systems are non-ideal, each having its own unique set of problems that affect measurements. The purpose of calibration is to sharpen the SAR imagery where we compensate for non-ideal transmitter and receiver transfer functions (e.g., the effect of the various transmitter and receiver components in the signal chain on what is transmitted and what is received).

In this book a point target is used as a calibration (cal) target because it is easier to work with and less expensive. Our objective is to sharpen the imagery to a sufficient degree to demonstrate imaging principles. This point target is usually a metal rod or a pole of tall enough dimension to provide a high SNR.

The cal target is placed at a known location down range from the radar. A range profile is acquired of the pole represented by $s_{pole}\big(\omega(t)\big)$. The pole is then removed and a background range profile is acquired; the result is represented by $s_{calback}\big(\omega(t)\big)$. The background is subtracted from the pole range profile resulting in a clean range profile of the pole only,

$$s_{cal}\big(\omega(t)\big) = s_{pole}\big(\omega(t)\big) - s_{calback}\big(\omega(t)\big). \tag{4.11}$$

The cal data is referenced to a theoretical point scatterer

$$s_{caltheory}\big(\omega(t)\big) = e^{-j2k_r R_{pole}}, \tag{4.12}$$

where R_{pole} is the range to the cal pole and $k_r = \omega(t)/c$. The cal factor is calculated,

$$s_{calfactor}\big(\omega(t)\big) = \frac{s_{caltheory}\big(\omega(t)\big)}{s_{cal}\big(\omega(t)\big)}. \tag{4.13}$$

After the SAR data is acquired this cal factor is multiplied by each range profile before the results are fed to the SAR imaging algorithm.

4.5.2 Coherent Background Subtraction and Coherent Change Detection (CCD)

CCD is useful when the user wants to observe changes to a target scene, where the user acquires an image of a target scene then re-visits that target scene sometime later. The newest image is subtracted from the previous one to show what has changed. If done coherently (using both magnitude and phase data) this process is very sensitive to change, showing what is new or what is missing from the first image. This can also be done incoherently (with magnitude only data) to a lesser degree of effectiveness.

Coherent background subtraction is typically used in the laboratory setting to reduce static clutter due to imperfect target scenes and often used in small rail SAR systems like those discussed in this book. Background subtraction is implemented by first measuring the target scene without the target of interest present $s_{back}\big(x(n), \omega(t)\big)$. Next, place the target of interest in the scene and re-measure $s_{scene}\big(x(n), \omega(t)\big)$. The resulting background-subtracted data set is the difference between the target scene with and without the target placed:

$$s_{targets}\big(x(n), \omega(t)\big) = s_{scene}\big(x(n), \omega(t)\big) - s_{back}\big(x(n), \omega(t)\big), \tag{4.14}$$

where $s_{targets}\big(x(n), \omega(t)\big)$ is fed into the SAR imaging algorithm producing an image with significantly less background clutter.

We have shown coherent change detection (CCD) applied to range profiles acquired by an FMCW radar (Sec 3.1.3.3), where the current range profile is subtracted from the previous one revealing only moving targets. CCD can also be applied to SAR imagery and coherent background subtraction can also be considered one example of CCD. More sophisticated approaches to CCD exist that are more robust to phase errors, but in this book, the method discussed above will be used for all CCD and background subtracted data. The reader is encouraged to implement more robust CCD approaches using the data sets provided by this book.

4.5.3 Motion Compensation

Motion compensation is a technique used to account for the non-ideal motion of a moving SAR imaging sensor (ideal motion can be easily accounted for, such as constant platform velocity). This is a practical consideration, where almost all aircraft (or any moving vehicle) tend to bounce around adding error to the SAR measurement geometry. This motion must be accounted for either prior to image formation or during image formation. In this book only rail SAR and near-field phased array imaging sensors are considered where motion compensation is unnecessary because the location of each range profile is well known. However, for any application that requires a moving airborne or ground vehicle to acquire the aperture of data, motion compensation must be considered and for these applications the reader is encouraged to study further [2].

4.6 Summary

SAR is a method of synthesizing an aperture across the known path of a moving radar sensor. This aperture can be very large and is in effect a phased array radar equal to the length of the path of the radar sensor. High resolution can be achieved by synthesizing this large aperture. SAR technology is a valuable ground mapping tool but it can also be implemented on a small scale by moving a UWB radar sensor along a mechanized rail. To this end, a SAR imaging algorithm was described for rail SAR and near-field imaging sensors. Such a SAR imaging algorithm would be useful for RCS measurement, observing nature, studying scattering phenomenology, through-wall imaging, or any application where near-field beamforming is valuable.

Bibliography

[1] G.W. Stimson, *Introduction to Airborne Radar,* Hughes Aircraft Company, El Segundo, California, 1983.

[2] W.G. Carrara, R.S. Goodman, and R.M. Majewski, *Spotlight Synthetic Aperture Radar Signal Processing Algorithms,* Artech House, Boston, MA, 1995.

[3] G.L. Charvat, "Low-Cost, High Resolution X-Band Laboratory Radar System for Synthetic Aperture Radar Applications," Antenna Measurement Techniques Association Conference, Austin, Texas, October 2006.

[4] W.G. Carrara, R.S. Goodman, and R.M. Majewski, *Spotlight Synthetic Aperture Radar Signal Processing Algorithms,* Artech House, Boston MA, 1995, eq. 2.16 p. 36.

5

Practical Examples of Small Synthetic Aperture Radar Imaging Systems

SAR imaging sensors are useful in imaging or mapping a scene from a stand-off range, where the image provided is a bird's eye view. Small SAR sensors are used for numerous imaging applications including measuring target scattering, imaging through walls of a building, medical imaging, or proving the concept of a much more complicated array-based system without significant investment. Small SAR imaging sensors provide range versus range imagery of near-field target scenes.

The least complex SAR imaging systems can be implemented by placing a small radar sensor onto a linear rail and moved (manually or automatically) along a known path while acquiring range profiles. In this chapter, working examples of rail SAR imaging systems will be shown including a UWB X-band FMCW rail SAR (Sec. 5.1), the MIT coffee can radar in SAR mode (Sec. 5.2), and two examples of the range-gated FMCW radar configured to SAR image at S- and X-bands (Sec. 5.3).

5.1 UWB FMCW X-Band Rail SAR Imaging System

A high resolution UWB rail SAR at X-band was created for the purposes of learning how to develop SAR imaging algorithms and technology [1]-[4]. This radar produces high resolution SAR imagery of small target scenes including model airplanes, vehicles, groups of point targets, and other items.

The radar front end (Sec. 3.3.1) is mounted to a 96 inch long linear rail where a stepper motor controller within the radar CTRL chassis moves the radar down the length of the rail at pre-defined increments (Fig. 5.1) of either 1 or 0.5 inch.

This system was deployed in the author's back yard (Fig. 5.2). To prepare the range, the lawn was mowed to reduce in-scene clutter. Targets were placed down range on a styrofoam table about 20 feet from the rail.

The linear rail is made from an old Genie garage door opener with a carrier plate that is pulled by a lead screw inside an aluminum extrusion. The radar front end is bolted to an aluminum plate that is screwed down to a carrier that mates with the lead screw (Fig. 5.3a). The lead screw is lubricated with axial grease. The stepper motor does not provide enough torque to pull the lead

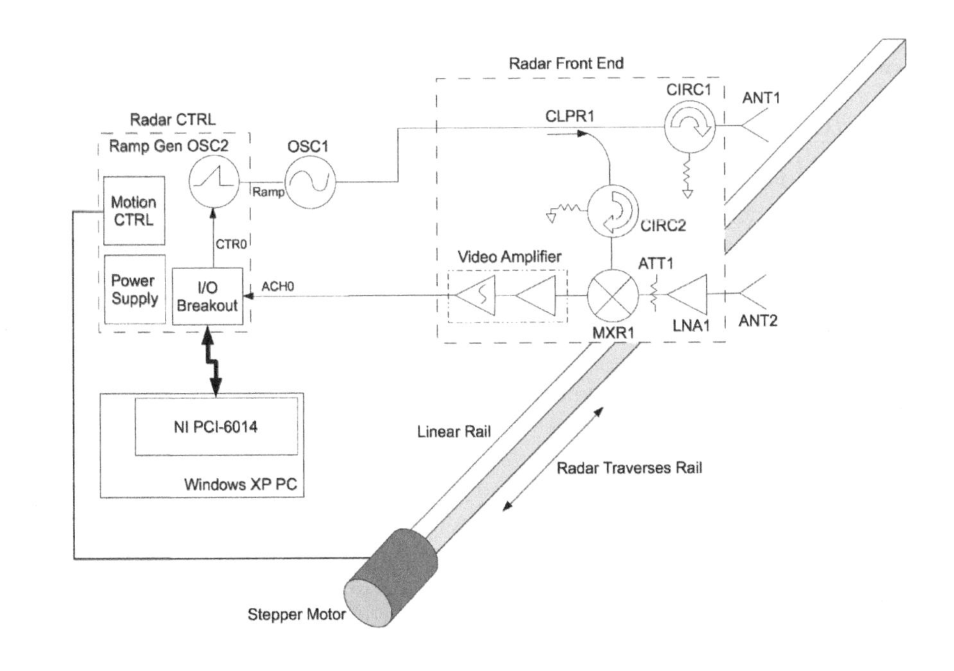

FIGURE 5.1
Block diagram of the UWB FMCW X-band rail SAR imaging system.

screw by itself; for this reason a 6:1 planetary gear transmission from an old cordless drill is used to gear down the stepper to drive the lead screw (Fig. 5.3b through d). These assemblies can be fabricated with basic machining skills. Precision of the linear rail is sufficiently good to support rail SAR imaging.

The X-band YiG Oscillator OSC1 chirps from 7.835 GHz to 12.817 GHz over the duration of the up-ramp providing a chirp bandwidth (BW) of 4.982 GHz. The radar CTRL chassis and a Windows XP computer are mounted in a small 4 foot tall rack. The RF output of OSC1 is fed through a low-loss flexible microwave cable to the radar front end. The rail moves very slowly, requiring approximately 20 minutes to acquire one image. Radar operators sit in lawn chairs behind the rail. A technical description of this UWB X-band FMCW radar was detailed previously (Sec. 3.3.1).

5.1.1 Expected Performance

The maximum range and minimum target RCS will be discussed (Sec. 5.1.1.1) and range resolution estimates will be made (Sec. 5.1.1.2).

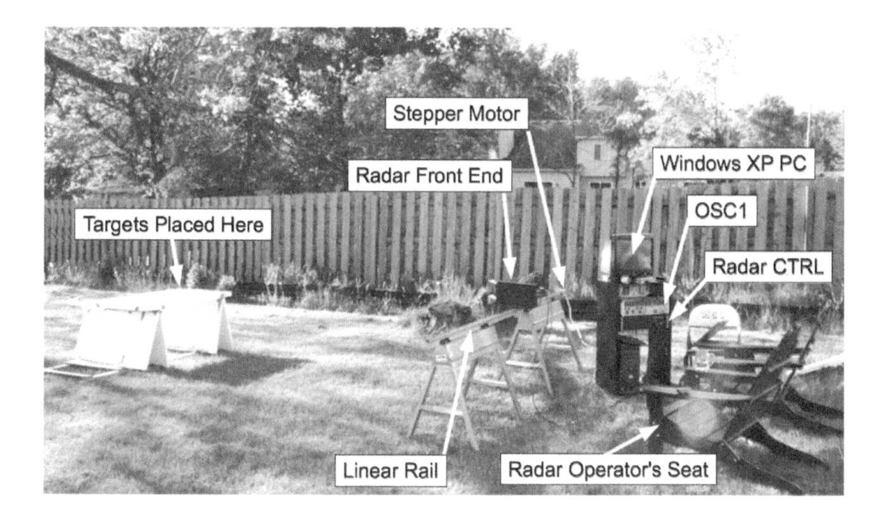

FIGURE 5.2
The UWB FMCW X-band rail SAR imaging system deployed in the author's backyard with call-outs showing the major subsystems.

TABLE 5.1
UWB FMCW X-band rail SAR specifications.

$$
\begin{aligned}
P_{ave} =& \; 31 \cdot 10^{-3} \text{ (watts)}\\
G_{tx} =& \; 17 \text{ dBi antenna gain}\\
G_{rx} =& \; 17 \text{ dBi antenna gain}\\
A_{rx} =& \; G_{rx}\lambda_c^2/(4\pi) \text{ (m}^2\text{)}\\
\lambda_c =& \; c/10 \cdot 10^9 \text{ (m) wavelength of carrier frequency}\\
f_c =& \; 10 \text{ GHz radar center frequency}\\
\rho_{rx} =& \; 1 \text{ because antenna efficiency is accounted for in antenna gain}\\
\sigma =& \; 10 \text{ (m}^2\text{) for automobile at 10 GHz}\\
L_s =& \; 6 \text{ dB miscellaneous system losses}\\
\alpha =& \; 0 \text{ attenuation constant of propagation medium}\\
F_n =& \; 4 \text{ dB receiver noise figure}\\
B_n =& \; 2/t_{sample} \text{ system noise bandwidth (Hz) where } t_{sample} = 10 \text{ ms}\\
(SNR)_1 =& \; 13.4 \text{ dB}\\
N =& \; 96 \text{ number of range profiles used in synthesizing the}\\
& \; \text{aperture for 1 inch spacing across the rail}
\end{aligned}
$$

5.1.1.1 Maximum Range and Minimum Target RCS

Substituting specifications (Table 3.10) into the radar range equation (4.8), the maximum range is estimated to be 3.7 km for a 10 dBsm automobile target using the MATLAB® script [5].

The noise figure is an educated guess based on the apparent age of the surplus microwave amplifier LNA2, and the IDFT is applied over 10 ms

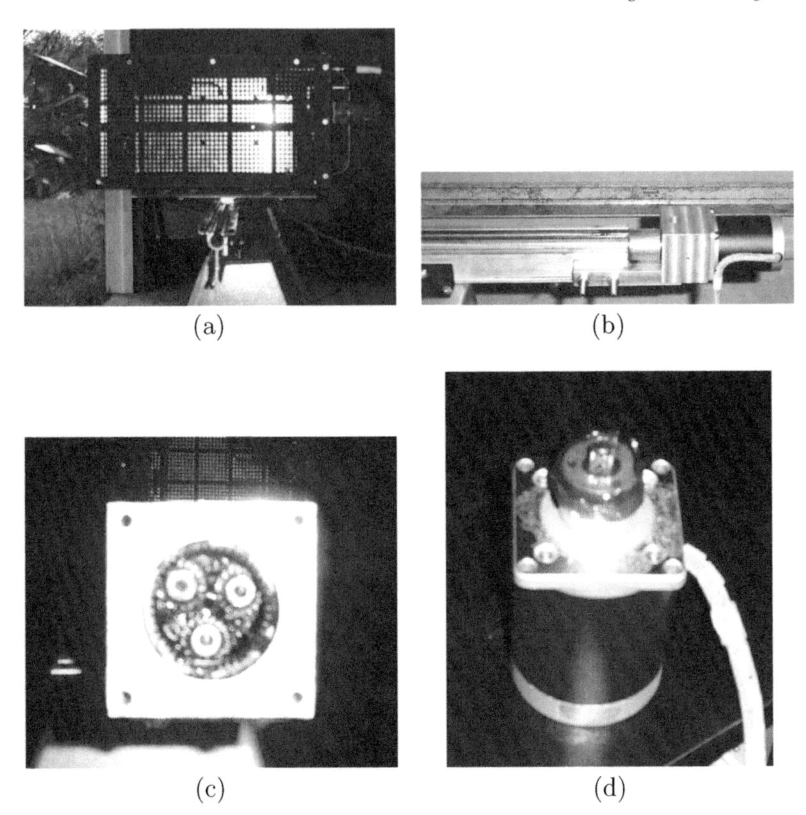

FIGURE 5.3
Radar front end mounted on rail (a), stepper motor and transmission assembly at opposite end of rail (b), inside view of transmission assembly with motor removed (c), stepper motor with pinion gear attached as it would mount into the transmission assembly (d).

up-chirps with a direct conversion receive architecture resulting in a 200 Hz effective noise bandwidth. The number of range profiles required to process a SAR image is 96 ($= N$).

Unfortunately the video amplifier (Fig. 2.13) limits the maximum range to $f_{-3db}c/c_r = 48.2$ m, where $f_{-3db} = 80$ KHz and $c_r = BW/t_{sample} = 498.2$ GHz/s. Therefore, the radar range equation can be solved for σ by substituting $R_{max} = 48.2$ m, thereby showing that the minimum radar cross section that can be detected within the 48.2 m maximum range is −65.5 dBsm. This can be estimated using the same MATLAB script [5].

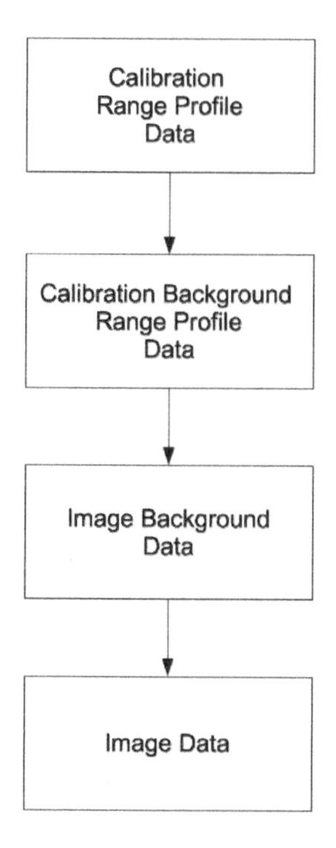

FIGURE 5.4
Rail SAR data collection follows this block diagram where each rail SAR data acquisition follows these steps.

5.1.1.2 Range Resolution Estimate

Expected range resolution is 2.7 cm and cross range with no weighting ($K_r = 0.89$). The cross range resolution is expected to be 2.8 cm when the targets are located 5 m down range, centered with respect to the rail with a length of 96 inches, with no weighting ($K_r = 0.89$). This too can be estimated using the same MATLAB® script [5].

5.1.2 Implementation

The experimental setup is shown (Fig. 5.2). To acquire an image, four data sets are collected to calibrate and subtract background clutter following the block digram (Fig. 5.4).

A 3/8 inch diameter 24 inch long aluminum rod was chosen as the point calibration target for this radar system because it is significantly smaller than the expected range resolution and it provides a high SNR return. The calibration range profile is acquired by measuring the rod as it is held vertical in Styrofoam. The calibration background range profile is acquired by measuring the Styrofoam without the rod present. An image background is acquired without targets present by scanning the radar across the 96 inch aperture at 0.5 or 1 inch increments acquiring a range profile for each with no targets present. Finally, the target that is to be measured is placed 20 feet in front of the radar and the radar is again scanned across the rail providing the image data set.

Data is processed into an image following the block diagram (Fig. 5.5). Calibration and calibration background range profiles are used to calculate the calibration coefficients providing an amplitude scale and phase shift for each data point in the range profile (Sec. 4.5.1). Calibration coefficients are multiplied by each range profile for both the image and image background data sets.

If a background image data set was acquired, then coherent background subtraction can be used to subtract out the static (non-moving) background clutter and the transmit-to-receive antenna coupling. If there is no background subtracted data, or if the user prefers to image without background subtraction, then the transmit-to-receive antenna coupling will dominate the image data. One method to remove this coupling, the average range profile is computed across all range profiles in the image data set. This mean range profile can be subtracted from each individual range profile in the data set, thereby removing the transmit-to-receive coupling because the only signal that is always present and dominating the mean is the transmit-to-receive coupling. The result is fed into the image formation algorithm described previously (Chapter 4) and displayed.

5.1.3 Measured Results

Resolution of the UWB X-band rail SAR imaging sensor is evaluated followed by measuring imagery of a variety of target scenes.

5.1.3.1 Resolution

A group of point targets were imaged to test resolution. When imaging point targets, it helps to locate each in a known configuration that is recognizable in the image. This group of targets spells out GO STATE in push pins (thumb tacks) which are convenient point targets at X-band. The measured image clearly shows each target (Fig. 5.6).

A bright point target is selected that is some distance away from nearby targets, the point target at the middle top of G is approximately 290 cm down range and 25 cm cross range. Down range resolution measured 4.2 cm which is close to the theoretical best possible range resolution of 2.7 cm. Cross range resolution measured 4 cm, which is somewhat close to the expected 1.7 cm cross range resolution (this data is available for the reader to process [6]).

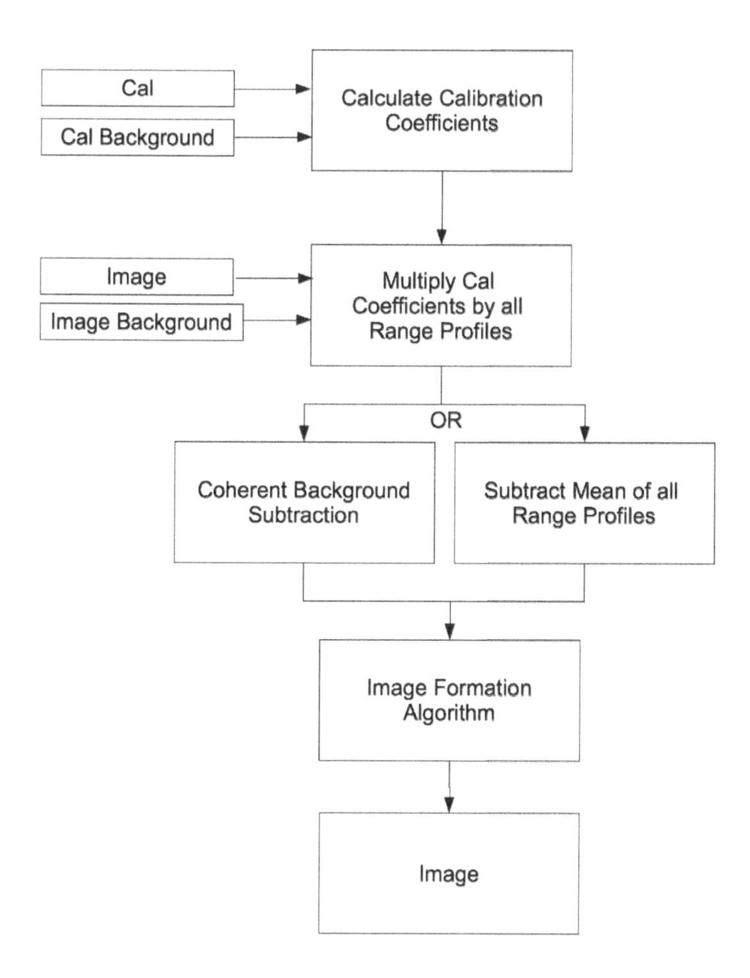

FIGURE 5.5
Data processing block diagram: calculate calibration coefficients, multiply calibration coefficients by all background and image data, perform coherent background subtraction or subtract the means of all range profiles which have the persistent TX-RX coupling, image formation, display image.

5.1.3.2 Sensitivity

A group of −55 dBsm spheres were imaged to test sensitivity. These spheres are steel BB's found in sporting good stores and each has a diameter of 4.3 mm. Spheres were placed in a recognizable configuration GO STATE so their locations would be observable in close proximity to clutter at

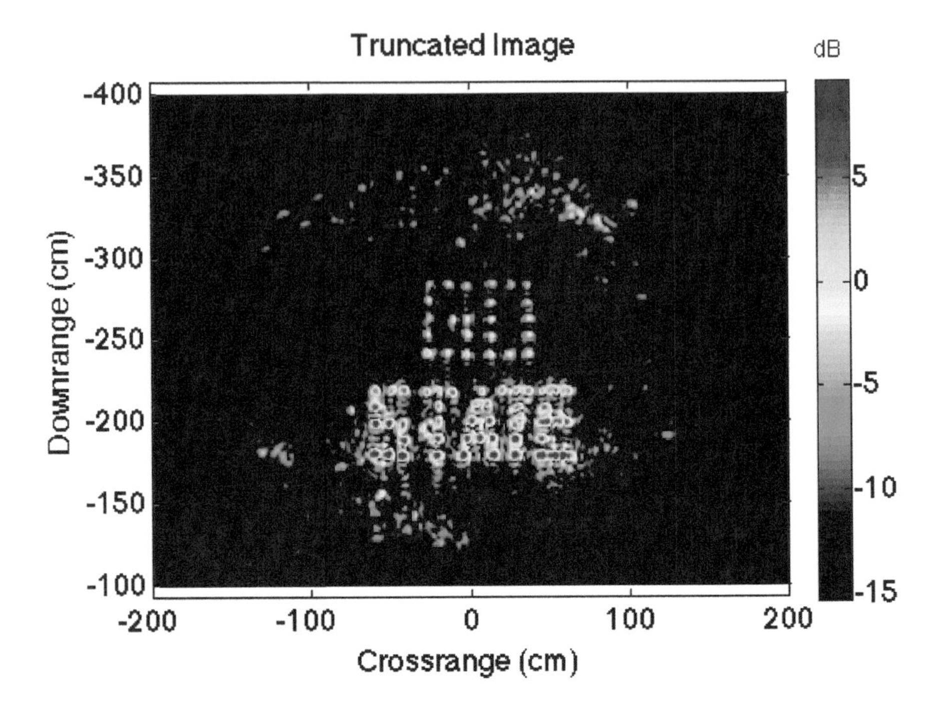

FIGURE 5.6
GO STATE in push pins (plastic thumbtacks); imaging groups of point targets allow the range resolution to be assessed.

or near their scattered magnitudes. Nearly every sphere was imaged (Fig. 5.7) demonstrating the radar's sensitivity is at least −55 dBsm at 175 cm down range.

5.1.3.3 Imagery

Now that the performance of this system has been verified, other more interesting targets can be imaged. As an example of a small target, a scale model (1:32 scale) F14 was imaged (Fig. 5.8a), showing the nose, inlets, wings, and tail wings. The outline of the aircraft model is clearly shown.

A scale mode (1:48) B52 was imaged (Fig. 5.8b), clearly showing the engine inlets, the nose, outline of the wings, and the tail. The engine inlets are the brightest targets in this image.

The author's Cannondale M300 mountain bike was imaged (Fig. 5.9a) showing the frame, handle bars, saddle, chain, and a faint outline of the wheels. The data and a MATLAB processing script are available [7].

The author's 5.0 Mustang was imaged (Fig. 5.9b). The front and side outline of the Mustang is clearly visible. The boundary between the windshield

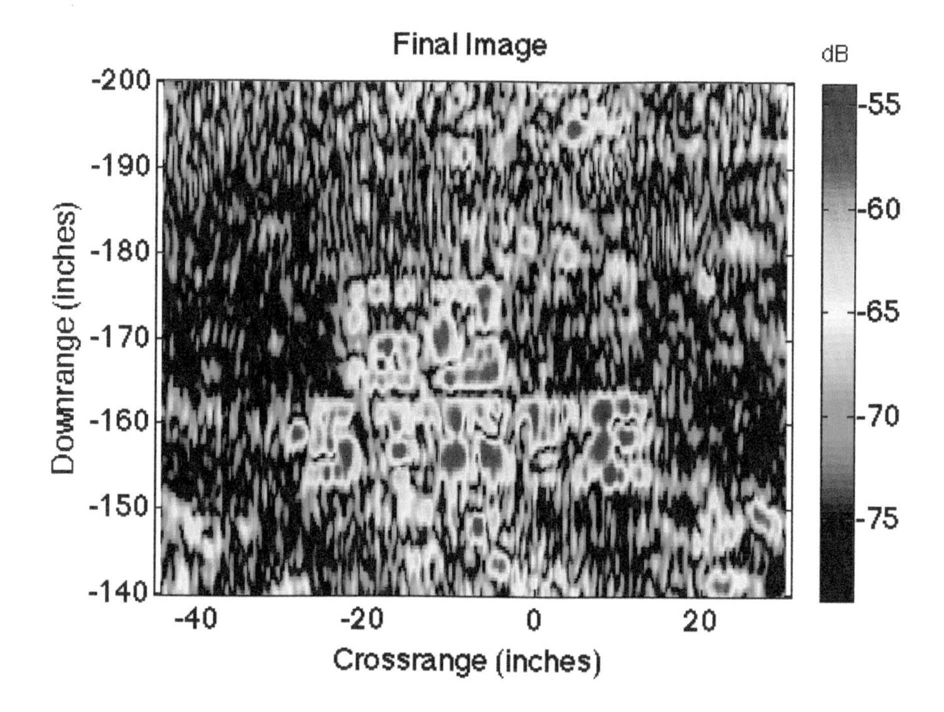

FIGURE 5.7
GO STATE in 4.3 mm diameter metal spheres that have an RCS of approximately −55 dBsm at X-band; SAR sensitivity can be measured by imaging low RCS targets and groups of low RCS targets are preferred to a single target because they become easier to locate when clutter is competing with the desired target returns.

and hood of the car is also clearly shown. Both the right-side mirror and rear-view mirror are also shown. The data and a MATLAB processing script are available [8].

A video demo was made to show this system in operation [9], where a ship's wheel and a 2012 Ford Focus were imaged (Fig. 5.10). This data is available to the reader including MATLAB scripts to process the data [10].

5.2 MIT Coffee Can Radar in Imaging Mode

The coffee can radar was developed to be an instructional tool to show students how radar systems work by building and using it in a series of field experiments [11]–[13]. In addition to Doppler and FMCW ranging, this radar is capable of crude SAR imaging.

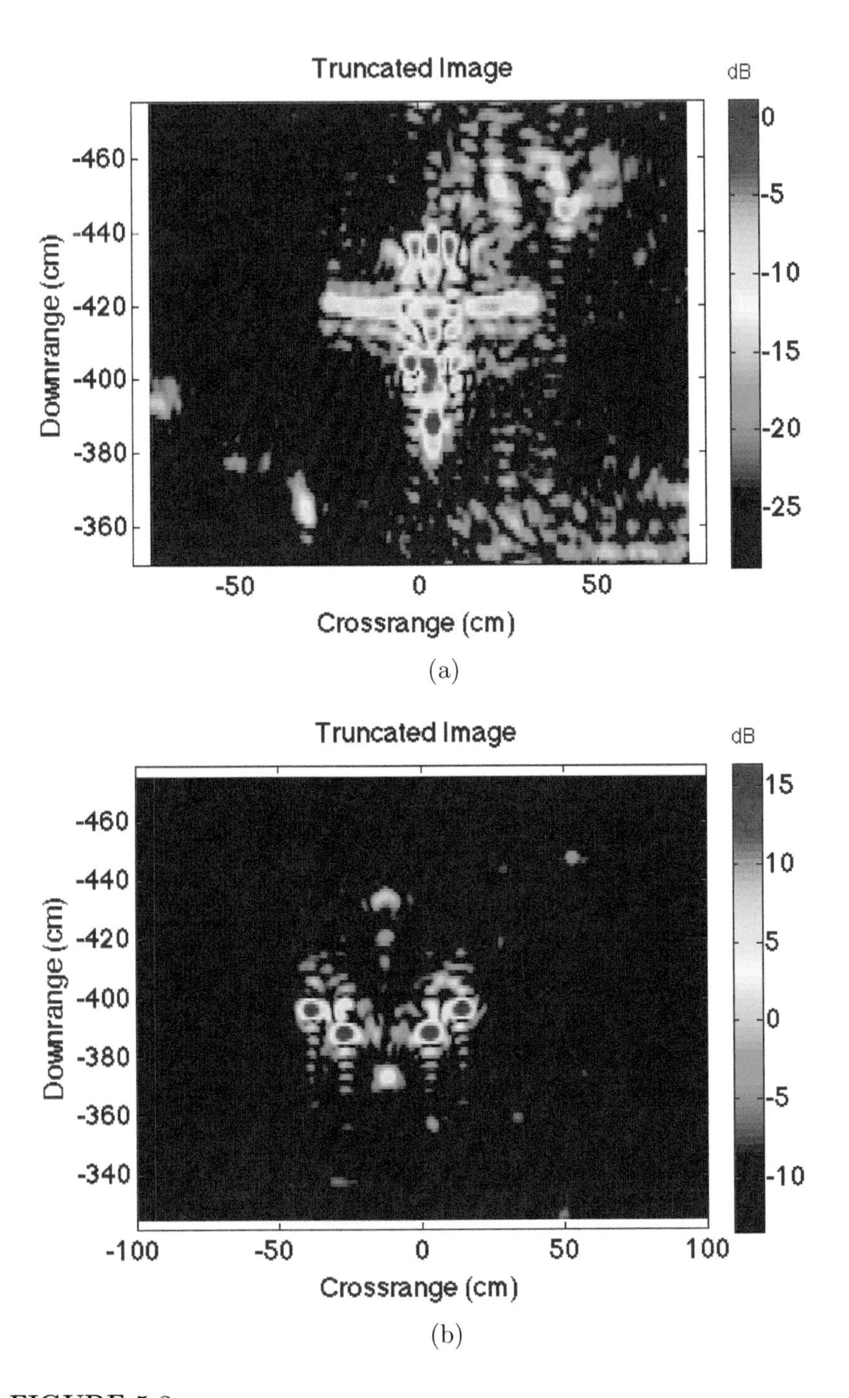

FIGURE 5.8
X-band SAR imagery of model aircraft including a 1:32 scale F14 (a) and 1:48 scale B52.

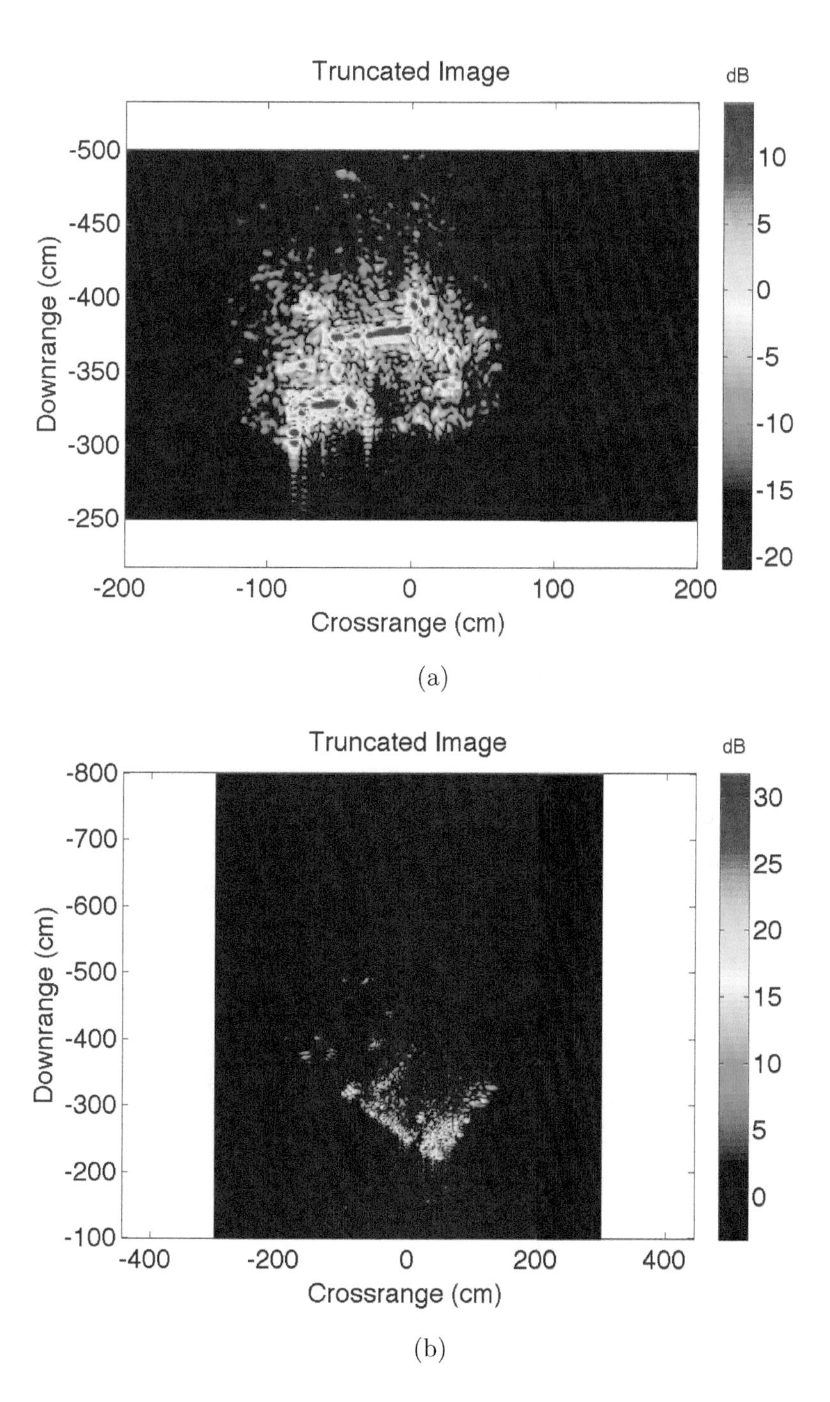

FIGURE 5.9
X-band SAR imagery of a mountain bike (a) and the author's car in grad school, a 5.0 Mustang (b).

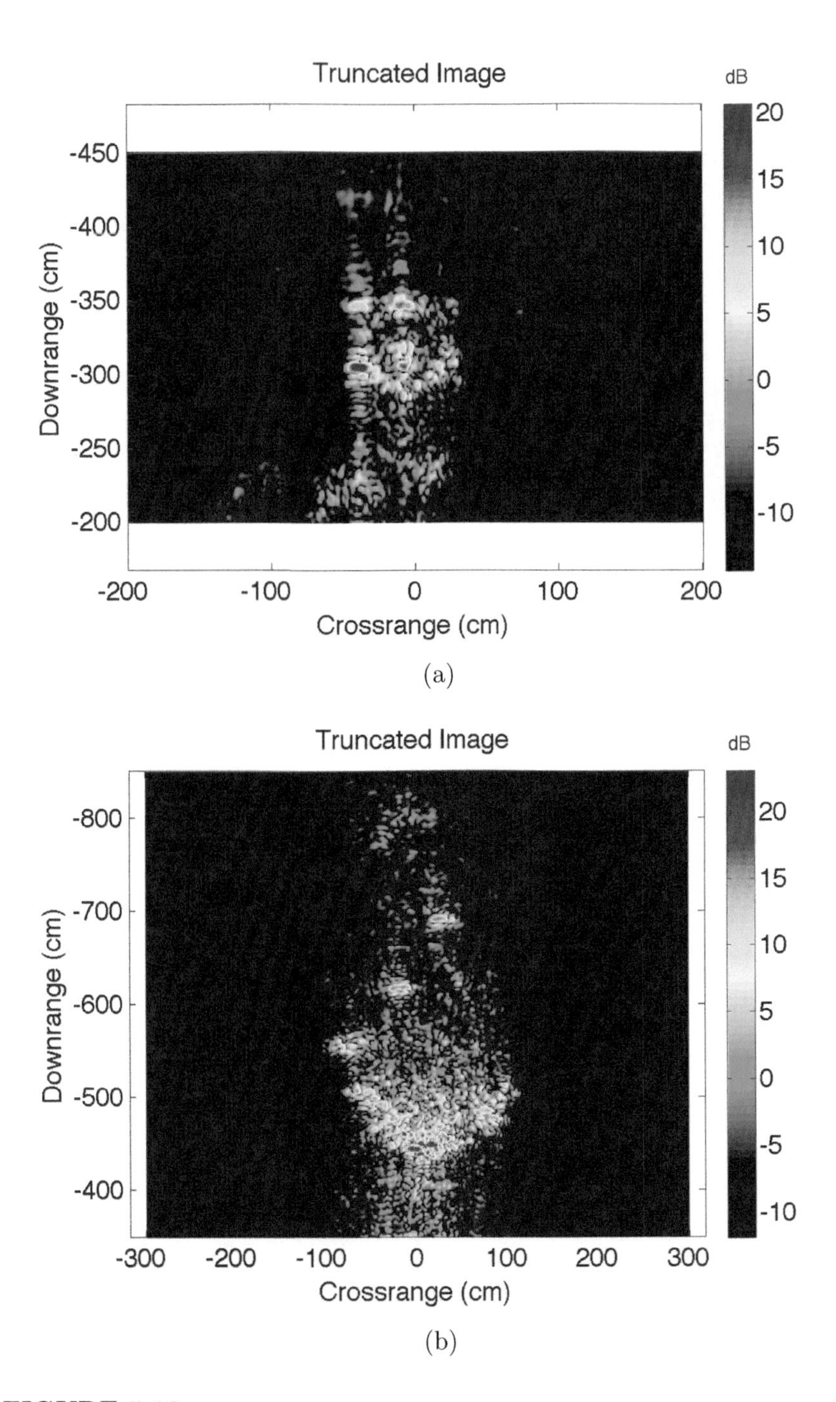

FIGURE 5.10
X-band SAR imagery of a ship's wheel (a) and the author's car, a 2012 Ford Focus (b).

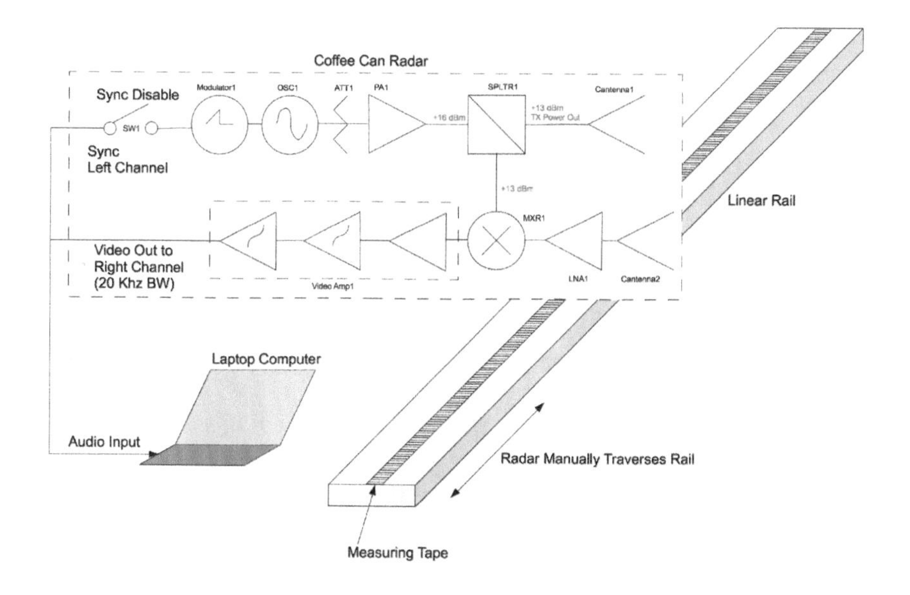

FIGURE 5.11
SAR imaging using the MIT coffee can radar system, where a measuring tape is placed across a long flat surface and the radar is manually moved every 2 inches and the left channel is interrupted to indicate to the computer where a new range profile should be processed (call-outs in Fig 2.5b).

This radar follows a coherent FMCW architecture (Sec. 3.3.2); it is built on a wood board with coaxial microwave components for ease of construction, analog components built on a solderless breadboard, and coffee cans are used as open-ended waveguide antennas (Fig. 2.4). This radar uses the audio input port of a laptop computer to digitize the video and synchronization pulses which correspond to the up-ramp FMCW modulation of the radar.

To acquire a SAR image data set the radar must record range profiles at approximately $\lambda/2$ spacing across a linear rail. To reduce costs the student is expected to manually move the radar every 2 inches and record range profiles. A tape measurer is supplied with the radar kit (Fig. 5.11).

The MATLAB script must know where the radar is at each increment in the recording. Right and left channel data is recorded continuously by a wav ($*$.wav) recorder on a laptop computer. To show the MATLAB script where the radar is located, synchronization pulses on the left audio input channel are muted to denote change in position. This is achieved by opening the toggle switch SW1 when the radar is moved and closing it when the radar is lined up with the desired 2-inch location on the measuring tape (Fig 5.12).

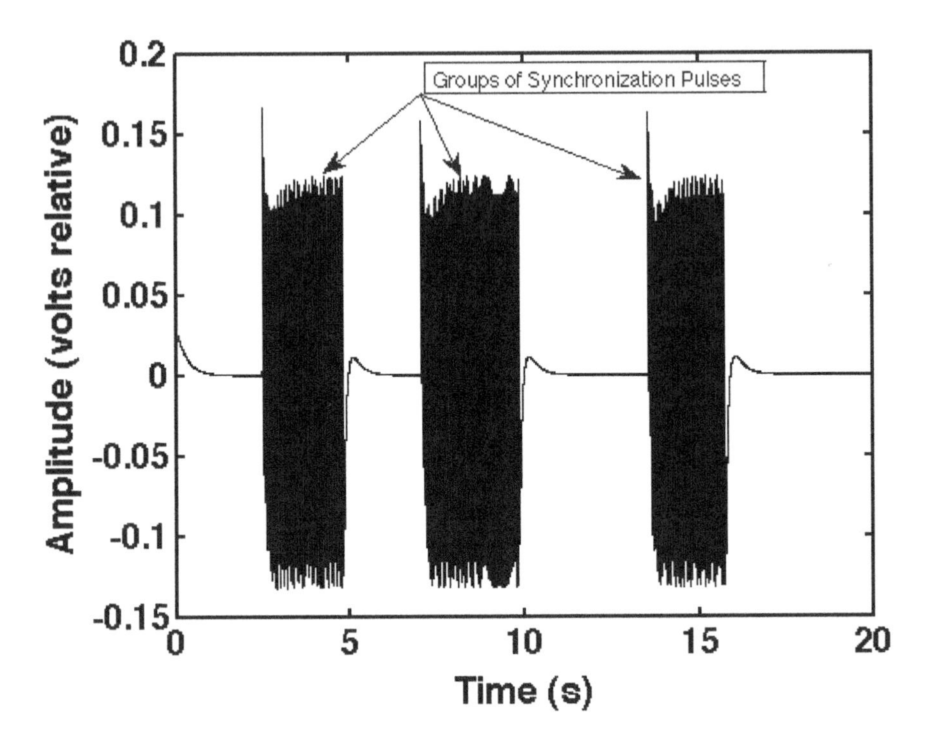

FIGURE 5.12
Synchronization pulses on the left channel are interrupted by toggle switch SW1 whenever the radar is moved between 2 inch increments, thereby showing the MATLAB script the position of the radar along the linear rail.

The MATLAB script parses through the left channel first, searching for groups of synchronization pulses and saving both the group of synchronization pulses and the corresponding right-channel video data to an array. It then parses through each of these arrays, coherently integrating all video data lined up with their synchronization pulses within the pulse group. These video data for each two inch position are then fed to an RMA SAR imaging algorithm which produces the SAR image.

5.2.1 Expected Performance

The maximum range and minimum RCS will be estimated (Sec. 5.2.1.1) followed by an estimate of the range resolution (Sec. 5.2.1.2).

TABLE 5.2
MIT coffee can radar in imaging mode specifications.

$P_{ave} =$ 10^{-3} (watts)
$G_{tx} =$ 7.2 dBi measured antenna gain
$G_{rx} =$ 7.2 dBi measured antenna gain
$A_{rx} =$ $G_{rx}\lambda_c^2/(4\pi)$ (m^2)
$\lambda_c =$ c/f_c (m) wavelength of carrier frequency
$f_c =$ 2.4 GHz center frequency of radar
$\rho_{rx} =$ 1 because antenna efficiency is accounted for in measured antenna gain
$\sigma =$ 10 (m^2) for target of interest
$L_s =$ 6 dB miscellaneous system losses
$\alpha =$ 0 attenuation constant of propagation medium
$F_n =$ 1.2 dB receiver noise figure
$B_n =$ $2/t_{sample}$ system noise bandwidth (Hz) where $t_{sample} = 20$ ms
$(SNR)_1 =$ 13.4 dB
$N =$ 48 number of range profiles used in synthesizing the aperture for 2 inch spacing across the rail

5.2.1.1 Maximum Range and Minimum Target RCS

Substituting specifications (Table 5.2) into the radar range equation (4.8), the maximum range is estimated to be 2.2 km for a 10 dBsm automobile target using the MATLAB script [5]

The IDFT is applied over 20 ms up-chirps with a direct conversion receive architecture resulting in a 100 Hz effective noise bandwidth. The number of range profiles required to process a SAR image is 48 ($= N$).

The video amplifier (Fig. 2.13) limits the maximum range to $f_{-3db}c/c_r =$ 272 m, where $f_{-3db} = 15$ KHz and $c_r = BW/t_{sample} = 16.5$ GHz/s. With this, the radar range equation can be solved for σ by substituting $R_{max} = 272$ m, thereby showing that the minimum radar cross section that can be detected within the 272 m maximum range is -26.1 dBsm. This can be estimated using the same MATLAB script [5].

5.2.1.2 Range Resolution Estimate

Expected range resolution is 40.5 cm with no weighting ($K_r = 0.89$). The cross range resolution is expected to be 34.9 cm when the targets are located 50 feet down range, centered with respect to the rail with a length of 96 inches, with no weighting ($K_r = 0.89$). This too can be estimated using the same MATLAB script [5].

5.2.2 Implementation

Neither calibration nor background subtraction is used when acquiring a coffee can SAR image; therefore all but the last step to collecting SAR data are followed (Fig. 5.4). For processing, (Fig. 5.5), only the image data is used

FIGURE 5.13
Example of a SAR image acquired with the MIT coffee can radar showing the contours of the tree line at an outdoor site.

where the means of all range profiles are subtracted from the image data which is then fed through the image formation algorithm.

5.2.3 Measured Results

The coffee can SAR was deployed in an open field in Westford Massachusetts. A SAR image was acquired of the surrounding terrain overlaid on a satellite image for reference (Fig. 5.13). The tree line is clearly shown on the left side of the image. Two large metal structures are bright on the right side of the image. The corridor in the middle is open and free of clutter as expected. This image is a good example of the quality that can be expected from the coffee can SAR; it is not great but demonstrates the principles of SAR imaging. Data from this experiment can be downloaded [11].

5.3 Range-Gated FMCW Rail SAR Imaging Systems

Each range-gated FMCW radar front end (Sec. 3.3.3) is bolted to the linear rail and motion control used previously (Sec. 5.1) thereby making a range-gated linear rail SAR imaging system which is capable of imaging target areas deliberately constrained by the range gate circuitry. Examples of this will be shown at X-band (Sec. 5.3.1) and S-band (Sec. 5.3.2).

TABLE 5.3
X-band range-gated FMCW SAR specifications.

$$
\begin{aligned}
P_{ave} &= 10 \cdot 10^{-3} \text{ (watts)} \\
G_{tx} &= 17 \text{ dBi antenna gain (estimated)} \\
G_{rx} &= 17 \text{ dBi antenna gain (estimated)} \\
A_{rx} &= G_{rx}\lambda_c^2/(4\pi) \text{ (m}^2\text{)} \\
\lambda_c &= c/f_c \text{ (m) wavelength of carrier frequency} \\
f_c &= 10 \text{ GHz center frequency of radar} \\
\rho_{rx} &= 1 \text{ because antenna efficiency is accounted for in antenna gain} \\
\sigma &= 10 \text{ (m}^2\text{) for automobile at 10 GHz} \\
L_s &= 6 \text{ dB miscellaneous system losses} \\
\alpha &= 0 \text{ attenuation constant of propagation medium} \\
F_n &= 4 \text{ dB receiver noise figure} \\
B_n &= 2/t_{sample} \text{ system noise bandwidth (Hz) where } t_{sample} = 10 \text{ ms} \\
(SNR)_1 &= 13.4 \text{ dB} \\
N &= 96 \text{ number of range profiles used in synthesizing the} \\
&\quad \text{aperture for 2 inch spacing across the rail}
\end{aligned}
$$

5.3.1 X-Band

A range gated X-band rail SAR is useful for acquiring imagery in a high clutter environment, such as inside a laboratory space.

5.3.1.1 Implementation

The range gated X-band FMCW radar system is mounted to a linear rail, where the radar controller moves the stepper motor so that the radar acquires data at every 0.5 or 1 inch increment (Fig. 5.14).

This rail SAR [14] and [15] is located in the author's garage and directed out toward the target platform 20 feet down range (Fig. 5.15). An absorber is located on the right side of the garage door over the metal track to reduce high-intensity in-scene clutter. The X-band front-end is bolted to the linear rail. Numerous cables are fed to and from the front end. Additional test gear used while developing this sensor is also shown.

5.3.1.2 Expected Performance

Substituting specifications (Table 5.3) into the radar range equation (4.8), the maximum range is estimated to be 2.8 km for a 10 dBsm automobile target using the MATLAB script [5].

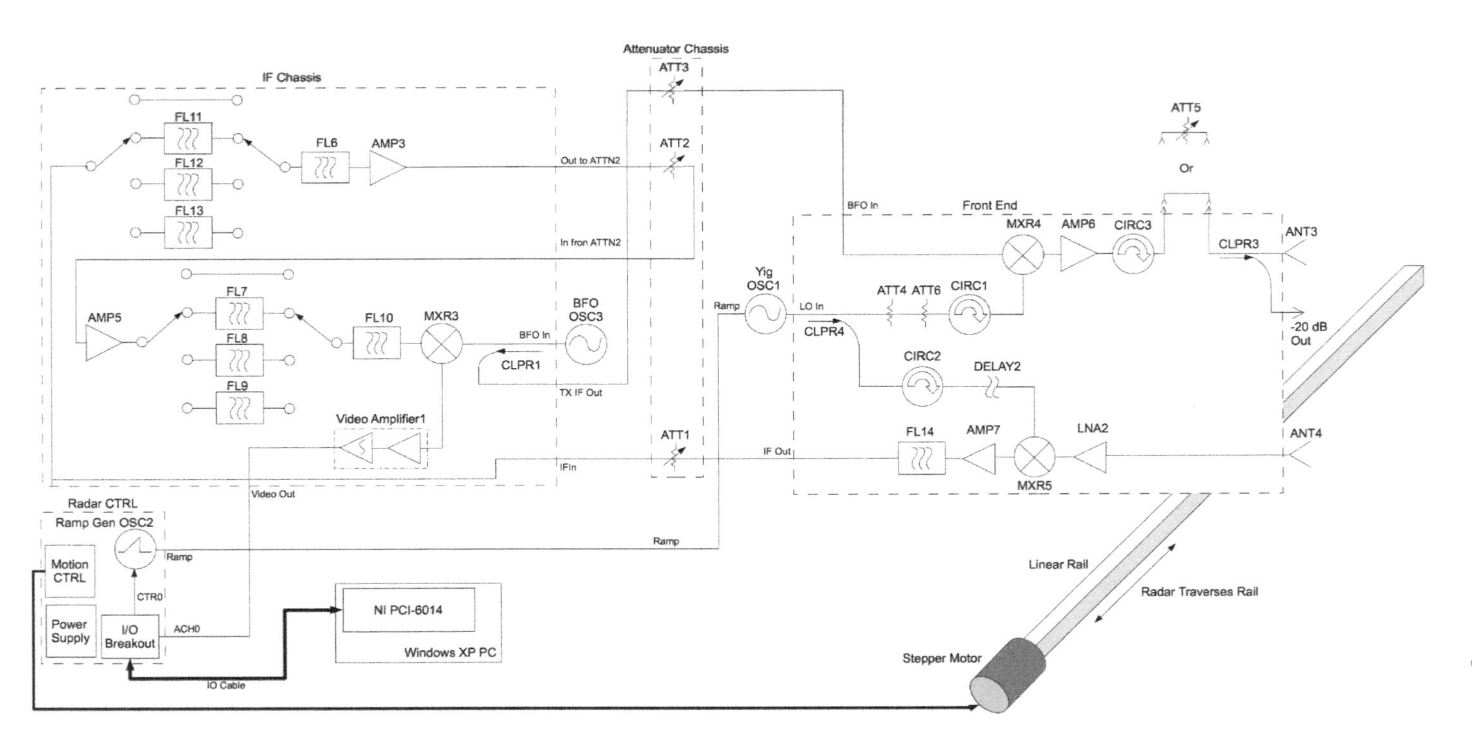

FIGURE 5.14
Block diagram of the X-band range-gated FMCW radar used as a rail SAR imaging system.

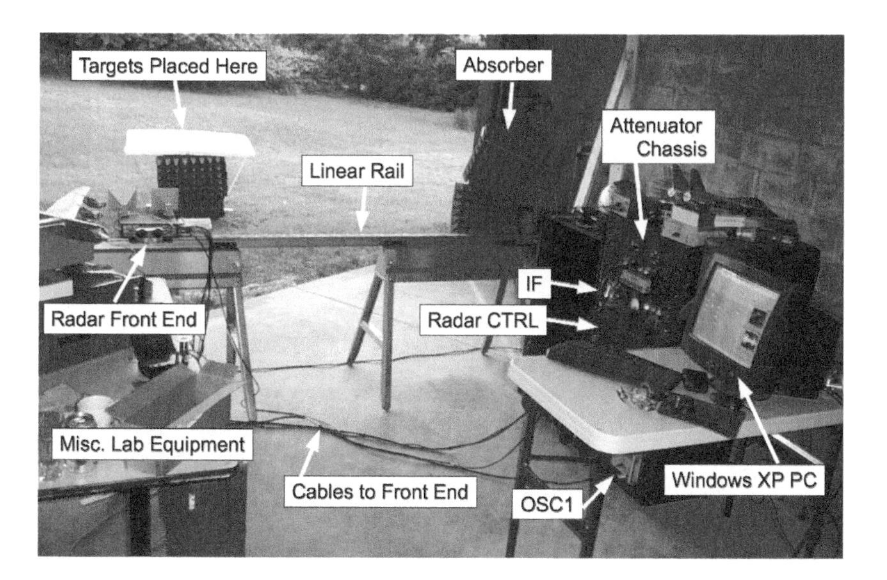

FIGURE 5.15
Experimental setup and call-out diagram for the X-band range-gated FMCW radar used as a rail SAR imaging system.

The antenna gains G_{rx} and G_{tx} were estimated based on aperture size, the noise figure is an educated guess based on the apparent age of the surplus microwave amplifier LNA2, and the IDFT is applied over 10 ms up-chirps with a direct conversion receive architecture resulting in a 200 Hz effective noise bandwidth. The number of range profiles required to process a SAR image is 96 $(= N)$.

The video amplifier (Fig. 2.13) limits the maximum range to $f_{-3db}c/c_r =$ 48 m, where $f_{-3db} = 80$ KHz and $c_r = BW/t_{sample} = 498$ GHz/s. With this, the radar range equation can be solved for σ by substituting $R_{max} = 48$ m, thereby showing that the minimum radar cross section that can be detected within the 48 m maximum range is -60 dBsm. This can be estimated using the same MATLAB script [5].

Expected range resolution is 2.7 cm with no weighting $(K_r = 0.89)$. The cross range resolution is expected to be 2.8 cm when the targets are located 5 m down range, centered with respect to the rail with a length of 96 inches, with no weighting $(K_r = 0.89)$. This too can be estimated using the same MATLAB script [5].

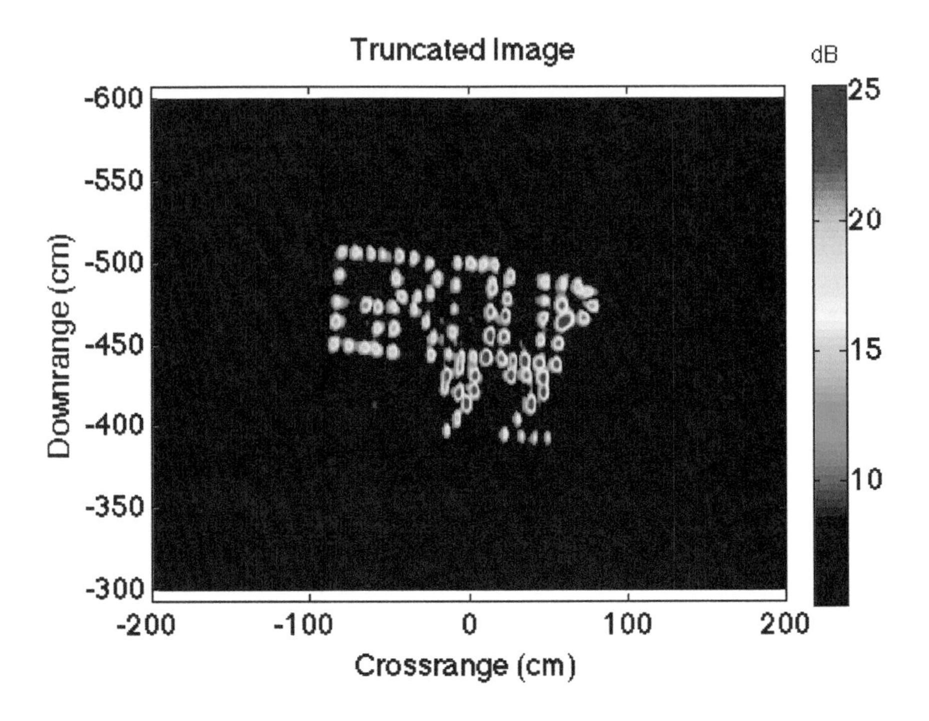

FIGURE 5.16
SAR image of pushpins (convenient point targets) to measure resolution.

5.3.1.3 Measured Results

Targets are placed 20 ft down range and in front of the linear rail (Fig. 5.15). To acquire a SAR image, calibration, calibration background, image background, and image data is acquired as described previously (Fig. 5.4). The range gate is set to gate in the length of the foam table where the targets are placed. To process a SAR image, the data acquired for calibration and imaging is fed into the processing chain as described previously (Fig. 5.5).

To measure range resolution a group of point targets (pushpins or thumbtacks) were imaged (Fig. 5.16. This data and a MATLAB processing script are available for the reader to process [16]. Each pin is clearly shown in this image spelling out GROUP 92.

Using the point target at the bottom of the 9, which is approximately 3.8 m down range and 20 cm cross range, the measured down range resolution with Hanning weighting was measured to be 5.75 cm which was very close to theoretical best possible down range resolution of 4.3 cm. The cross range resolution without weighting was measured to be 4.25 cm which was somewhat

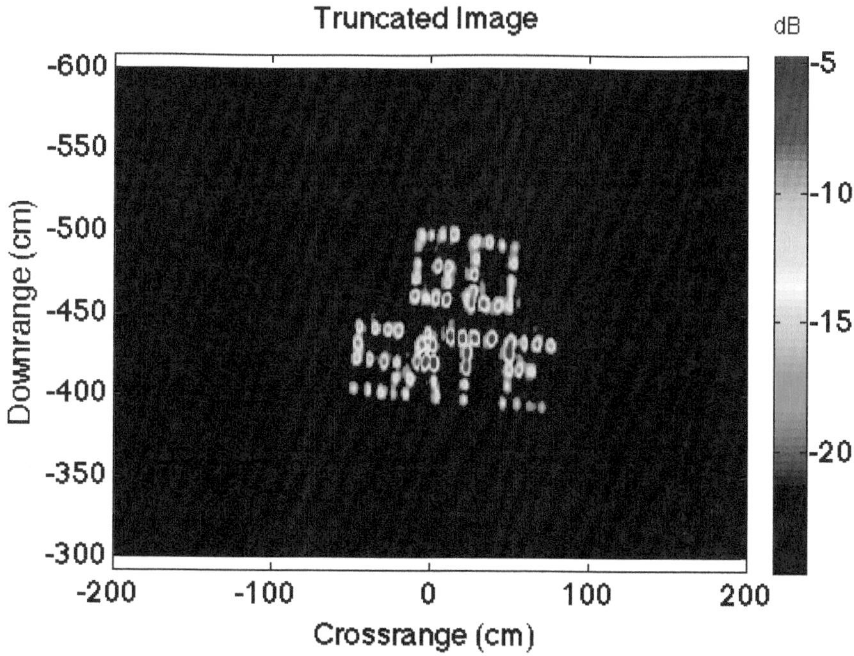

FIGURE 5.17
SAR image of pushpins (convenient point targets) at a low transmit power of 100 nw demonstrating the SAR's ability to image using low transmit power.

close to the theoretical best cross range resolution of 2.2 cm. Resolution was estimated using the MATLAB script [5].

Using attenuator ATTN5 the transmitter power was dialed from 10 mw down to 100 nw to test the SAR's ability to image pushpins with low transmit power. Using lower transmit power is advantageous because it allows the radar to be used for applications where transmit power is limited. In this image, Hanning weighting was used down range and no weighting in cross range (Fig. 5.17). This data and a MATLAB processing script are available for you to process [17]. The location of each pin is clearly shown, spelling out GO SATE and the resolution is comparable to the image at full power 5.16.

This image was re-processed without Hanning weighting (Fig. 5.18) resulting in a fairly clear GO STATE image where the location of each pushpin is clear, the apparent range resolution is higher, but there is additional clutter in the image.

To measure the limits of this SAR's performance, the transmit power was dialed down further to 1 nw using ATTN5. A group of pushpins was imaged

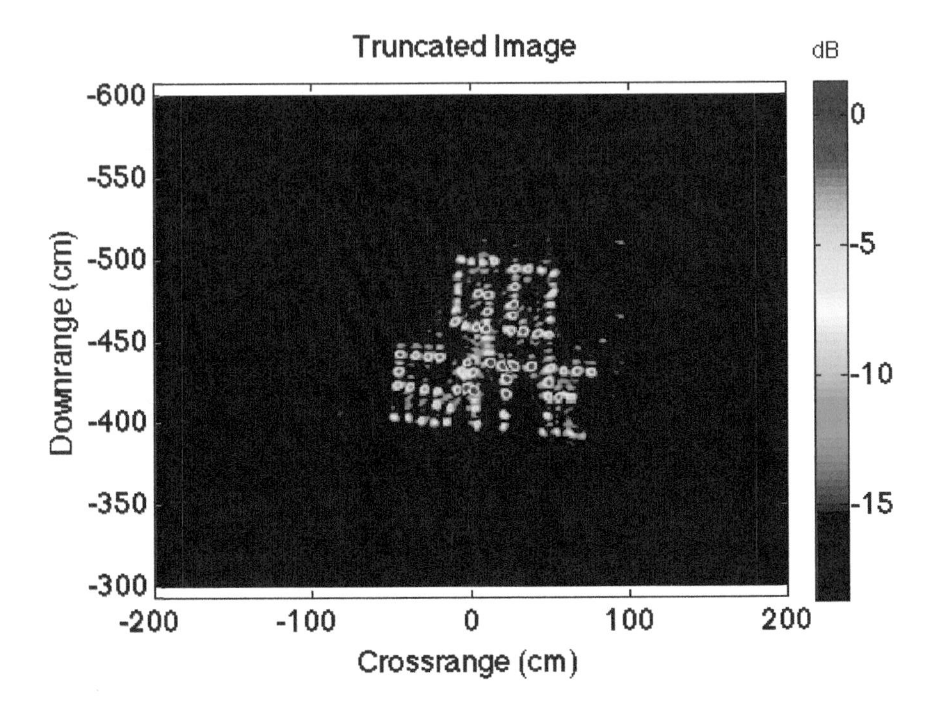

FIGURE 5.18
SAR image of pushpins (convenient point targets) at a low transmit power of 100 nw without Hanning weighting.

which spells out MSU (Fig. 5.19), where Hanning weighting was used down range and no weighting was used in cross range. The location of each pin is shown but some are fading into the clutter. The clutter could also be noise floor because of the low transmit power. This is the lowest transmit power that can be used to image targets as small as pushpins. This data and a MATLAB processing script are available for you to process [18].

This radar system is demonstrated in a video [19] where the author's bike is imaged (Fig. 5.20). Data from this demonstration is available for the reader including MATLAB scripts to process the data [20].

5.3.2 S-Band

An S-band range-gated rail SAR was developed for use in through-wall radar imaging and phased array radar experiments. This system could also be used for ground penetrating radar applications or for any application where radar imagery must be acquired in a high clutter environment.

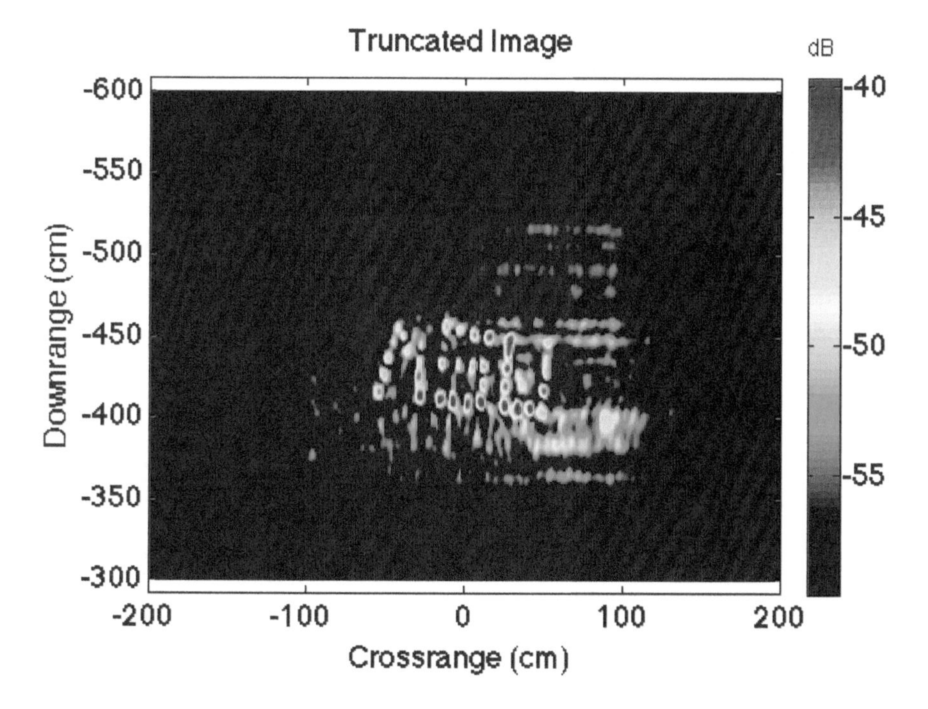

FIGURE 5.19
SAR image of pushpins (convenient point targets) at a low transmit power of 1 nw demonstrating the SAR's ability to image using extremely low transmit power.

5.3.2.1 Implementation

The range-gated S-band front end is mounted onto a linear rail (Fig. 5.21), where the radar control chassis commands the radar down the rail at 2 inch increments as it acquires range profiles.

The S-band front ends are configured for rail SAR imaging [21] by placing two UWB antenna elements onto an aluminum slab (Fig. 5.22a). To preserve noise figure the S-band front-end is co-located with the antenna elements (Fig. 5.22b).

5.3.2.2 Expected Performance

Substituting specifications (Table 5.4) into the radar range equation (4.8), the maximum range is estimated to be 2.5 km for a 10 dBsm automobile target using the MATLAB script [5]

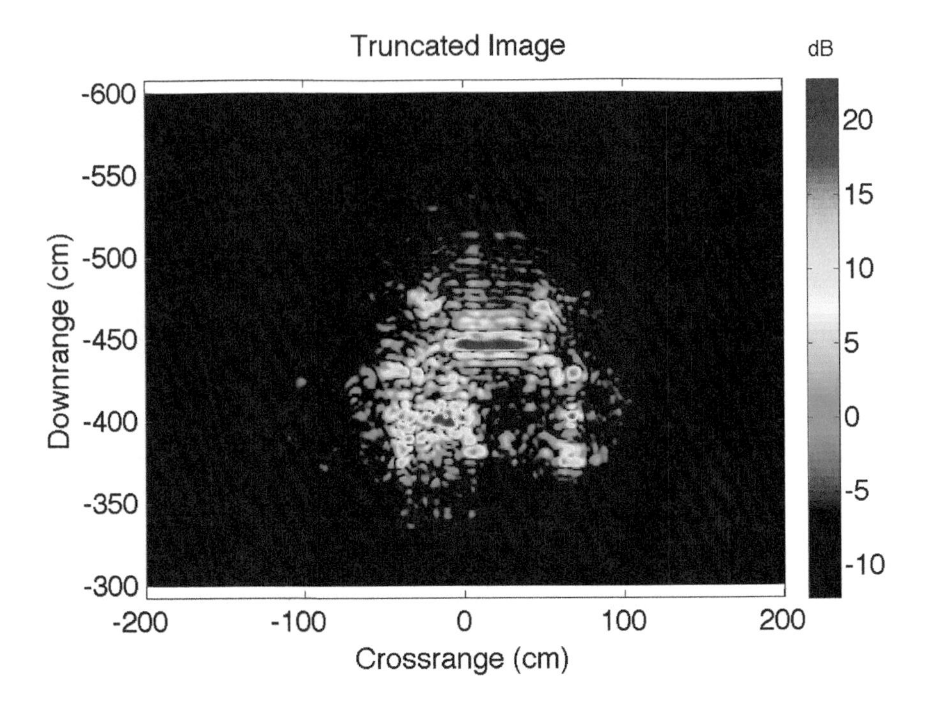

FIGURE 5.20
SAR image of the author's bike, a Cannondale M300, using the range-gated X-band FMCW rail SAR imaging system.

The IDFT is applied over 10 ms up-chirps with a direct conversion receive architecture resulting in a 200 Hz effective noise bandwidth. The number of range profiles required to process a SAR image is 48 ($= N$).

The video amplifier (Fig. 2.13) limits the maximum range to $f_{-3db}c/c_r =$ 112 m, where $f_{-3db} = 80$ KHz and $c_r = BW/t_{sample} = 214$ GHz/s. With this, the radar range equation can be solved for σ by substituting $R_{max} = 112$ m, thereby showing that the minimum radar cross section that can be detected within the 112 m maximum range is -44 dBsm. This can be estimated using the same MATLAB script [5].

This radar chirps from 1.926 to 2.069 GHz. Expected range resolution is 6.2 cm with no weighting ($K_r = 0.89$). The cross range resolution is expected to be 7.6 cm when the targets are located 4 m down range, centered with respect to the rail with a length of 96 inches, with no weighting ($K_r = 0.89$). This too can be estimated using the same MATLAB script [5].

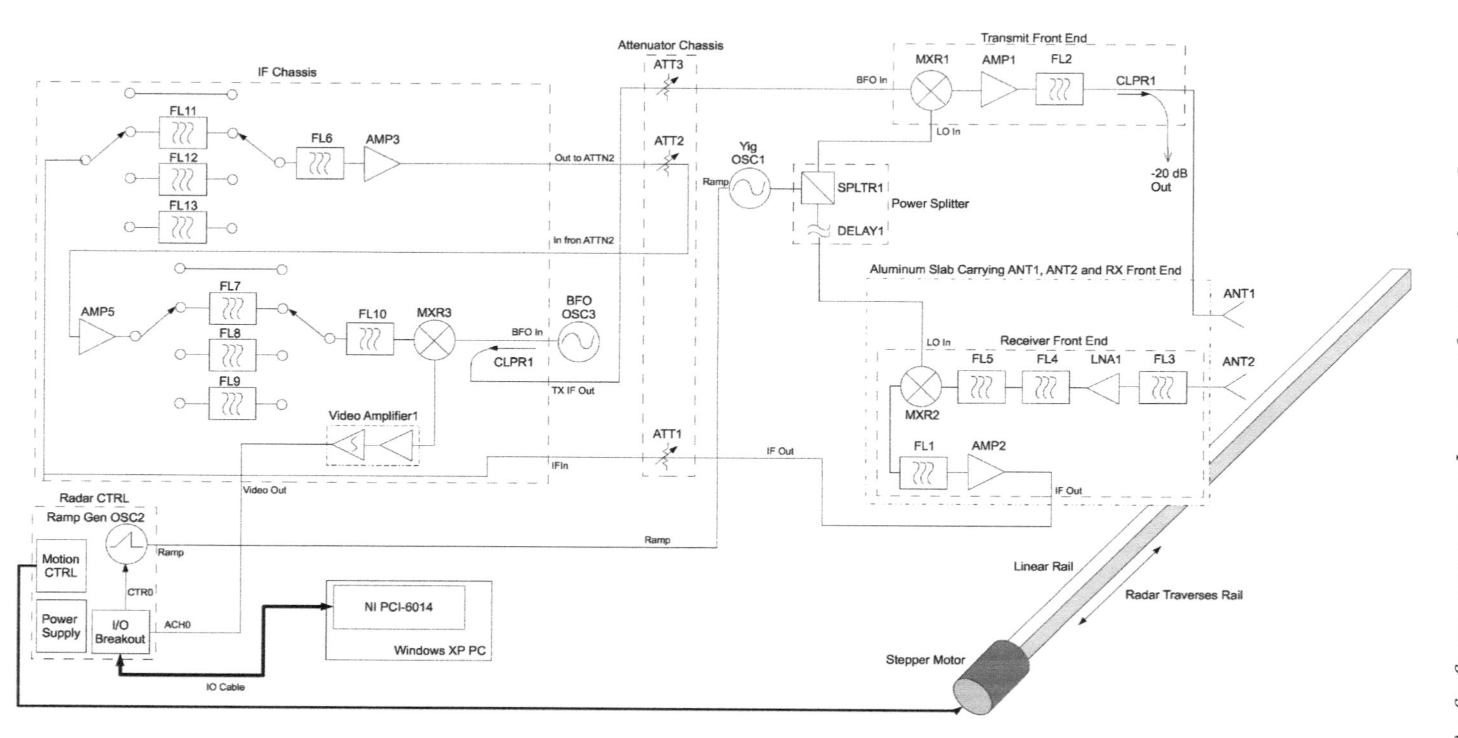

FIGURE 5.21
Block diagram of the X-band range-gated FMCW radar used as a rail SAR imaging system.

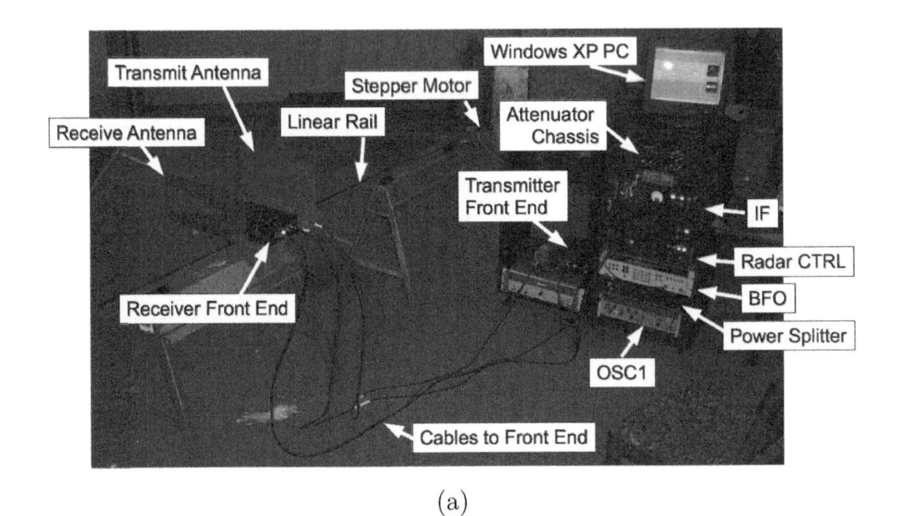

(a)

(b)

FIGURE 5.22
S-band front end (a) and inside view with call-outs (b).

TABLE 5.4
Range-gated S-band FMCW rail SAR specifications.

$$
\begin{aligned}
P_{ave} &= 10 \cdot 10^{-3} \text{ (watts)} \\
G_{tx} &= 12 \text{ dBi antenna gain (estimated)} \\
G_{rx} &= 12 \text{ dBi antenna gain (estimated)} \\
A_{rx} &= G_{rx}\lambda_c^2/(4\pi) \text{ (m}^2\text{)} \\
\lambda_c &= c/f_c \text{ (m) wavelength of carrier frequency} \\
f_c &= 3 \text{ GHz center frequency of radar} \\
\rho_{rx} &= 1 \text{ because antenna efficiency is accounted for in antenna gain} \\
\sigma &= 10 \text{ (m}^2\text{) for automobile at 10 GHz} \\
L_s &= 6 \text{ dB miscellaneous system losses} \\
\alpha &= 0 \text{ attenuation constant of propagation medium} \\
F_n &= 3.5 \text{ dB receiver noise figure} \\
B_n &= 2/t_{sample} \text{ system noise bandwidth (Hz) where } t_{sample} = 10 \text{ ms} \\
(SNR)_1 &= 13.4 \text{ dB} \\
N &= 48 \text{ number of range profiles used in synthesizing the} \\
& \text{aperture for 2 inch spacing across the rail}
\end{aligned}
$$

5.3.2.3 Measurements

Targets are placed 20 ft down range and in front of the linear rail on a Styrofoam table. The radar traverses the linear rail, acquiring range profiles at 2 inch increments (Fig. 5.22b).

To acquire a SAR image, calibration, calibration background, image background, and image data is acquired as described previously (Fig. 5.4). The range gate is set to gate in the length of the foam table where the targets are placed. The calibration rod is 5 feet tall and pounded into the foreground for calibration. To process a SAR image, the data acquired for calibration and imaging is fed into the processing chain as described previously (Fig. 5.5).

Range resolution was measured by imaging a group of point targets in an S configuration (Fig. 5.23a) where each point target is a 6 inch tall carriage bolt. The range resolution was measured for the target in the lower left of the S to be 8.8 cm and the best possible range resolution was estimated to be 6.2 cm. The cross range resolution was measured to be 11.7 cm and the best possible cross range resolution was estimated to be 9.2 cm. Range resolution estimates were calculated using the MATLAB script [5]. This data can be downloaded and processed [22].

This target scene was measured repeatedly with lower power levels to test the SAR's sensitivity at 10 nW, 100 pW, and 5 pW (Figs. 5.23b through d). This data can be downloaded and processed [23]–[25]. For transmit power levels of 10 nW and 100 pW the group of point targets appears to have nearly the same signal-to-clutter as the full-power (10 mW) image (Fig. 5.23a), showing this radar's ability to image target scenes with low power. When using 5 pW

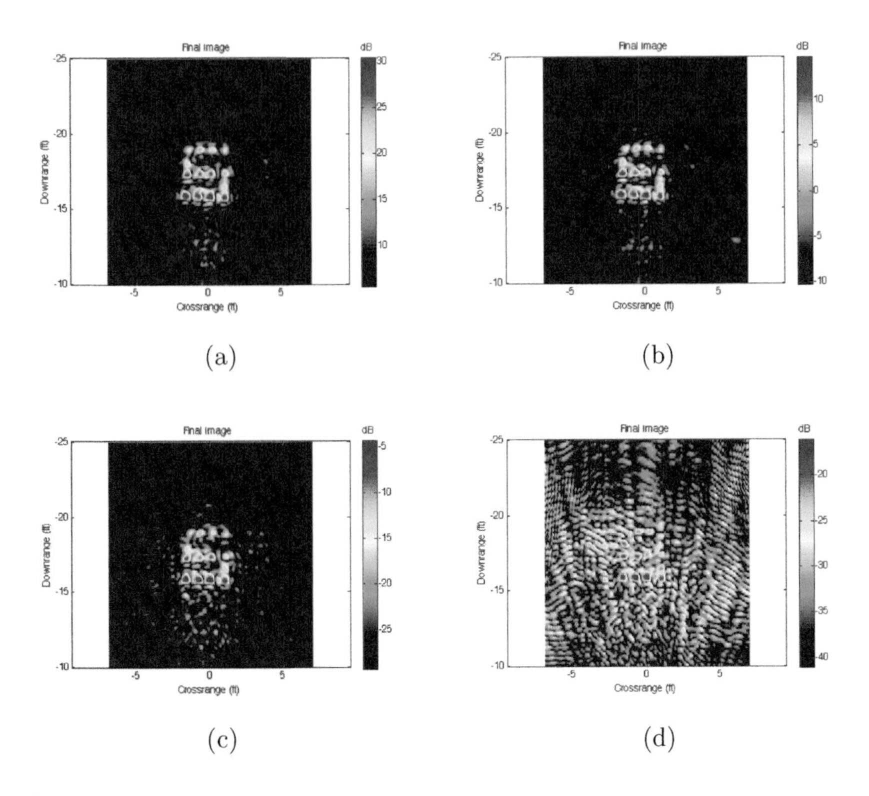

FIGURE 5.23
Imagery of a group of 6 inch carriage bolts using transmit power levels of 10 mW (a), 10 nW (b), 100 pW (c), and 5 pW (d).

of power the image fades into the clutter which is likely due to thermal noise of the receiver.

A demonstration video was made of this radar imaging system [26] where two copper cylinders (6 inches and 12 inches in diameter) and a group of five vertical cables are imaged (Fig. 5.24). This data is available for the reader including MATLAB scripts to process the data [27].

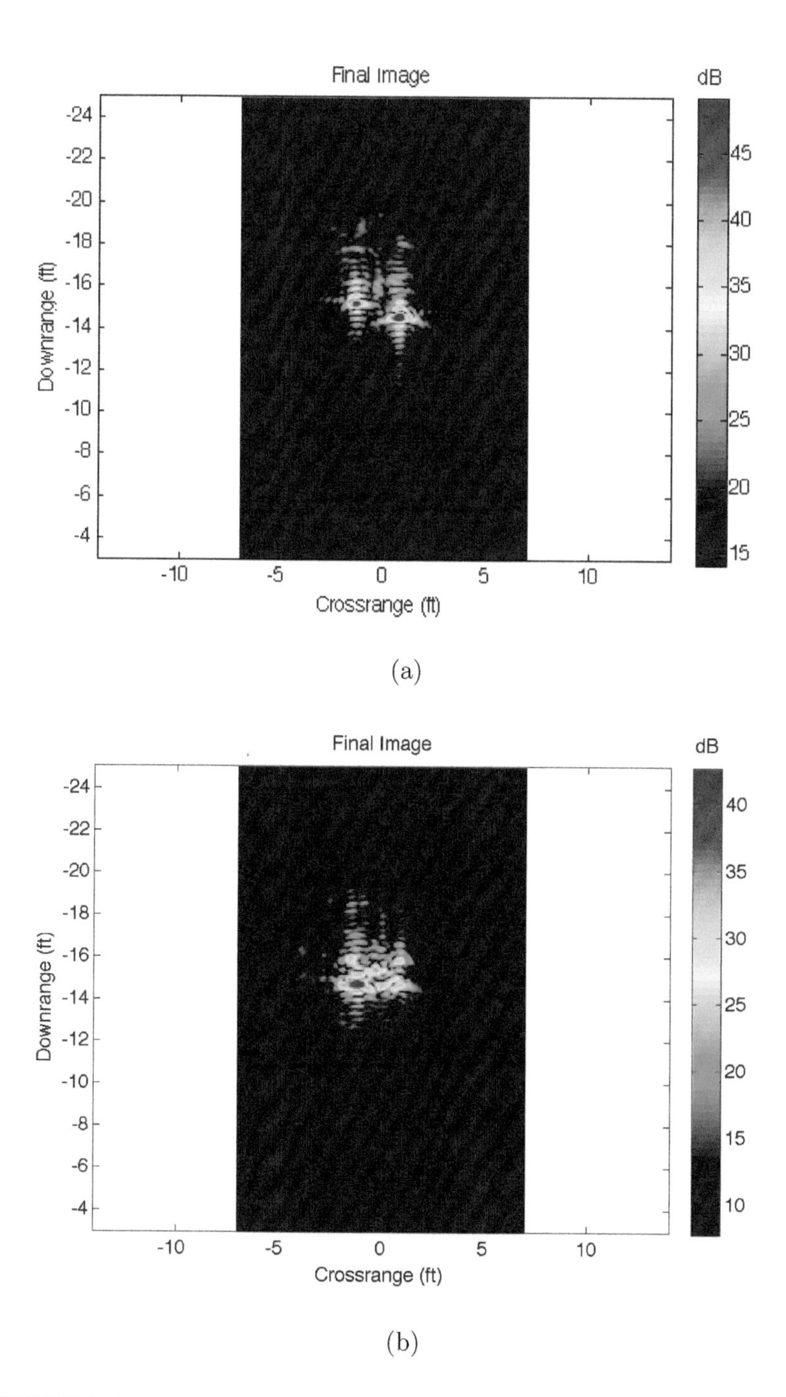

FIGURE 5.24
S-band SAR imagery of two metal cylinders (a) and five vertical cables.

5.4 Summary

Rail SAR imaging systems are used to learn SAR imaging algorithms, experiment with phased array techniques, image small target scenes, measure radar targets, and even image through walls. Rail SAR imaging equipment can be implemented by mounting a small radar system to a linear rail and acquiring range profiles at known locations. Imagery is formed by feeding this data into a SAR algorithm. High resolution imagery can be achieved with low-power UWB radar front ends. Numerous examples were shown and measured imagery of various target scenes processed. In the next chapter, the use of SAR imaging algorithms will be extended for use in a phased array radar system for near-field digital beamforming.

Bibliography

[1] G. L. Charvat, L. C. Kempel. "Low-Cost, High Resolution X-Band Laboratory Radar System for Synthetic Aperture Radar Applications," East Lansing, MI, IEEE Electro/Information Technology Conference, May 2006.

[2] G. L. Charvat, "Low-Cost, High Resolution X-Band Laboratory Radar System for Synthetic Aperture Radar Applications," Antenna Measurement Techniques Association Conference, Austin, Texas, October 2006.

[3] G. L. Charvat. "Build a high resolution synthetic aperture radar imaging system in your backyard," MIT Haystack Observatory, May 12, 2010.

[4] T. S. Ralston, G. L. Charvat, S. G. Adie, B. J. Davis, S. Carney, S. A. Boppart. "Interferometric synthetic aperture microscopy, microscopic laser radar," *Optics and Photonics News*, Vol. 21, No. 6, pp. 32–38, 2010.

[5] radar_range_eq_SAR.m, `http://glcharvat.com/shortrange/practical-examples-of-small-synthetic-aperture-radar-imaging-systems/`

[6] Backyard_SAR_GOSTATE.zip, `http://glcharvat.com/shortrange/practical-examples-of-small-synthetic-aperture-radar-imaging-systems/`

[7] Backyard_SAR_Bike.zip, `http://glcharvat.com/shortrange/practical-examples-of-small-synthetic-aperture-radar-imaging-systems/`

[8] Backyard_SAR_5_0_Mustang.zip, http://glcharvat.com/shortrange/practical-examples-of-small-synthetic-aperture-radar-imaging-systems/

[9] X Band Rail SAR Imaging System, http://glcharvat.com/shortrange/practical-examples-of-small-synthetic-aperture-radar-imaging-systems/

[10] xband_fmcw_railsar_demo.zip, http://glcharvat.com/shortrange/practical-examples-of-small-synthetic-aperture-radar-imaging-systems/

[11] G. L. Charvat, J. H. Williams, A. J. Fenn, S. M. Kogon, J. S. Herd, "RES.LL-003 Build a Small Radar System Capable of Sensing Range, Doppler, and Synthetic Aperture Radar Imaging, January IAP 2011," (Massachusetts Institute of Technology: MIT OpenCourseWare), http://ocw.mit.edu (Accessed 02 Sep, 2012). License: Creative Commons BY-NC-SA.

[12] G. L. Charvat, J. H. Williams, A. J. Fenn, S. M. Kogon, J. S. Herd. "The MIT IAP 2011 radar course: build a small radar system capable of sensing range, Doppler, and SAR," The Boston Chapters of the IEEE Life Members, AES, and GRSS, May 24, 2011.

[13] G. L. Charvat, A. J. Fenn, B. T. Perry, "The MIT IAP radar course: build a small radar system capable of sensing range, doppler and synthetic aperture radar (SAR) imaging," Atlanta, GA: IEEE Radar Conference, May 2012.

[14] G. L. Charvat, "A Low-Power Radar Imaging System," Ph.D. dissertation, Department of Electrical and Computer Engineering, Michigan State University, East Lansing, MI, 2007.

[15] G. L. Charvat, L. C. Kempel, and C. Coleman, "A low-power high-sensitivity X-band rail SAR imaging system." *IEEE Antennas and Propagation Magazine*, June 2008, pp. 108-115.

[16] xband_rangegated_SAR_10mw.zip, http://glcharvat.com/shortrange/practical-examples-of-small-synthetic-aperture-radar-imaging-systems/

[17] xband_rangegated_SAR_100nw.zip, http://glcharvat.com/shortrange/practical-examples-of-small-synthetic-aperture-radar-imaging-systems/

[18] xband_rg_SAR_1nw.zip, http://glcharvat.com/shortrange/practical-examples-of-small-synthetic-aperture-radar-imaging-systems/

[19] X Band Range Gated FMCW SAR Imaging System, http://glcharvat.com/shortrange/practical-examples-of-small-synthetic-aperture-radar-imaging-systems/

[20] xband_rg_fmcw_railsar_demo.zip, http://glcharvat.com/shortrange/practical-examples-of-small-synthetic-aperture-radar-imaging-systems/

[21] G. L. Charvat, L. C. Kempel, E. J. Rothwell, C. Coleman, and E. L. Mokole, "A through-dielectric radar imaging system," *IEEE Transactions on Antennas and Propagation*, Vol. 58, No. 8, pp. 2594–2603, 2010.

[22] S_band_free_space.zip, http://glcharvat.com/shortrange/practical-examples-of-small-synthetic-aperture-radar-imaging-systems/

[23] S_band_rg_bolts_10nw.zip, http://glcharvat.com/shortrange/practical-examples-of-small-synthetic-aperture-radar-imaging-systems/

[24] S_band_rg_bolts_100pw.zip, http://glcharvat.com/shortrange/practical-examples-of-small-synthetic-aperture-radar-imaging-systems/

[25] S_band_rg_bolts_5pw.zip, http://glcharvat.com/shortrange/practical-examples-of-small-synthetic-aperture-radar-imaging-systems/

[26] S Band Rangegated Rail SAR, http://glcharvat.com/shortrange/practical-examples-of-small-synthetic-aperture-radar-imaging-systems/

[27] sband_rg_fmcw_railsar_demo.zip, http://glcharvat.com/shortrange/practical-examples-of-small-synthetic-aperture-radar-imaging-systems/

6

Phased Array Radar

The larger your antenna, the further your signal will go. One way to make a large antenna at microwave frequencies is to build a parabolic dish and place a point radiator at its focus (Fig. 6.1). With this, spherical field from the point radiator spreads out as it travels toward the dish. When it hits the metal dish, it scatters (bounces) off the dish and radiates out toward the opening of the dish. The wave front is in phase by the time it reaches the opening of the dish, thereby providing a columnated plane wave [1]. Parabolic dish antennas can provide a considerable gain for applications from radar to satellite television reception.

Another way to achieve high gain and plane wave radiation is to build an array of small antenna elements (Fig. 6.2). Behind each is a phase shifter. All phase shifters are fed from a single power combiner (large power splitter). The combined phase-shifted antenna feeds are fed to a radar system or signal source. Assuming N elements and that all phase shifters are set to 0, then a plane wave would radiate out of this array just as it does from a parabolic dish. The advantage of a phased array is that when the phase is shifted linearly across all phase shifters then the radiated plane wave can be directed (or steered) up and down, at some angle off the normal of the antenna [2]. Steering this plane wave is known as beamforming, where each different plane wave projects a beam in its respective angular direction.

This phased array radar can transmit a plane wave and likewise can also receive a plane wave when the target is in the far field because the scattering spherical wavefront from the target will become planer before it is collected by the array elements (Fig. 6.3). A target is in the far field when its distance from the antenna $R = 2D^2/\lambda$, where D is the largest dimension of the antenna array (either the length or width, whichever is greatest).

6.1 Near Field Phased Array Radar

For small radar systems the target is often in the near field which makes it difficult to use traditional phased array technology because the scattered wavefront has curvature as it is being collected by the array elements (Fig. 6.4). When signals from each element are fed back through the phase shifters

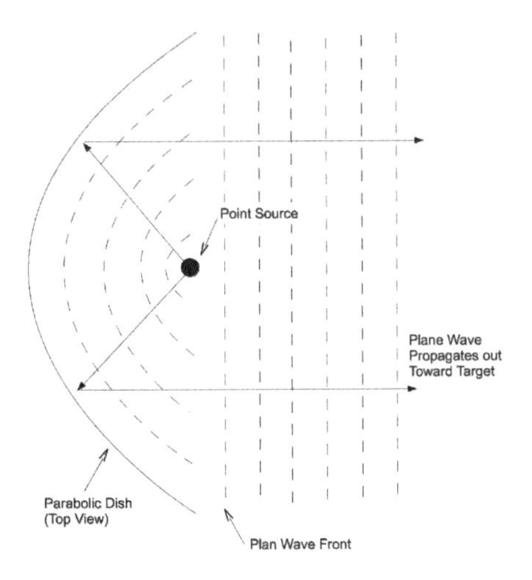

FIGURE 6.1
A real aperture antenna using a parabolic dish radiates a plane wave front.

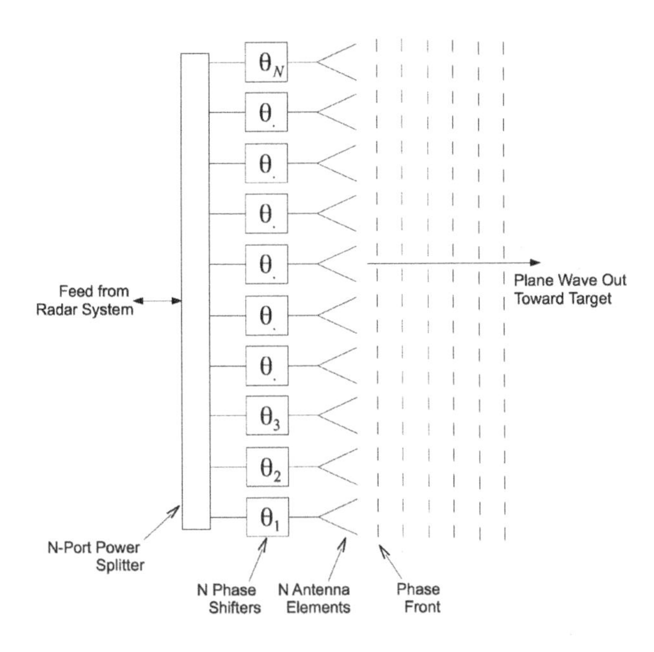

FIGURE 6.2
A phased array antenna radiates a plane wave front.

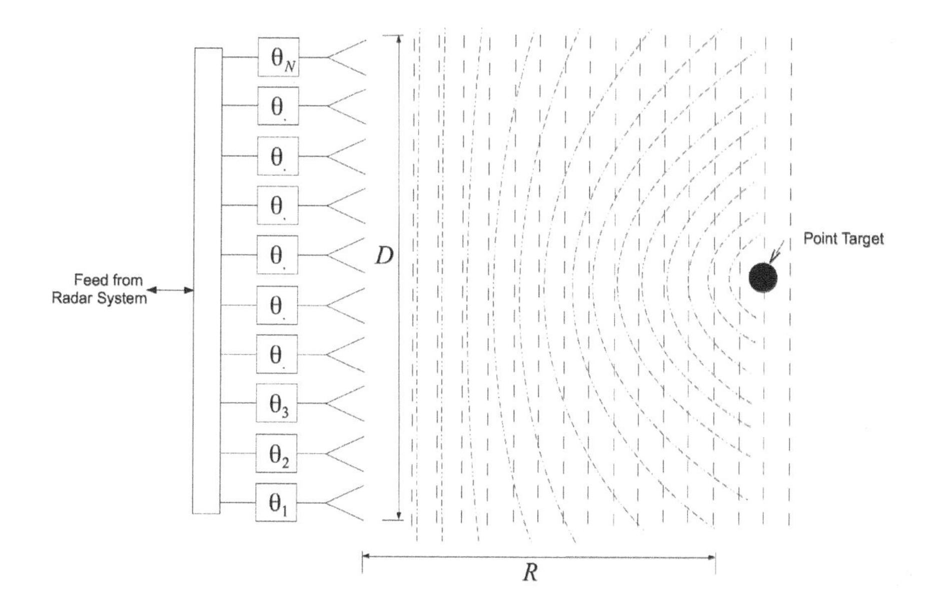

FIGURE 6.3
When a point target is in the far field, it scatters a plane wave front back to the phased array.

and into the power combiner they will not add up coherently, thereby providing significantly less gain and a poor time-domain response. For this reason small phased array radar systems must use more sophisticated methods of beamforming.

One approach is to apply a parabolic phase compensation across the phase shifters to flatten the spherical wave front before it enters the power combiner. This would make the array focused only at the point where the scatterer is located. This technique is used for many imaging modalities, including early versions of ultrasound, but it is limited because it focuses the array to only one point and must be re-set for each location in the field of view to form an image of the target scene.

Another approach is to transmit a plane wave but rather than using phase shifters on receive, digitize each antenna element and electronically focus the array for all positions in space. Still another method would be to use digital transmitters at each element. Various combinations of the above are also relevant. There are as many solutions to near field phased array imaging as there are radar engineers.

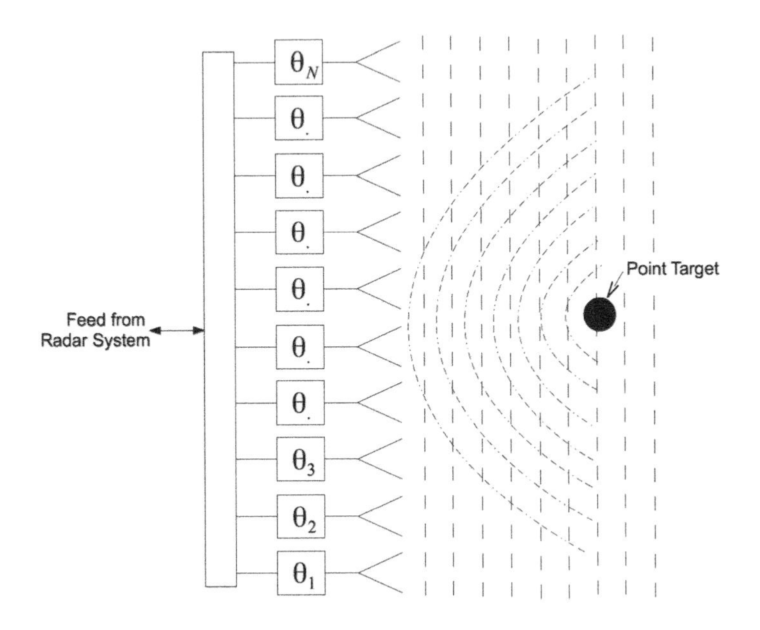

FIGURE 6.4
When a point target is in the near field it scatters a spherical wave back toward the array with a non-linear phase front incident on the surface of the elements which will not coherently add up in the power combiner.

6.2 Near Field Beamforming Using SAR Imaging Algorithms

One method to beamform in the near field is to use a SAR imaging algorithm providing a focused image in the full field of view of the array. In this book the range migration SAR imaging algorithm (Chapter 4) is used to beamform small short-range phased array radar systems. This can be achieved by electronically switching a radar system's antenna port across a linear array of antenna elements using a large microwave switch. The resulting switching action is the equivalent of moving a radar front end across a linear rail (Fig. 6.5).

Typically, when using FMCW radar, two separate transmit and receive elements must be used to reduce coupling from the transmitter to the receiver. To achieve a switched array equivalent to the S-band FMCW rail SAR previously shown (Sec. 5.3.2) would require an array with 48 separate transmit and receive elements with a 48 port microwave switch for a total of 96 elements (Fig. 6.6). This array would provide 48 phase centers spaced at $\lambda/2$ because given a separate transmitter and receiver, the location of the effective

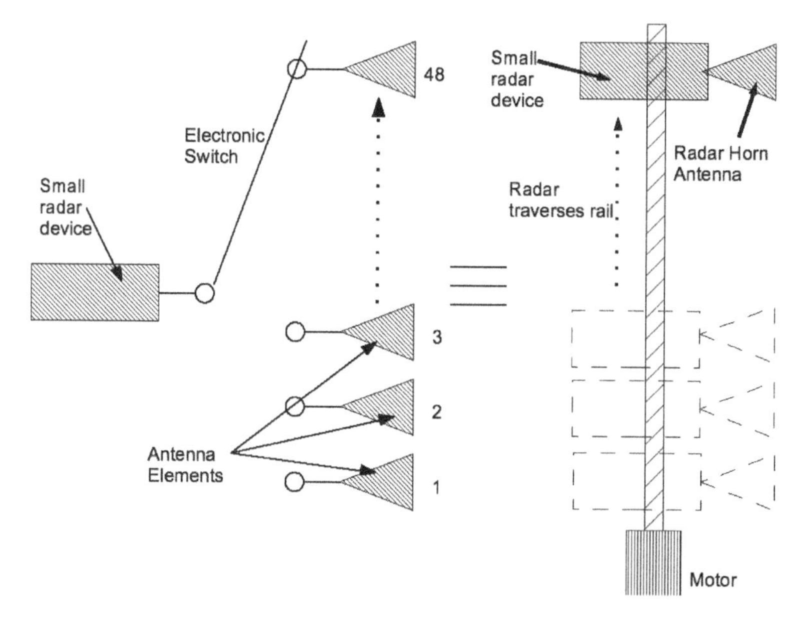

FIGURE 6.5
SAR imaging can be accomplished by electronically switching a radar front end across an array of elements.

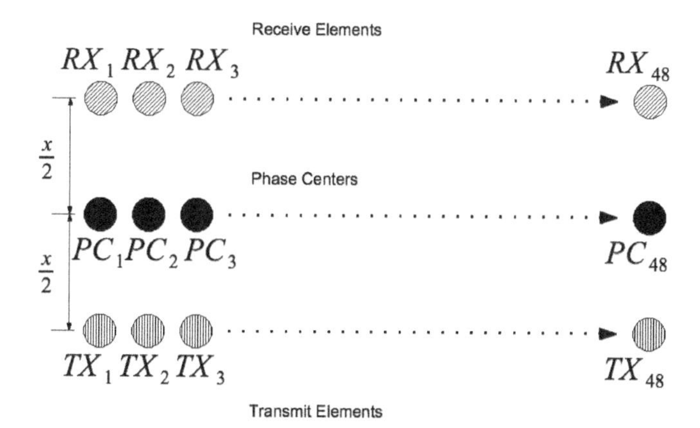

FIGURE 6.6
A switched array using an FMCW radar that must transmit while simultaneously receive, resulting in two separate linear arrays (transmit TX and receive RX) that are stacked on top of each other to provide 48 phase centers (PC).

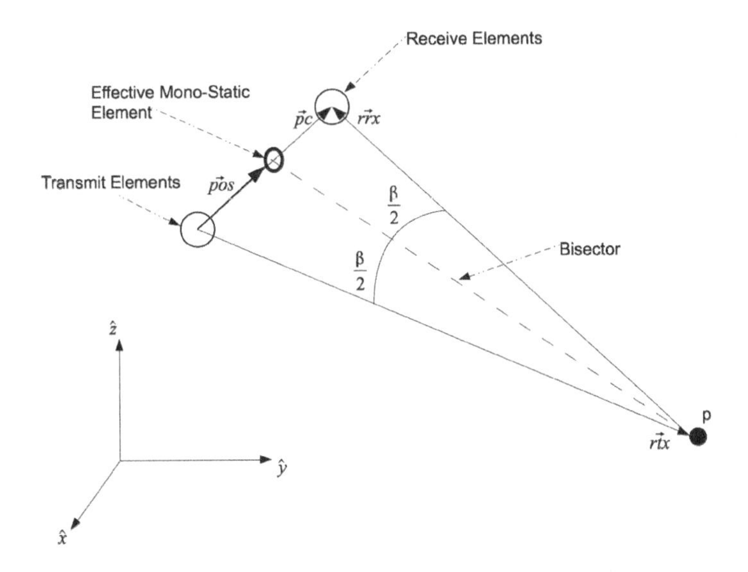

FIGURE 6.7
Location of the effective mono-static element, where $\vec{rtx}$ is the vector from the transmit element to the point target, $\vec{rrx}$ is the vector from the point target to the receive target representing the scattered field, $\vec{pc}$ is the vector from the transmit element to the receive element, and $\vec{pos}$ is the vector from the transmit element to the effective mono-static element.

mono-static element lies along the angle bisector of the triangle made up of the bi-static radar geometry of transmitter, receiver, and target [3].

Leveraging this, an array with fewer antenna elements but just as many phase centers can be achieved by staggering the receive and transmit elements. Each bi-static antenna combination makes up a triangle with a point target somewhere down range in the scene represented by point p in Fig. 6.7. Three points make up each triangle: the transmitter element, receive element, and p, where $\vec{rtx}$ is the vector from the transmit element to the point target, $\vec{rrx}$ is the vector from the point target to the receive target representing the scattered field, $\vec{pc}$ is the vector from the transmit element to the receive element, and $\vec{pos}$ is the vector from the transmit element to the effective mono-static element. The location of the effective mono-static element along $\vec{pc}$ is the position vector $\vec{pos}$ which depends on the length of $\vec{rrx}$ and $\vec{rtx}$ in the direction of $\vec{pc}$ as determined by the angle bisector theorem [4],

$$\vec{pos} = \frac{\vec{pc}}{\left(\frac{|\vec{rrx}|}{|\vec{rtx}|} + 1\right)}. \tag{6.1}$$

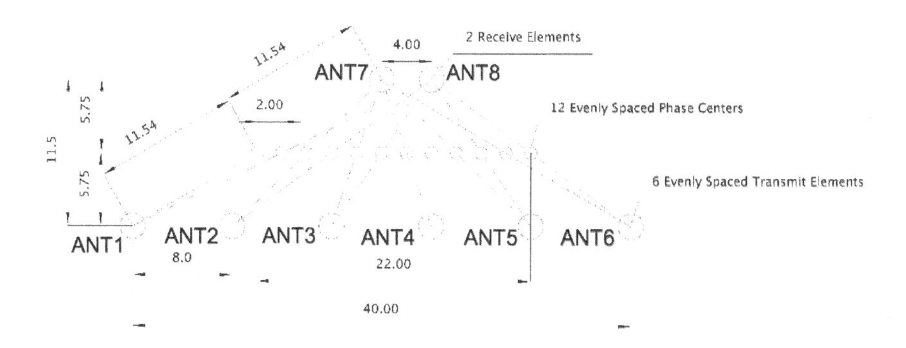

FIGURE 6.8
A switched antenna array showing a more efficient use of antenna elements when separate transmit and receive elements are needed; 12 evenly spaced phased centers are achieved with only 8 antenna elements (units relative).

An example array is shown where 12 phase centers are created using only 8 antenna elements (Fig. 6.8). This design can be extended to any size array at any wavelength. By replicating this pattern four times, a switched array equivalent to the SAR previously shown (Sec. 5.3.2) can be achieved using only 21 elements.

To capture a radar image using this array, range profiles are acquired at each phase center location by addressing the appropriate transmit and receive elements. These recorded range profiles are fed to the SAR imaging algorithm discussed in Chapter 4. In summary, when we must image or beamform in the near field it is advantageous to make a SAR imaging device with a switched array because we can image much faster with an array than by waiting for a rail SAR to complete a data acquisition.

6.3 Performance of Small Phased Array Radar Systems

Two key performance parameters of small phased array radar systems are the maximum range (Sec. 6.3.1) and resolution (Sec. 6.3.2).

6.3.1 The Radar Range Equation for Phased Array Radar Systems

A maximum range estimate for a small phased array imaging system like those discussed in this book is simply a matter of multiplying the performance of an FMCW radar by N, the number of phase centers provided by the switched array.

The maximum range to target for a small SAR is given by

$$R_{max}^4 = \frac{NP_{ave}G_{tx}A_{rx}\rho_{rx}\sigma e^{(2\alpha R_{max})}}{(4\pi)^2 kT_oF_nB_n\tau F_r(SNR)_1L_s}, \tag{6.2}$$

where:

R_{max} = maximum range of radar system (m)
P_{ave} = average transmit power (watts)
G_{tx} = transmit antenna gain for one single transmit element including transmit antenna efficiency
A_{rx} = receive antenna effective aperture (m^2) for one receive element
ρ_{rx} = receiver antenna efficiency for one receiver element
σ = radar cross section (m^2) for target of interest
L_s = miscellaneous system losses
α = attenuation constant of propagation medium
F_n = receiver noise figure (derived from procedure outlined in Sec 1.1.5.4)
k = $1.38 \cdot 10^{-23}$ (joul/deg) Boltzmann's constant
T_o = 290°K standard temperature
B_n = system noise bandwidth (Hz)
τF_r = the radar duty cycle which is equal to 1 for an FMCW radar
τ = pulse width (s)
F_r = pulse repetition frequency (Hz)
$(SNR)_1$ = single-pulse signal-to-noise ratio requirement
N = number of range phase centers in switched array

Switched array techniques can be applied to several types of radar. For FMCW radar, the noise bandwidth is inversely proportional to the discrete sample length $B_n = 1/t_{sample}$. For an impulse radar, the noise bandwidth is simply the -3 dB roll-off frequency (f_{-3db}) of the anti-aliasing filter.

For a direct conversion radar the noise bandwidth is twice as wide as a radar following an image rejection architecture because the noise image folds over onto itself, where $B_n = 2/t_{sample}$ for an FMCW radar or simply f_{-3db} for an impulse radar.

6.3.2 Resolution of Near Field Phased Array Imagery

The switched array radar devices in this chapter use a SAR imaging algorithm to process imagery in the near field. For this reason the procedure to estimate expected resolution is identical to that for SAR imaging previously discussed (Sec. 4.4.2).

6.4 Processing

The SAR imaging algorithm is applied to near field array data, but unlike the SAR radars shown previously the array radar acquires data at a very high rate. The SAR imaging algorithm must be applied repeatedly and in real time to each data set as it is acquired. For the small phased array radar system discussed in this chapter, calibration background is acquired first followed by calibration target data (Fig. 6.9). Imagery is processed by calculating the calibration coefficients and multiplying these by acquired image data. Image data can be either coherently subtracted from an image background or CCD can be used on a frame-to-frame basis by subtracting the previous image set

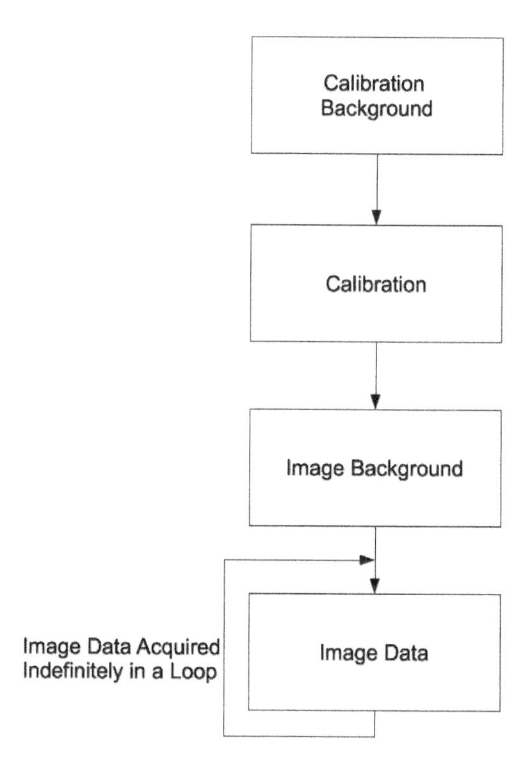

FIGURE 6.9
Data acquisition block diagram, where calibration background is first recorded, followed by calibration data, image background, and image data, all of which is looped indefinitely.

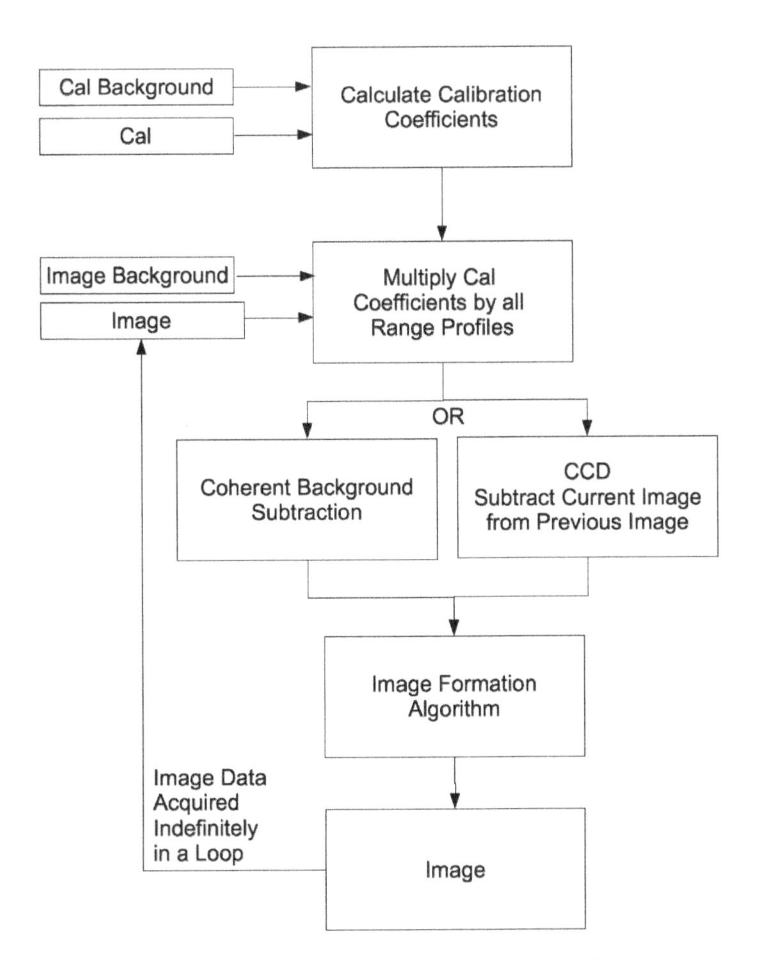

FIGURE 6.10
Array processing block diagram, where calibration (cal) data is acquired first, then applied to all image data and image data is acquired, processed, then displayed indefinitely in a while loop.

from the current one (Fig. 6.10). This result is fed into the image formation algorithm, which is the RMA SAR algorithm from Chapter 4. The final image is displayed to the user and the entire process is repeated over and over again in a while loop until the user stops the radar.

Calibration processing is used to sharpen imagery (Sec. 6.4.1) and coherent background or change detection is used to show only those targets in the scene that have changed between image frames (Sec. 6.4.2).

6.4.1 Calibration

Calibration improves imagery to its best possible resolution by accounting for phase and amplitude deviations across each phase center and through the radar's signal chains. For near field phased array antenna systems, calibration also compensates for the variation in round-trip delays due to the diverse baselines of the bi-static combinations shown (Fig. 6.8). One example would be comparing ANT1 and ANT8 to ANT3 and ANT7. The baseline between ANT1 and ANT8 is much longer than that between ANT3 and ANT7. This causes the round-trip delay to target from ANT1 and ANT7 to be greater than from ANT3 and ANT8.

A large point target is used; usually a metal pole or rod is convenient. The calibration target is placed at a known down range (d_{pole}) and centered to the middle of the array. Range profile data is acquired across the array at each of the phase centers. This data is represented by $s_{pole}\big(x_n, \omega(t)\big)$ where x_n is the cross range phase center position on the array. The pole is then removed and a background 2D range profile data array is acquired. The result is represented by $s_{calback}\big(x_n, \omega(t)\big)$. The background is subtracted from the pole range profile data resulting in a 2D range profile array of the pole only

$$s_{cal}\big(x_n, \omega(t)\big) = s_{pole}\big(x_n, \omega(t)\big) - s_{calback}\big(x_n, \omega(t)\big). \tag{6.3}$$

The calibration (cal) data is referenced to a point scatterer which is represented by

$$s_{caltheory}\big(x_n, \omega(t)\big) = e^{-j2k_r R_{pole}}, \tag{6.4}$$

where the R_{pole} is a 2D range to pole across the array represented by the

$$R_{pole} = \sqrt{x_n^2 + d_{pole}^2}, \tag{6.5}$$

and $k_r = \omega(t)/c$. The calibration factor is calculated by the equation

$$s_{calfactor}\big(x_n, \omega(t)\big) = \frac{s_{caltheory}\big(x_n, \omega(t)\big)}{s_{cal}\big(x_n, \omega(t)\big)}. \tag{6.6}$$

This 2D cal factor is multiplied by each range profile acquired at each phase center across the array.

6.4.2 Coherent Background Subtraction and Coherent Change Detection (CCD)

Coherent change detection can be used to image changes in a target scene or moving targets on a frame-to-frame basis. First, the background target scene is measured without the target of interest present $s_{back}\big(x(n), \omega(t)\big)$. Next, place the target of interest in the scene and re-measure $s_{scene}\big(x(n), \omega(t)\big)$. The resulting background-subtracted data set is the difference between the target scene with and without the target placed:

$$s_{targets}\big(x(n), \omega(t)\big) = s_{scene}\big(x(n), \omega(t)\big) - s_{back}\big(x(n), \omega(t)\big). \tag{6.7}$$

The difference is fed to the SAR imaging algorithm where only changes will be shown in the image.

To show moving targets in real time, subtract the previous $s(x(n), \omega(t))$ from the current one, thereby coherently subtracting all stationary clutter from frame to frame. Pass the result to the SAR imaging algorithm and image what has changed from the previous frame. This is very effective for imaging moving targets. Extremely slow moving targets such as stationary breathing humans can be imaged using this technique. This method is also referred to as moving targer indication (MTI).

6.5 An S-Band Switched Array Radar Imaging System

A switched array radar system was developed that provides equivalent performance to a SAR of the same size but yet acquires its image in a small fraction of the time [5]–[7]. In some circles this type of radar system can also be considered a multiple-input multiple-output (MIMO) device, because it has multiple input and output ports (e.g., array elements). Unlike conventional MIMO systems that transmit simultaneously and use orthogonal waveforms, this radar transmitter and receiver is time division multiplexed across the array elements by use of microwave switches. This switching action provides orthogonal excitations that are multiplexed in time rather than transmitted simultaneously with orthogonally coded waveforms.

6.5.1 System Implementation

In this example, the S-band range-gated FMCW radar discussed previously is connected to a switched antenna array system as described (Sec. 6.2). This radar is capable of chirping from 1.926 GHz to 4.069 GHz at 2.5 ms, 5 ms, and 10 ms providing chirp rates (c_r) of 857 GHz/s, 428 GHz/s, and 214 GHz/s. The reader is referred back to Sec. 5.3.2 for details on this range gate architecture for FMCW radar.

FMCW radar uses separate transmit and receive antenna elements to minimize transmitter-to-receiver coupling. This technique provides better isolation than using a circulator with a single element because the circulator's performance depends on the reflected power of the element to which it is fed. For example, if the S_{11} of the element were -20 dB this would offer only 20 dB of isolation. The measured S11s of elements used in this radar vary over the wide bandwidth, but in general each element provides a better than -10 dB S11 between 2 to 4 GHz, dipping as low as -17 dB S11 between 3 and 4 GHz. The coupling between elements was measured at the radar center frequency of 3 GHz to be -46 dB from ANT19 to ANT9 and from ANT19 to ANT11, providing less coupling than a circulator feeding a well matched antenna. It is important to note that this technique may not be as effective if the antenna elements had a wide beamwidth, such as dipoles or monopoles.

As discussed previously in this chapter, by using switched antenna array techniques fewer transmit and receive elements are required to provide $\lambda/2$ spacing across the array. This comes at the expense of time required to time division multiplex (TDM) the radar transmit and receive ports to the appropriate antenna elements. For this array, only 8 receive elements and 13 transmit elements are required. Fortunately the practical application of imaging the location of human targets behind concrete walls does not require instantaneous beamforming, allowing for the use of switched array techniques because human targets are slow moving compared to the switching speed of the array.

The transmit and receive ports of the FMCW radar are fed to two fan-out switch matrices (Fig. 6.11) that route transmit signals to transmit elements (ANT1-13) and received signals from receive elements (ANT14-21).

The radar transmitter provides a peak output power of 10 mW, but the transmit fan-out switch matrix (SW1-4) has approximately 10 dB of insertion loss, resulting in a 1 mW peak power at the transmit antenna elements. The receive fan-out switch matrix (SW5-7) also has insertion loss; therefore in order to preserve the noise figure, each receive element has an LNA mounted directly on it (LNA1-8). The resulting cascaded noise figure is estimated to be approximately 3.3 dB.

The transmit and receive elements are physically separated into two subarrays made up of ANT1-13 and ANT14-21. Each element is a linear tapered slot antenna etched on FR-4 substrate (Fig. 6.14). The physical location of each element is represented by a large circle in Fig. 6.13, where the receive elements are on the top row and the transmit elements are on the bottom row.

Antenna switches are controlled digitally by the programming word defined (Table 6.2). Each phase center is dialed into the radar's transmit and receive ports by the hexadecimal (HEX) look-up table shown (Table 6.3).

At any given time the transmitter and receiver ports are routed to only one antenna pair. This pair represents a bi-static radar baseline. Only 44 pairs (or baselines) are used to synthesize a $\lambda/2$ aperture, where each of the pairs is represented by straight lines drawn between elements (Fig. 6.13). Each bi-static pair of transmit and receive elements functions like a mono-static element approximately located half way along the line drawn between them [3]. The middle row of small circles in Fig. 6.13 indicates where the effective mono-static elements are located.

The beamforming algorithm is the SAR imaging algorithm discussed in Chapter 4. This algorithm assumes a uniformly sampled $\lambda/2$ aperture. These assumptions are an approximation to the actual effective mono-static element position which, according to analysis [6], holds true for targets at practical stand-off ranges of 4.5 m or greater. Due to the relatively small size of this SAR, the target scene of interest is within the main beam of each element.

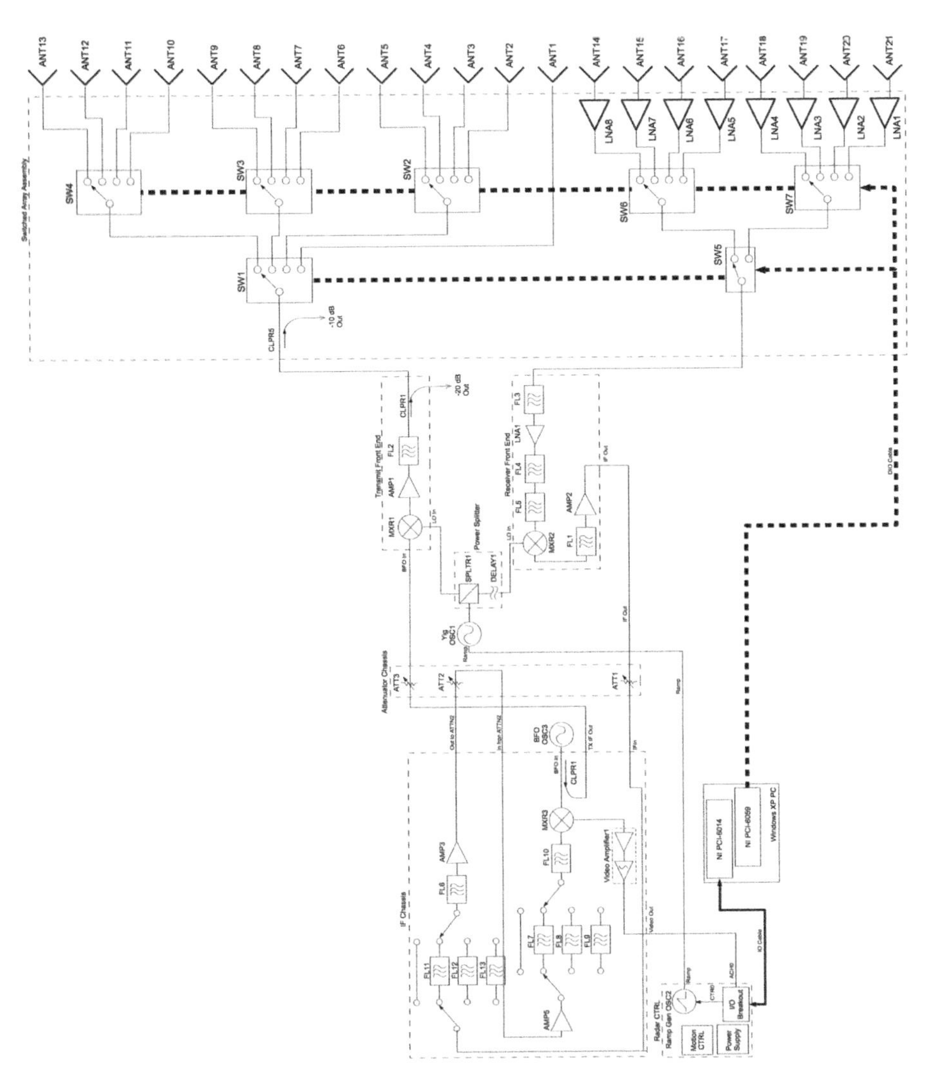
ANT13
ANT12
ANT11
ANT10
ANT9
ANT8
ANT7
ANT6
ANT5
ANT4
ANT3
ANT2
ANT1
ANT14
ANT18
ANT19
ANT21
LNA8
LNA7
LNA6
LNA5
LNA4
LNA3
LNA2
LNA1
Switched Array Assembly
SW4
SW3
SW2
SW6
SW7
SW1
SW5
CLPR5
-10 dB Out
CLPR1
-20 dB Out
Transmit Front End
MXR1
AMP1
FL2
Receiver Front End
FL3
LNA1
FL4
FL5
MXR2
AMP2
FL1
IF Out
Power Splitter
SPLTR1
DELAY1
Yig OSC1
Attenuator Chassis
ATT3
ATT2
ATT1
DIO Cable
BFO OSC3
CLPR1
MXR3
FL10
Video Amplifier1
IF Chassis
AMP3
FL6
FL11
FL12
FL13
FL7
FL8
FL9
AMP5
Radar CTRL
Ramp Gen OSC2
I/O Breakout
Motion CTRL
Power Supply
IO Cable
NI PCI-6014
NI PCI-6059
Windows XP PC

FIGURE 6.11

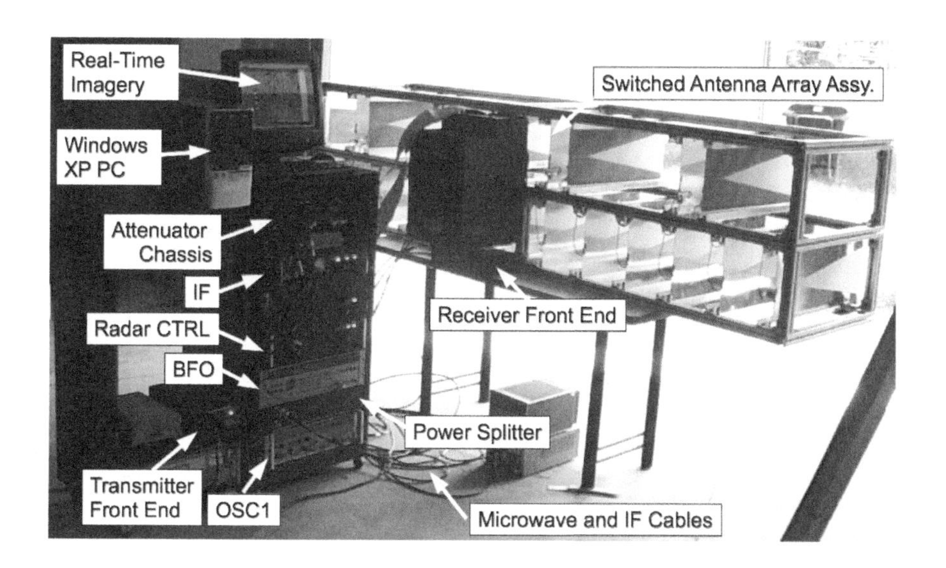

FIGURE 6.12
S-band real-time near-field phased array radar system with call-outs.

TABLE 6.1
List of materials for the S-band switched array assembly.

Component	Description
ANT1-21	LTSA built on FR4 (Fig. 3.25)
CLPR1	Narda −10 dB Directional Coupler
LNA1-8	Mini-Circuits ZX60-6013E, 20 MHz to 6 GHz, gain = 14 dB, NF = 3.3 dB
SW1-4, 6, 7	Mini-Circuits ZSWA-4-30DR, DC-3 GHz 4-way GaAs Switch
SW5	Mini-Circuits ZSDR-230, DC-3 GHz PiN Diode Switch

The switched antenna-array radar system is shown in Fig. 6.12 and the front of the array is shown in Fig. 6.15. A Labview graphical user interface (GUI) controls the switch matrices, pulses the transmitter, digitizes the de-chirped video signal, and computes then displays the SAR image, providing a pulse rate frequency (PRF) of approximately 22 Hz and an image rate of approximately 0.5 Hz.

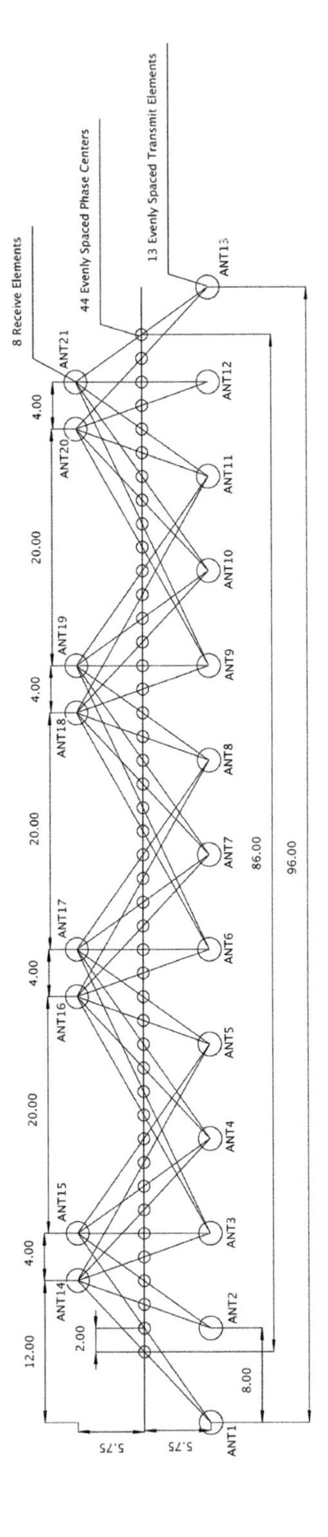

FIGURE 6.13
Array layout (units in inches) showing antenna element locations (large circles), bi-static baselines, and effective mono-static phase centers (small circles).

FIGURE 6.14
Photo of the production antenna element on FR-4 substrate; these antennas were produced at a board fabrication house to maintain consistency across the array (Fig. 3.25).

Each antenna pair is calibrated to a 1.52 m tall 1.9 cm diameter copper pole exactly 3.35 m down range and centered to the middle of the array in free space. This pole is treated as if it were a point target. Calibration coefficients are applied to each frame of data before the SAR image is processed.

6.5.2 Performance Estimate

Substituting specifications (Table 6.4) into the radar range equation (6.2), the maximum range is estimated to be 971 m for a 10 dBsm automobile target using the MATLAB® script [5]

The IDFT is applied over 2.5 ms up-chirps with a direct conversion receive architecture resulting in a 800 Hz effective noise bandwidth. The number of range profiles required to process a SAR image is 44 ($= N$).

The video amplifier (Fig. 2.13) limits the maximum range to $f_{-3db}c/c_r =$ 28 m, where $f_{-3db} = 80$ KHz and $c_r = BW/t_{sample} = 856$ GHz/s. With this, the radar range equation can be solved for σ by substituting $R_{max} = 28$ m, thereby showing that the minimum radar cross section that can be detected within the 28 m maximum range is −51 dBsm. This can be estimated using the same MATLAB script [5].

This radar chirps from 1.926 to 4.069 GHz. With this transmit bandwidth, expected range resolution is 6.2 cm (Equation (3.6)) with no weighting ($K_r =$ 0.89). The cross range resolution is expected to be 7.6 cm (Equation (4.9)) when the targets are located 4 m down range, centered with respect to the array with an aperture length of 2.24m, and no weighting ($K_r = 0.89$). It is important to note that cross range resolution of near field phased array radar devices, like this one, is better than the resolution of far field phased array devices when targets are close to the array because a near field radar device focuses the radar beam (like a lens) in the near field, accounting for wavefront curvature.

TABLE 6.2
Digital wiring for the S-band switched-array antenna switches and mapping to a 32 bit hex word for communicating with the PCI-6059 digital IO card.

Receiver Switch Matrix									Transmitter Switch Matrix															
Port 3	Port 2								Port 1								Port 0							
0	7	6	5	4	3	2	1	0	6	5	4	3	2	1	0	7	6	5	4	3	2	1	0	
SW5	SW7				SW6				SW4				SW3				SW2				SW1			
TTL	C5	C3	C4	C6	C5	C3	C4	C6	C5	C3	C4	C6	C5	C3	C4	C6	C5	C3	C4	C6	C5	C3	C4	C6

TABLE 6.3
Array control hex look-up table.

Phase Center	Receive Element	Transmit Element	Hex Code
1	14	1	1010001
2	15	1	1020001
3	14	2	1010012
4	15	2	1020012
5	14	3	1010022
6	15	3	1020022
7	14	4	1010042
8	15	4	1020042
9	14	5	1010082
10	15	5	1020082
11	16	3	1040022
12	17	3	1080022
13	16	4	1040042
14	17	4	1080042
15	16	5	1040082
16	17	5	1080082
17	16	6	1040104
18	17	6	1080104
19	16	7	1040204
20	17	7	1080204
21	16	8	1040404
22	17	8	1080404
23	18	6	0100104
24	19	6	0200104
25	18	7	0100204
26	19	7	0200204
27	18	8	0100404
28	19	8	0200404
29	18	9	0100804
30	19	9	0200804
31	18	10	0101008
32	19	10	0201008
33	18	11	0102008
34	19	11	0202008
35	20	9	0400804
36	21	9	0800804
37	20	10	0401008
38	21	10	0801008
39	20	11	0402008
40	21	11	0802008
41	20	12	0404008
42	21	12	0804008
43	20	13	0408008
44	21	13	0808008

FIGURE 6.15
Antenna array.

TABLE 6.4
Range gated S-band FMCW rail SAR specifications.

$P_{ave} =$ $10 \cdot 10^{-3}$ (watts)
$G_{tx} =$ 12 dBi antenna gain (estimated)
$G_{rx} =$ 12 dBi antenna gain (estimated)
$A_{rx} =$ $G_{rx}\lambda_c^2/(4\pi)$ (m^2)
$\lambda_c =$ c/f_c (m) wavelength of carrier frequency
$f_c =$ 3 GHz center frequency of radar
$\rho_{rx} =$ 1 because antenna efficiency is accounted for in antenna gain
$\sigma =$ 10 (m^2) for automobile at 10 GHz
$L_s =$ 6 dB miscellaneous system losses
$\alpha =$ 0 attenuation constant of propagation medium
$F_n =$ 3.5 dB receiver noise figure
$B_n =$ $2/t_{sample}$ system noise bandwidth (Hz) where $t_{sample} = 10$ ms
$(SNR)_1 =$ 13.4 dB
$N =$ 44 range profiles used in synthesizing the aperture for 2 inch spacing across the rail

This is achieved by using the RMA SAR imaging algorithm described in Chapter 4. Resolution for this imaging device can be estimated using the same MATLAB script used to estimate SAR resolution [5].

6.5.3 Free Space Results

Target scenes are simulated and measured in free space to assess the radar's range resolution, range and cross range sidelobes, and ability to measure low RCS imagery.

6.5.3.1 Simulated Sidelobes

To compare the effect of the array layout compared to a linear rail SAR, imagery of a point target at $p = (-484.2, 2.5, 0)$ cm was simulated. One data set was simulated assuming an error-free uniformly spaced linear array (Fig. 6.16a) and the other was simulated using the actual array layout (Fig. 6.16b), where the effective mono-static elements are determined by Equation (6.1).

A down range cut of these images at the point target location shown in Fig. 6.17, shows that the errors are negligible between the switched array and a mono-static linear array with uniform element spacing.

A cross range cut of these images at the point target location is shown in Fig. 6.18, showing that the peak magnitude of the point scatterer is 0.5 dB lower for the switched array than for the uniformly spaced linear array. This simulation also shows that the first cross range sidelobes of the switched array are approximately 2 dB lower than for a uniformly spaced linear array.

This demonstrates that the switched array is a close approximation to a linear uniformly spaced array. Furthermore, it shows that the switched array has superior cross range sidelobe performance compared to a uniformly spaced linear array. This is likely due to a slight randomization of phase centers.

6.5.3.2 Measured Sidelobes

A measured free space image of a soda can located 484.2 cm down range and 2.5 cm cross range is shown in Fig. 6.19b. For the analysis presented here it is assumed that this is a point target because of the range resolution, operational frequency, and length of this phased array radar. The simulated image of a point scatterer is shown in Fig. 6.19a.

Measured and simulated down range responses are shown in Fig. 6.20. The measured location of the soda can is slightly shifted toward the array. This could be due to errors in the expected physical down range location of the calibration pole. Some increases in range sidelobe levels are indicated at 540, 600, 650, and 675 cm. These are likely due to clutter in the target scene (measurements were acquired outdoors in the author's backyard). From this it is clear that the measured down range resolution is in agreement with simulation.

Measured and simulated cross range responses are shown in Fig. 6.21. The further out sidelobes starting around ± 50 cm do not agree with simulation. This could be due to mutual coupling between elements, feed lines, and antenna switches and radar errors in assumed phase center locations due to measurement geometry on target with respect to array elements. The switches

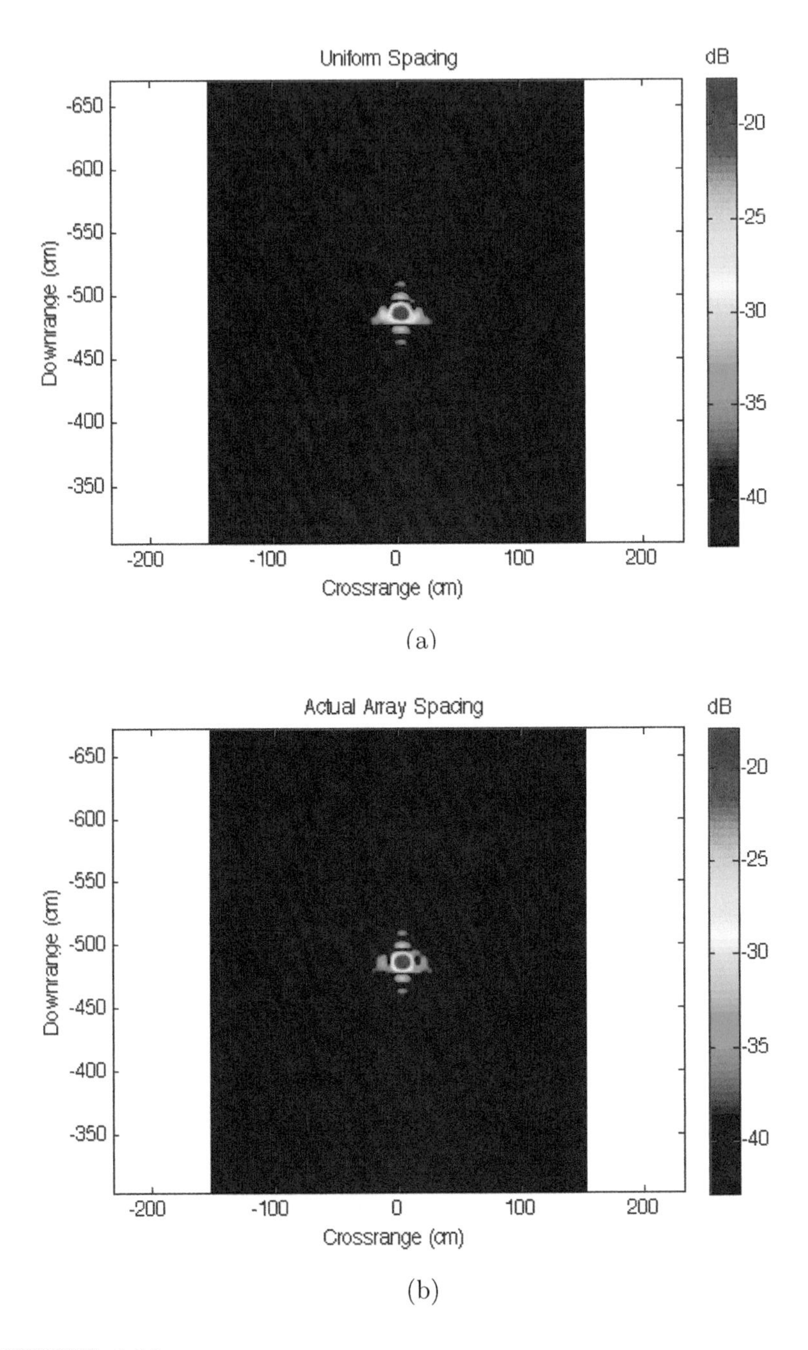

FIGURE 6.16
Simulated imagery of a point target with a uniform linear array (a) and actual array layout (b).

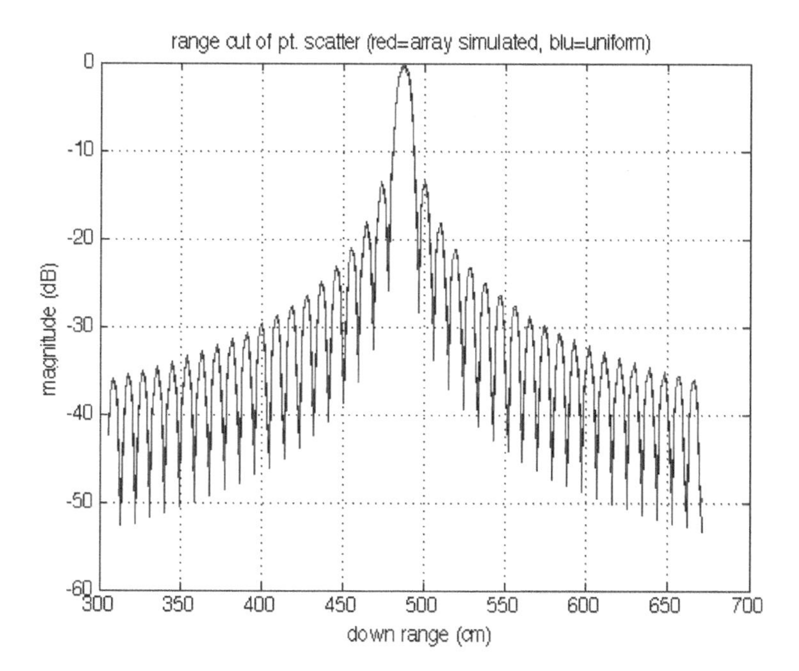

FIGURE 6.17
Down range cut of a simulated point target.

used in this system are designed to work up to 3 GHz with 40 dB of port-to-port isolation. It was decided to push the limit of operation of these switches to 4 GHz (often times microwave devices work well beyond their specified frequency limits). At 4 GHz the port-to-port isolation drops to about 35 dB, increasing mutual coupling due to the switches. Further coupling could be due to the fact that the element feed lines are bundled together on one wiring harness which is fed across the middle support member on the rear of the array. Regardless of this coupling, the close-in sidelobes and the −3 dB points agree with simulation; therefore the measured cross range resolution is in agreement with simulation.

6.5.3.3 Resolution

In this section, a free space image of a block-S configuration of 14 carriage bolts with equal spacing of approximately 0.305 m between adjacent bolts is discussed to characterize the radar's resolution. The bolts are mounted on a Styrofoam board that is parallel to the ground and approximately 5.25 m from the antenna array. Each bolt is 15.24 cm long with a diameter of 0.95 cm. Each bolt is mounted vertically, providing a small down range and cross range extent allowing the bolts to act like point scatterers. Imaging these bolts allows the radar's resolution to be tested.

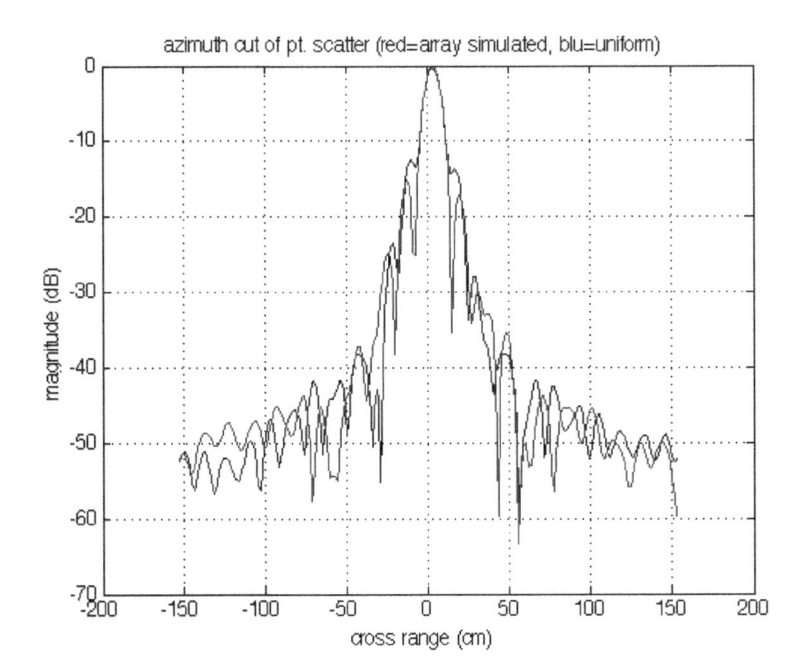

FIGURE 6.18
Azimuth cut of a simulated point target.

The range resolution was measured resulting in the image shown in Fig. 6.22. The expected down range resolution based on chirp bandwidth is 6.2 cm. The down range resolution measured from this image is 9.8 cm. These results show that the switched antenna-array radar is performing close to the smallest theoretical range resolution possible.

The expected cross range resolution for all targets shown at 546 cm is 11.1 cm. The measured cross range is 10.2 cm. The expected cross range resolution for all targets shown at 661 cm down range is 13.2 cm. The measured cross range is 12 cm. These results show that this switched antenna-array radar is performing close to the smallest theoretical cross range resolution possible.

6.5.3.4 Low RCS imagery

Additional free space SAR imagery was acquired to test the radar's sensitivity for target scenes with multiple low RCS point targets.

Figure 6.23 shows an image of a target scene consisting of 7.6 cm tall metal nails in a block S configuration. The location of each nail is clearly shown, although there does appear to be some degree of clutter at the bottom of the S. This is not surprising, considering the cross range sidelobe issues discussed previously.

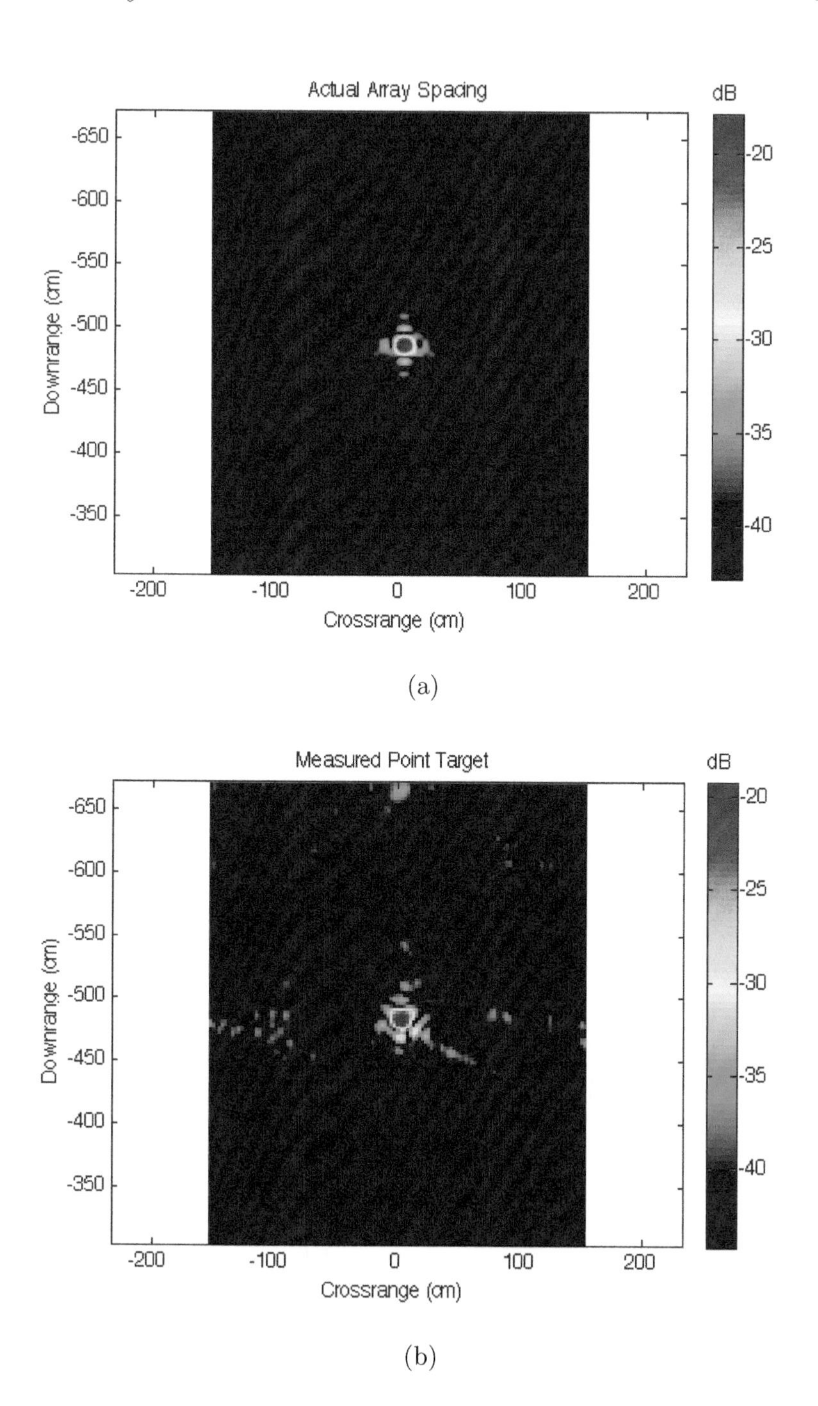

FIGURE 6.19
A near-real-time SAR image of a point target: simulated (a), measured (b).

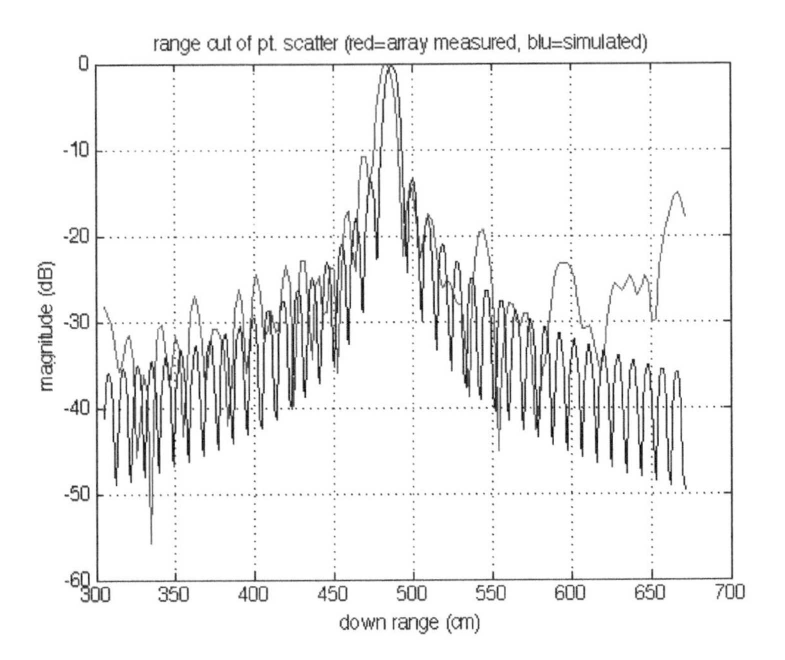

FIGURE 6.20
Simulated and measured down range cuts of a point target.

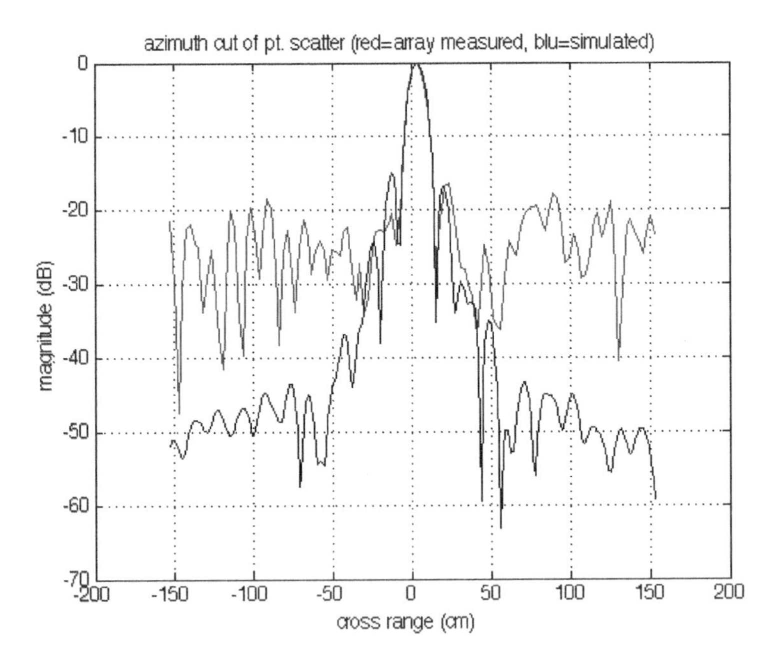

FIGURE 6.21
Simulated and measured cross range cuts of a point target.

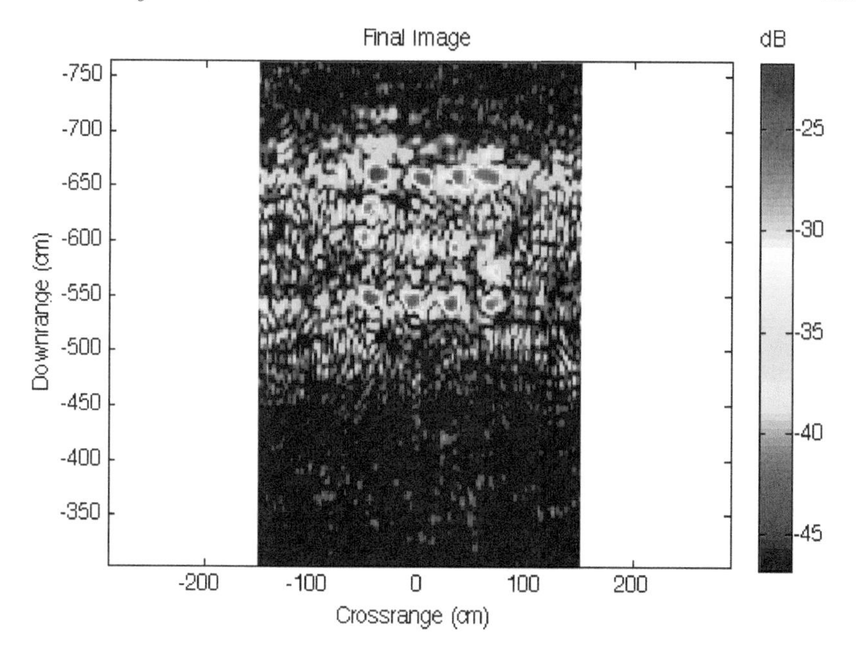

FIGURE 6.22
SAR image of a group of 15.24 cm long 0.95 cm diameter carriage bolts in free space.

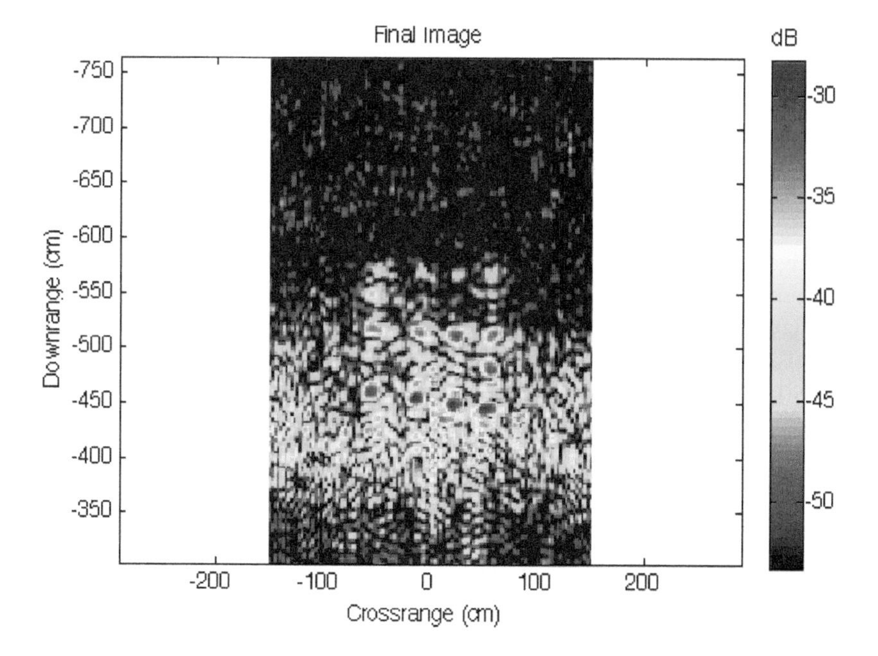

FIGURE 6.23
Image of a group of 7.6 cm tall nails in free space.

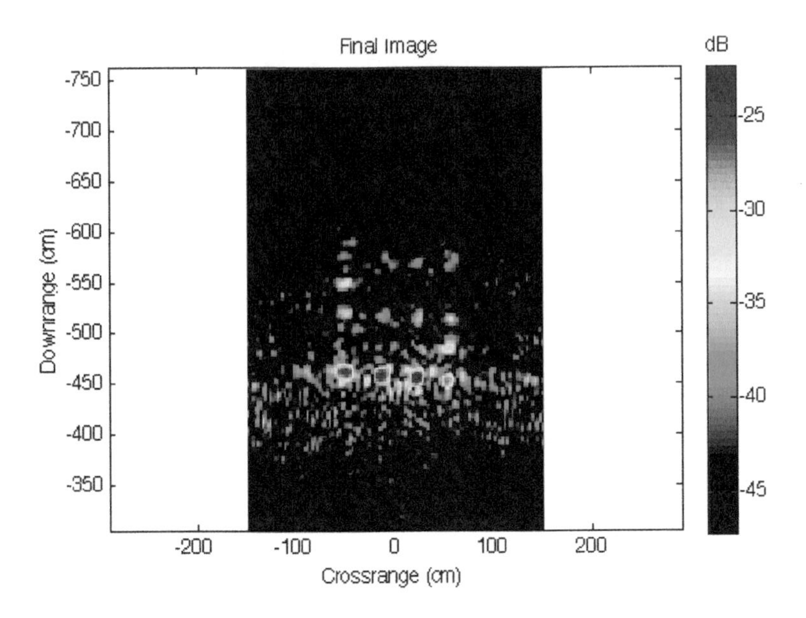

FIGURE 6.24
Image of a group of 5.08 cm tall nails in free space.

Figure 6.24 shows an image of a target scene made up of 5.08 cm tall metal nails in the same block S configuration. The position of each nail is clearly shown, where the last two rows are fading into the noise and clutter. The relative magnitude of each target in this image is 5 dB greater than for the image using 7.6 cm nails, likely due to the fact that 5.08 cm tall nails are half-wave resonant at the radar's center frequency of 3 GHz.

Figure 6.25 shows an image of a group of 3.18 cm tall nails in a block S configuration. The bottom two rows are clearly shown and a few of the nails at the top row are noticeable. The fact that these images could be acquired using 1 mW of peak transmit power demonstrates the radar's sensitivity.

6.5.3.5 Demonstrations

Demonstrations of this radar are shown, where a 6 inch diameter metal cylinder and a 12 oz soda can were imaged while moving in free space [9] and [10].

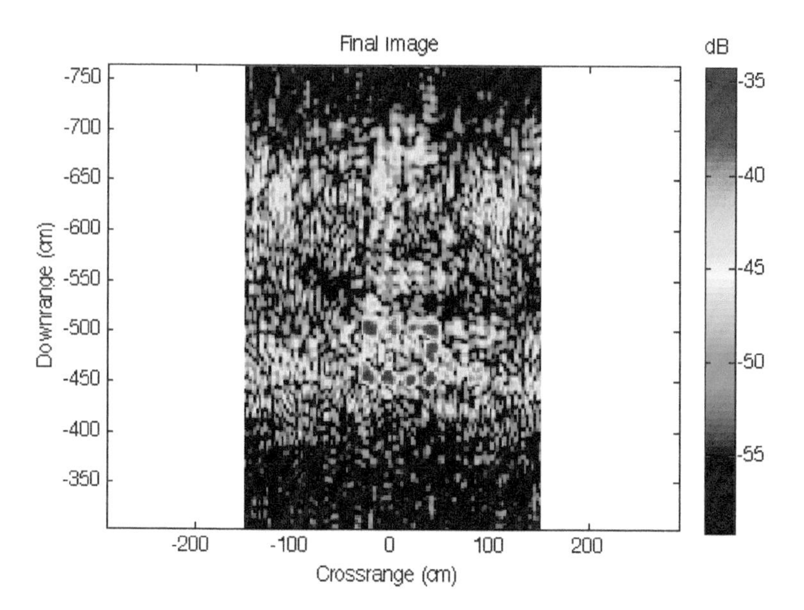

FIGURE 6.25
Image of a group of 3.2 cm tall nails in free space.

End-to-end operation of this radar system is demonstrated in a video [11] where all data from these experiments is available to the reader including MATLAB scripts to process the data [12].

6.6 MIT IAP Phased Array Radar Course

A low-cost switched array radar kit was developed based on the bi-static array techniques described above for a radar course at MIT to teach students the principles of phased array radar systems by actually building one [13] and [14]. In this the MIT IAP coffee can radar front end was plugged into a pair of four port microwave switches which fanned-out to four wi-fi antennas. These antennas were arranged on a pegboard following the bi-static antenna geometry described previously. Photos of this radar's implementation are shown (Fig. 6.26).

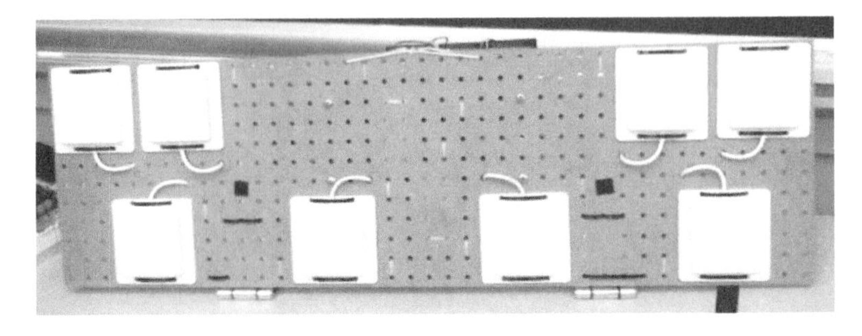

(a)

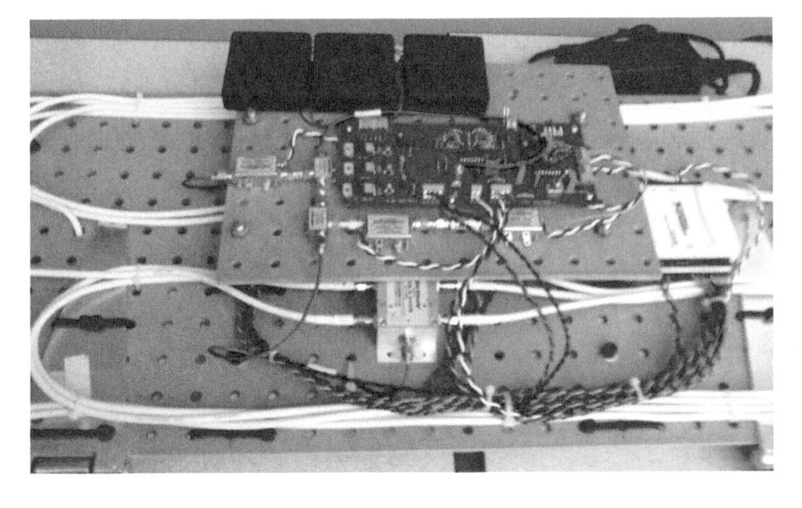

(b)

FIGURE 6.26
The MIT IAP phased array radar course radar system showing the array face (a) and electronics (b).

6.7 Summary

Phased array radar techniques for long range radar systems were shown to be impractical for short-range applications because targets are often in the near field for short range geometries. To address this issue, a near field phased array radar system was shown based on the fact that FMCW radar requires separate transmit and receive antennas, providing an array of 44 phase centers using only 21 antenna elements with comparable performance to a linear rail SAR of equal size. Numerous targets were imaged and simulated, quantifying performance. In addition to this, a short course on radar was developed at MIT based on this layout.

Bibliography

[1] J. D. Kraus, *Antennas, Second Edition,* McGraw Hill, Boston, MA, 1988, pp. 561–563.

[2] C.A. Balanis, *Antenna Theory Analysis and Design, Second Edition,* John Wiley & Sons, New York, NY, 1997.

[3] N. J. Willis, *Bistatic Radar,* Scitech Publishing, Inc., Raleigh, NC, 1995.

[4] E. W. Weisstein, Angle bisector theorem. From MathWorld–A Wolfram Web Resource. http://mathworld.wolfram.com/AngleBisectorTheorem.html.

[5] G. L. Charvat, "A Low-Power Radar Imaging System," Ph.D. dissertation, Department of Electrical and Computer Engineering, Michigan State University, East Lansing, MI, 2007.

[6] G. L. Charvat, L. C. Kempel, E. J. Rothwell, C. Coleman, and E. L. Mokole, "An ultrawideband (UWB) switched-antenna-array radar imaging system," 2010 International Symposium on Phased Array Systems and Technology, October 12–15, Waltham, MA.

[7] G. L. Charvat, L. C. Kempel, E. J. Rothwell, C. Coleman, and E. L. Mokole, "A through-dielectric ultrawideband (UWB) switched-antenna-array radar imaging system," *IEEE Transactions on Antennas and Propagation*, Vol. 60, No. 11, 2012, 5495–5500.

[8] G. L. Charvat, L. C. Kempel, E. J. Rothwell, C. Coleman, and E. L. Mokole, "A through-dielectric radar imaging system," *IEEE Transactions on Antennas and Propagation*, Vol. 58, No. 8, pp. 2594–2603, 2010.

[9] Radar imaging a cylinder in free space, `http://glcharvat.com/shortrange/phased-array-radar/`

[10] Radar imaging a 12 oz soda can in free space, `http://glcharvat.com/shortrange/phased-array-radar/`

[11] S Band Near-Field Phased Array Imaging System, `http://glcharvat.com/shortrange/phased-array-radar/`

[12] sband_array_demo_data.zip, `http://glcharvat.com/shortrange/phased-array-radar/`

[13] G. L. Charvat, B. T. Perry, and J. P. Kitchens, "The MIT IAP 2012 radar course: build a small phased array radar system capable of imaging moving targets," The Boston Chapters of the IEEE Life Members, AES, and Signal Processing Society, May 24, 2012.

[14] Bradley Perry, "Low-cost phased array radar for applications in engineering education," 2013 International Symposium on Phased Array Systems and Technology, October, Waltham, MA.

7

Ultrawideband (UWB) Impulse Radar

Impulse radar is valuable for small radar applications where range Doppler coupling is a problem (e.g., when using FMCW to range a fast moving target, if the target is fast enough its Doppler will cause a shift in the measured target position away from its actual position) such as in automotive radar. UWB radar is also useful when low spectral power density is desired for low probability of detection military applications or to comply with UWB spread spectrum regulations. Specific applications include but are not limited to automotive, ground penetrating, through-wall, laboratory radar systems, and longer range high power proof-of-concept systems [1].

Impulse radar is simply conventional radar implemented on a very small scale. As shown previously in a conventional radar (Sec. 1.2.1), a microwave pulse is generated and transmitted out of an antenna (Fig. 1.22). The pulse travels out to the target(s) and some of this pulse scatters off the target and travels back to the radar. Some of this scattered energy is collected by the antenna. The antenna is shared with the receiver where any scattered energy is directed through the circulator (additional switching circuitry protects the receiver when transmitting). Range to target is determined by measuring the time difference between transmitting the microwave pulse and receiving it. This can be done using an oscilloscope or a digitizer. Multiple simultaneous targets are superimposed on the oscilloscope and can be measured at the same time.

Impulse radar functions in the same way except that maximum range to targets is limited to typically less than 150 m and the range resolution requirements are very low. For this reason, this type of radar transmits an extremely short pulse (typical $Tp \leq 2$ nS) of microwave energy. This pulse is extremely short in duration and therefore it occupies a considerable instantaneous bandwidth $BW = 1/T_p \geq 500$ MHz. The difference between chirped radar devices previously discussed and impulse radar is that a chirped radar device takes time to sweep a given bandwidth but an impulse radar covers this bandwidth instantaneously in a single pulse. Additional theoretical analysis and laboratory experiments using impulse radar in a laboratory setting are investigated [2]–[6].

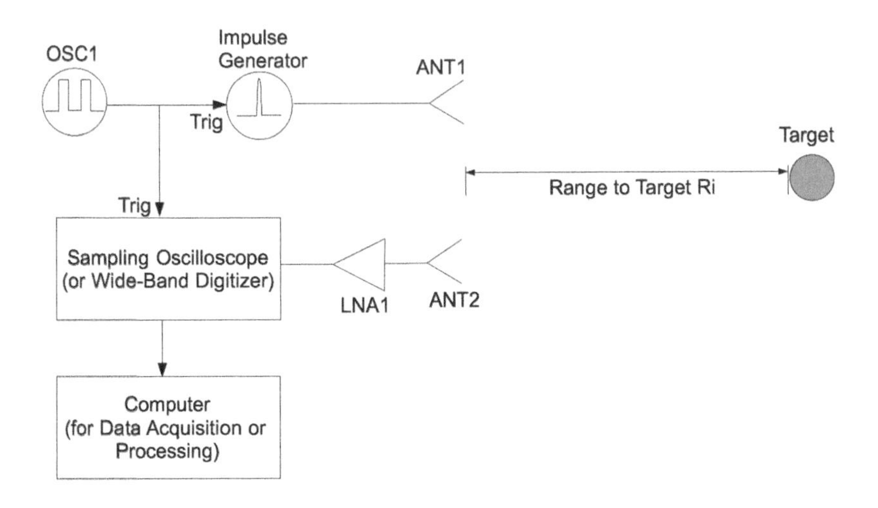

FIGURE 7.1
Basic UWB impulse radar system.

7.1 Architectures for UWB Impulse Radar

Unlike conventional radar, impulse radars often use two different antennas for transmit and receive or use low power transmitters with one antenna and a microwave circulator limited by the –20 dB isolation possible from a typical circulator. This is done because impulse radars cannot switch between transmit and receive due to the extremely short pulse times making it difficult to use the same antenna for both transmit and receive.

Two basic architectures will be described including directly driving and sampling from the antenna (Sec. 7.1.1) and the use of frequency conversion to increase the operational frequency of an impulse radar to a specified microwave spectrum (Sec. 7.1.2).

7.1.1 Basic UWB Impulse Radar System

A basic impulse radar is shown (Fig. 7.1). This system is typical of what would be found in an applied electromagnetics laboratory. Synchronization pulses are provided by OSC1, which is a square wave or pulse generator set to a rate of 100 to 500 KHz. The rising edge from OSC1 triggers the impulse generator to emit an impulse while simultaneously triggering the sampling oscilloscope to begin data acquisition.

A typical laboratory impulse generator is the Picosecond Pulse Labs 4015 which provides a pulse output with a fall-time of $T_f \leq 50$ pS. With the addition of an external pulse head and differentiator network the 4015 is capable of providing 50 pS impulses. The instantaneous bandwidth of a 50 pS pulse is $BW_{instantaneous} = 1/T_r = 20$ GHz. Such a pulse can be fed directly into a wide-band antenna, ANT1. ANT1 can be a Vivaldi, linear tapered slot antenna (LTSA), transverse electromagnetic (TEM) wave antenna, spiral, or other UWB antenna.

The impulse is radiated out of ANT1 toward the target scene, scattering off targets and some of the scattered impulse travels back toward the radar to be collected by ANT2 which is identical to ANT1. ANT2 is fed into a wide bandwidth low noise amplifier LNA1. The output of LNA1 is fed into the sampling scope where range-to-target information is acquired or displayed. Optionally, data can be fed into a computer for further analysis or signal processing.

Filters can be placed after the impulse generator to shape or band-limit the impulse. Similarly, filters can be placed after LNA1 to band-limit the input to the sampling scope, thereby reducing high frequency noise from aliasing into the desired acquisition bandwidth but usually the sampling scope has this filter built-in. With this architecture, a basic UWB impulse radar design has been shown.

7.1.2 UWB Impulse Radar Using Frequency Conversion

Frequency conversion techniques can be used to shift the impulse frequency up to a higher microwave band (Fig. 7.2), this is typically used to implement impulse radar systems in specific microwave bands (e.g., for automotive applications at 24 and 77 GHz).

OSC1 is a square wave generator that synchronizes the impulse generator and the sampling oscilloscope. On the rising edge of OSC1, the impulse generator produces an impulse and the sampling scope begins to acquire data.

The output of the impulse generator is fed to the IF port of frequency mixer MXR1. The LO port of MXR1 is fed by the microwave oscillator OSC2 through the power divider SPLTR1. The RF output of MXR1 is the product of the impulse and OSC2, shifting an image of the impulse above and below the frequency of OSC2. This product is amplified by PA1 and fed through FL1. FL1 can be placed before or after PA1 or ANT1 can function as FL1 given its bandwidth characteristics. The purpose of FL1 is to pass only one image from MXR1, either the upper or the lower image of the impulse. In most cases FL1 passes the upper image product from MXR1.

The output of PA1 feeds ANT1 where the microwave impulse is radiated out to the target scene. The microwave impulse scatters off target(s) and some of it propagates back toward the radar and collected by ANT2.

The output of ANT2 is fed into FL2 where only one of the receiver images is

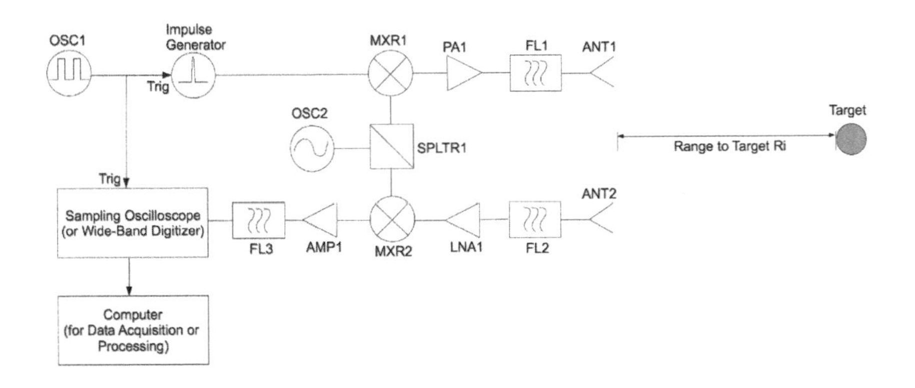

FIGURE 7.2
UWB impulse radar using frequency conversion.

passed through to low noise amplifier LNA1 and to the RF port of MXR2. FL2 can also be located after LNA1 or there may be two filters, before and after LNA1. FL2 can also represent the pass-band characteristics of the antenna ANT2.

The LO port of MXR2 is sourced by OSC2 through SPLTR1 and because OSC2 feeds the LO port of both MXR1 and MXR2, the frequency conversion maintains radar phase coherence. The IF output of MXR2 is the scattered return from the target scene at base band. In other words, waveforms sampled here will appear as though they are impulses from the impulse generator but spread in time, attenuated, and phase shifted according to target scene. The IF is amplified by AMP1 and fed through FL3 to the sampling oscilloscope for data acquisition, display, and recording. FL3 serves as an IF noise-limiting and anti-aliasing filter where FL3 is often built in to the sampling scope. Data can be fed to a computer for processing and additional analysis.

Additional filters can be added after the impulse generator to shape the pulse. With this architecture an impulse radar can be designed for any desired frequency.

7.2 Signal Processing for UWB Impulse Radar

Measuring range to target is greatly simplified because raw data is already in the time domain (Sec. 7.2.1). Calibration will be discussed (Sec. 7.2.2). Impulse radar data can also be used to create SAR imagery (Sec 7.2.3). Finally, coherent change detection using impulse radar systems will be outlined (Sec. 7.2.4).

7.2.1 Computing Range to Target

Measurements are already in the time domain when using an impulse radar system; therefore it is not always necessary to apply signal processing to the raw data if we assume the transmitted waveform sufficiently approximates a band-limited impulse. Sometimes it is convenient for a low-cost radar sensor to compute the absolute value of the raw data $S(t)$,

$$S_{display}(t) = |S(t)|, \tag{7.1}$$

where this function can be implemented with either an envelope detector circuit or in software. A threshold can be applied and a simple algorithm developed to find range to nearest or brightest target.

7.2.2 Calibration

Scattered impulses contain phase information which can facilitate Doppler measurements, moving target indication, coherent change detection, beamforming, and SAR imaging. Sometimes it is not accurate enough to assume that the transmitted impulse is close to a band-limited impulse or the range and cross range sidelobe specifications of the radar require a precision calibration of the pulse generation and radar signal chain.

To this end, calibration (cal) can be applied to impulse radar data just as it can be to FMCW frequency domain data. The first step is to apply the DFT of impulse data to convert this data into the frequency domain.

$$s\big(\omega(t)\big) = DFT(S(t)). \tag{7.2}$$

When in the frequency domain, the calibration procedure is the same as it was for FMCW radar, where a point cal target is placed at a known location down range from the radar sensor. In this case a metal pole is used and assumed to be a point target (for more accurate results a sphere should be used with the known scattering solution to a sphere applied [7]). A range profile is acquired of the pole represented by $s_{pole}\big(\omega(t)\big)$. The pole is then removed and a background range profile is acquired and the result is represented by $s_{calback}\big(\omega(t)\big)$. The background is subtracted from the pole range profile, resulting in a clean range profile of the pole only,

$$s_{cal}\big(\omega(t)\big) = s_{pole}\big(\omega(t)\big) - s_{calback}\big(\omega(t)\big). \tag{7.3}$$

The cal data is referenced to a theoretical point scatterer

$$s_{caltheory}\big(\omega(t)\big) = e^{-j2k_r R_{pole}}, \tag{7.4}$$

where R_{pole} is the range to the cal pole center and $k_r = \omega(t)/c$. The difficult task is to determine the frequency range of the sampled impulse signal. If a basic UWB impulse radar architecture is used (Sec. 7.1.1), then using the

Fourier transform relationships between time and frequency samples as previously described (Sec. 2.2) allows us to determine the frequency bandwidth of the impulse from the frequency domain plots providing the reader with needed values of k_r. If a frequency shifted UWB impulse radar architecture is used (Sec. 7.1.2) then the bandwidth as described for the frequency span of k_r is offset by the LO frequency of OSC2, where the term 2π fosc 2/c is added to k_r.

With the above understood, the cal factor

$$s_{calfactor}\big(\omega(t)\big) = \frac{s_{caltheory}\big(\omega(t)\big)}{s_{cal}\big(\omega(t)\big)}. \tag{7.5}$$

After data is acquired this cal factor is multiplied by each range profile before the profiles are fed to the beamforming or SAR imaging algorithm. Typically these algorithms operate in the frequency domain and for this reason calibrated data can be left as it is. If only high precision range profile is required, the IDFT can be applied to this data to transform back into the time domain.

7.2.3 Synthetic Aperture Radar

To apply the SAR imaging algorithm described previously in Chapter 4, the DFT must first be applied to all data to convert to frequency domain. The wavenumber (k_r) range must then be determined as described above. Calibration is optional. If it is not applied then the assumption is that the impulse sufficiently approximates a band-limited pulse and this may be enough to form an image. A practical example of SAR imaging using impulse radar will be described in Sec. 7.4.2.

7.2.4 Coherent Change Detection (CCD)

Just as it was applied to FMCW, CCD can also be applied to raw impulse data revealing anything that has moved or changed in the target scene since the last range profile or radar image was acquired. To implement CCD for impulse radar acquiring n pulses, one must subtract the previous pulse $S_{n-1}(t)$ from the current pulse $S_n(t)$

$$S_{CCD}(t) = S_n(t) - S_{n-1}(t) \tag{7.6}$$

to reveal the moving targets in a range profile or radar image. Some authors refer to this as moving target indication (MTI).

7.3 Expected Performance of UWB Impulse Radar Systems

Like all previous radar architectures the key performance specifications for impulse radar are the maximum range (Sec. 7.3.1) and the resolution (Sec. 7.3.2).

7.3.1 The Radar Range Equation for UWB Impulse Radar

The maximum range for a small UWB impulse radar system can be estimated using a similar procedure to other radar systems discussed in this book except that the noise bandwidth is inversely proportional to the impulse width and therefore is significantly higher than the noise bandwidth for a typical FMCW or CW radar system. Additionally, the average power of one transmit pulse, rather than the average power of the transmitter over many pulses, is used to make this estimate.

The maximum range to target for a small UWB impulse radar

$$R_{max}^4 = \frac{N P_t G_{tx} A_{rx} \rho_{rx} \sigma e^{(2\alpha R_{max})}}{(4\pi)^2 k T_o F_n B_n (SNR)_1 L_s}, \tag{7.7}$$

where:

$R_{max} =$ maximum range of radar system (m)
$P_t =$ RMS transmit power of the transmit pulse (watts)
$G_{tx} =$ transmit antenna gain
$A_{rx} =$ receive antenna effective aperture (m^2)
$\rho_{rx} =$ receiver antenna efficiency
$\sigma =$ radar cross section (m^2) for target of interest
$L_s =$ miscellaneous system losses
$\alpha =$ attenuation constant of propagation medium
$F_n =$ receiver noise figure (derived from procedure outlined in Sec 1.1.5.4)
$k =$ $1.38 \cdot 10^{-23}$ (joul/deg) Boltzmann's constant
$T_o =$ 290°K standard temperature
$B_n =$ system noise bandwidth (Hz)
$(SNR)_1 =$ single-pulse signal-to-noise ratio requirement
$N =$ number of range profiles used in synthesizing an aperture

For an impulse radar, the noise bandwidth is simply the -3 dB roll-off frequency (f_{-3db}) of the anti-aliasing filter. For a well-designed impulse radar where the pulse width matches the radar receiver bandwidth, the -3 dB roll-off frequency should correspond to the bandwidth of the transmitted impulse; therefore for a matched-filter radar the noise bandwidth $B_n = 1/T_p$.

Impulse radar will provide significantly lower maximum range for the same peak transmit power compared to FMCW radar. This is because FMCW radar transmits for a significantly longer duration, integrating scattered returns over the duration of the chirp by use of the IDFT. To match performance, an impulse radar would have to significantly increase the single-pulse transmit power P_t to difficult-to-achieve at low-cost levels.

For an impulse radar that is ranging only $N = 1$ and for an impulse radar that is used as a SAR imaging device, N is the number of range profiles acquired. Similarly, for phased array impulse radar N is the number of effective elements in the array.

7.3.2 Range Resolution for UWB Impulse Radar

Range resolution for any pulse radar is defined as the minimum detectable or observable difference between two targets. We consider the actual pulse width T_p to estimate the range resolution [8]

$$\rho_r = \frac{cT_p}{2}. \tag{7.8}$$

For SAR imaging using an impulse radar, cross range resolution is estimated the same as it would be for an FMCW SAR imaging system described previously in Equation (4.9).

7.4 UWB Impulse Radar Systems

Two examples of UWB impulse radar systems will be shown including a ranging system (Sec. 7.4.1) and a SAR imaging system (Sec. 7.4.2) at X-band.

7.4.1 X-Band UWB Impulse Radar System

A functioning example of a X-band UWB impulse radar system will be described, the performance estimated, and a demonstration shown. This design is an interesting case study because it uses frequency conversion architecture and therefore can be scaled to the automotive 24 GHz band or other microwave bands.

7.4.1.1 Implementation

A block diagram (Fig. 7.3) with photos and call-outs (Fig. 7.4 and 7.5) is shown. A complete bill of material is described (Table 7.1). This radar uses the front-end subsystem from the range-gated X-band FMCW radar previously discussed (Sec. 3.3.3.5) with extensive modification to operate as an impulse radar.

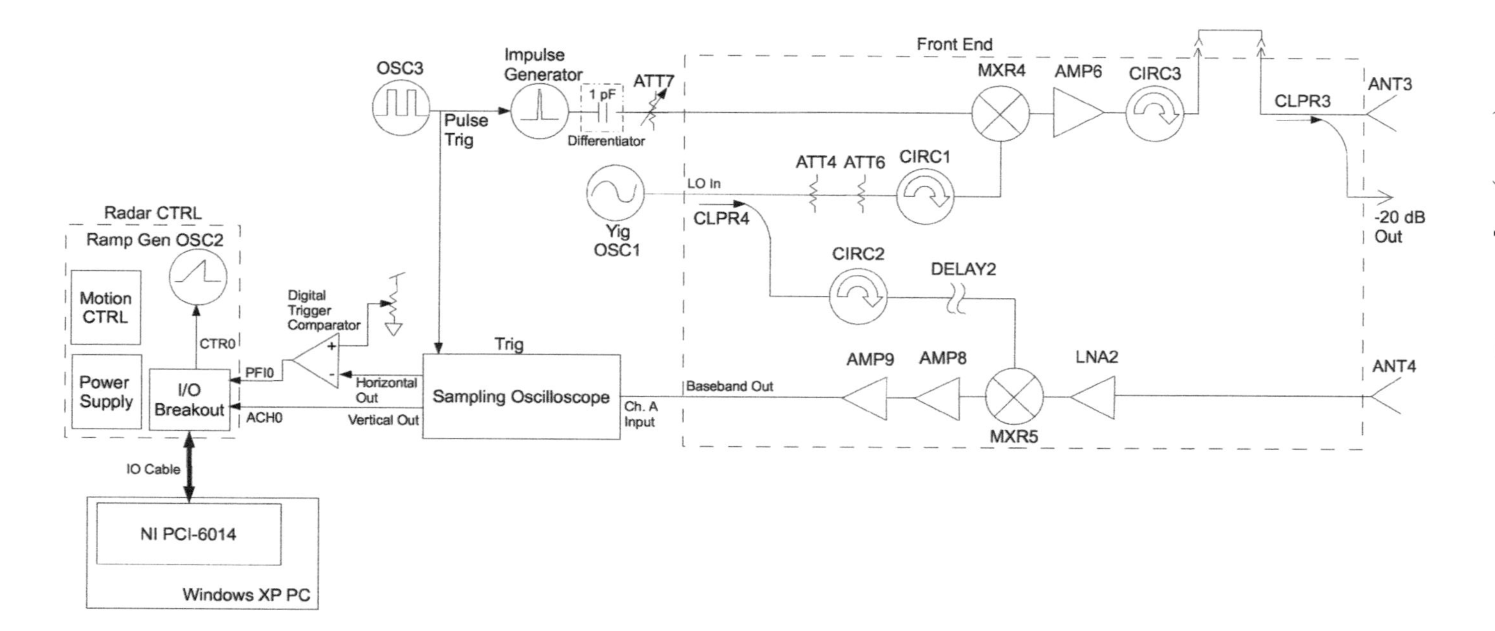

FIGURE 7.3
Block diagram of the X-band impulse radar system.

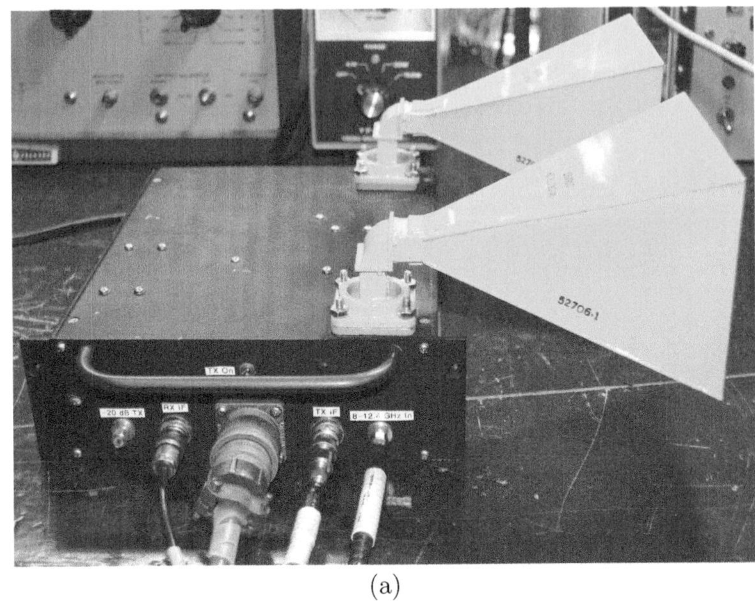

(a)

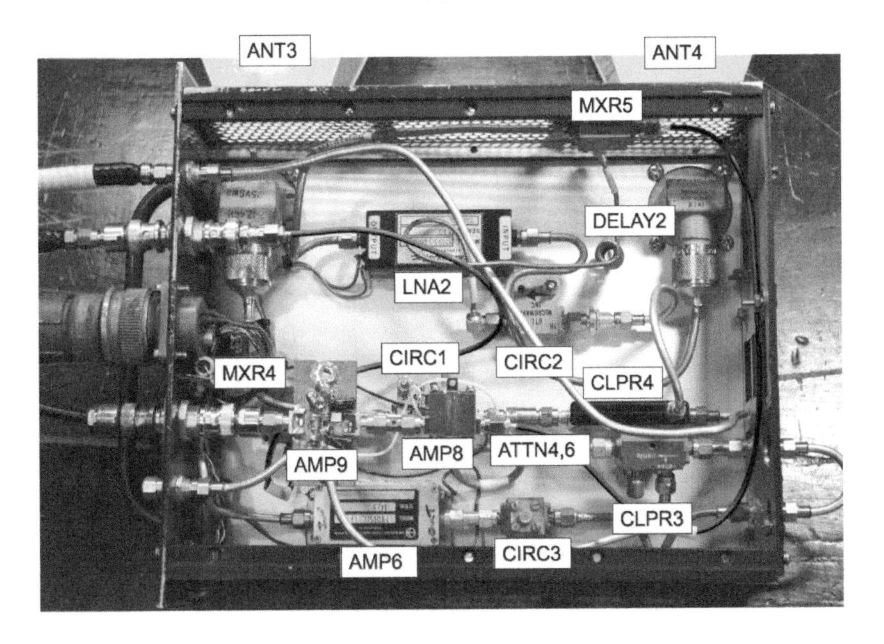

(b)

FIGURE 7.4
X-band impulse radar front end (a), front-end call-out diagram (b).

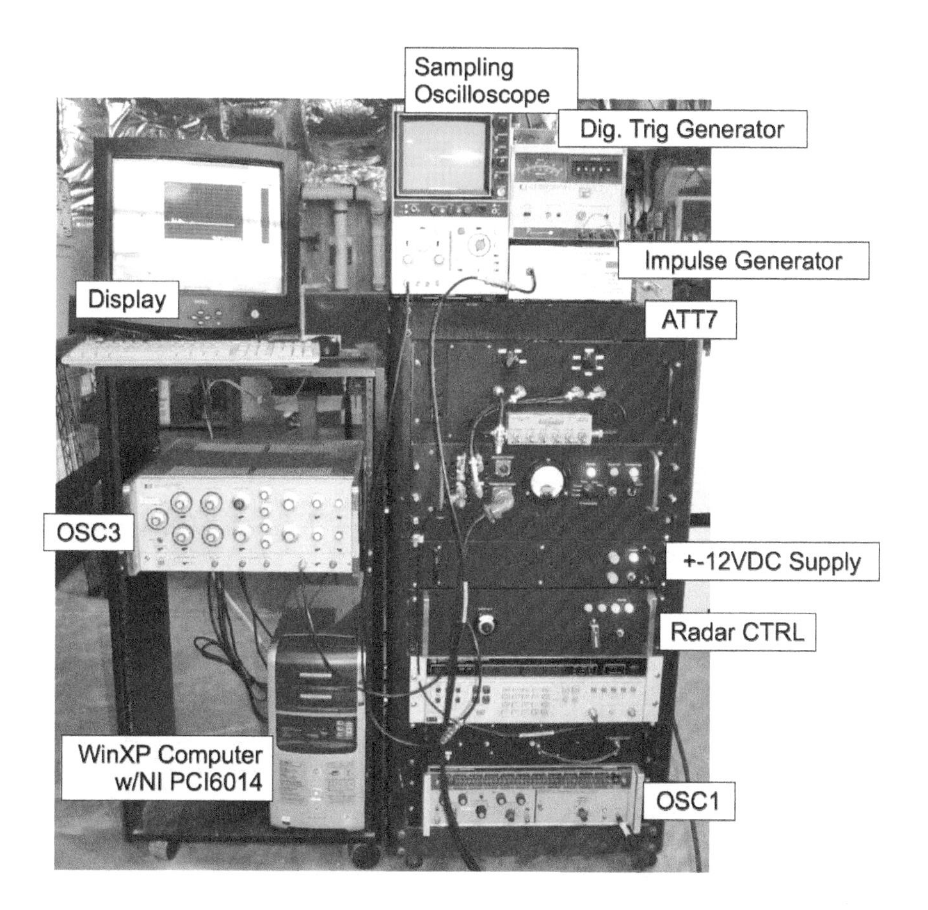

FIGURE 7.5
X-band impulse radar back-end call-out diagram, showing sampling scope, impulse generator, data acquisition computer, power supply, and radar control.

OSC3 is an HP8010A pulse generator that feeds a 500 KHz square wave as the trigger input for both the Picosecond Pulse Labs 4015 impulse generator and the Hewlett Packard 182C with 1810A plug-in sampling oscilloscope, causing the impulse generator to generate an impulse and the sampling scope to begin acquisition of a range profile at the same time (Fig. 7.6). In this timing system the impulse generator feeds a UWB impulse to the radar and the sampling scope acquires one sample in the time domain for every rising edge of the pulse generator. Before the next rising edge, the sampling scope delays the sample time on the order of 300 ps and on the next rising edge it

TABLE 7.1
X-band UWB impulse radar bill of material.

Component	Description
AMP6	Microwave Components Corporation MH858231, gain = 25 dB, IP1 = 23 dBm
AMP8	Mini-Circuits ZX60-6013E, 20 MHz to 6 GHz, gain = 14 dB, NF = 3.3 dB
AMP9	Mini-Circuits ERA-2 on microstrip line, DC to 6 GHz, gain = 16 dB, NF = 3.8 dB
ANT3	Microtech 205297, X-band horn with WR90 waveguide flange
ANT4	Microtech 205297, X-band horn with WR90 waveguide flange
ATTN4	3 dB attenuator
ATTN6	Midwest Microwave 6 dB in line attenuator
ATTN7	Narda Microline 705-69 step attenuator set to 8 dB
CIRC1	UTE Microwave X-band isolator
CIRC2	UTE Microwave X-band isolator
CIRC3	Unknown surplus X-band isolator
CLPR3	Unknown surplus X-band -20 dB directional coupler
CLPR4	Omni Spectra X-band -10 dB directional coupler
DELAY2	Coaxial delay line
Differentiator	1 pF microwave capacitor on a 50 Ω microstrip line
Digital trigger comparator	LM339 circuit in Fig. 7.7
Impulse generator	Picosecond Pulse Labs 4015, $T_r \leq 50$ pS
OSC1	Weinschel Engineering 430A Sweep Oscillator with 434A RF Unit set to 7.82 GHz CW
OSC3	Hewlett Packard HP8010A Pulse Generator
PC	PC running Labview, all system control software was written in Labview
PCI-6014	National Instruments PCI-6014 data acquisition and IO card
LNA1	Amplica, Inc. XM553403, gain = 20dB, IP1 = 25 dBm
MXR4	Watkins Johnson M31A
MXR5	TRW Microwave MX18533
Sampling oscilloscope	Hewlett Packard 182C with 1810A 1 GHz sampling plug-in

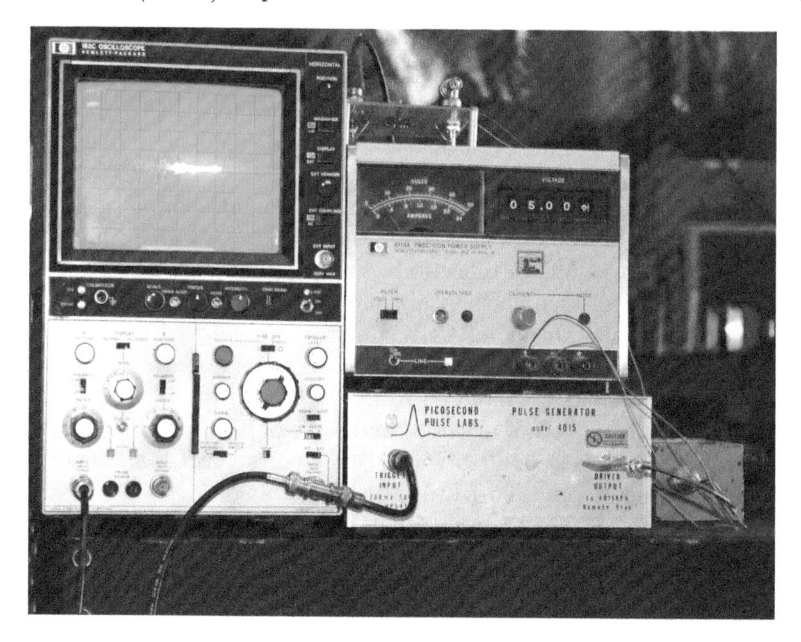

FIGURE 7.6
Three critical pieces of equipment: the sampling oscilloscope (left), the impulse generator (lower right), and the digital trigger circuit board (top right).

samples again [9] and [10]. This process is repeated at the rate of 500 KHz for every pulse from the HP8010A.

The output of the 4015 impulse generator is not actually an impulse; it is a 10 ns wide pulse with a fall-time of $\leq$ 50 ps. To generate an impulse with this waveform one must add a differentiator to the output. Picosecond Pulse Labs offers high-precision differentiators but the author decided to use a 1 pF microwave capacitor in series on a 50 Ω microstrip line. Results are sufficiently good for demonstration purposes.

The output of the differentiator is fed through a step attenuator ATT7 set to 8 dB and into the IF port of MXR4. The purpose of ATT7 is to reduce time domain-reflected energy from the mismatch between the differentiator and the IF port of MXR4.

The LO port of MXR4 is fed by the Yig Oscillator OSC1 through ATTN4, ATTN6, and CIRC1 to reduce the LO amplitude feeding MXR1 and to reduce reflected power from the LO port mismatch respectively. OSC1 is tuned to 7.82 GHz CW (unmodulated). When multiplying a UWB impulse by OSC1 within MXR4, the output of MXR4 will ideally produce a lower and upper image of the UWB impulse centered at 7.82 GHz. In impulse radar design, it is often desirable to only transmit and receive one of the two images, either the upper or the lower. In this case the feed structure of ANT3 and ANT4 is made from WR-90 waveguide which does not support frequencies below 8 GHz and thereby attenuates the lower impulse image passing only the upper. For this reason, the UWB impulse radar transmits an impulse with a bandwidth that starts at 7.82 GHz and ends at the waveguide's upper frequency bound of 12.4 GHz.

The RF port of MXR4 is the product of OSC1 and the impulse. This signal is amplified by AMP6 and fed to the antenna through CIRC3 and CLPR3. CIRC3 reduces time domain reflections due to antenna mismatches. CLPR3 provides a convenient trouble-shooting and calibration port if necessary. It is possible to access the output of CIRC3 or the input to CLPR3 through an external loop-back for trouble shooting purposes.

A microwave impulse is radiated from ANT3 and propagates toward the target scene. It scatters off of target(s) and some of it propagates back toward the radar. Some of this scattered field is collected by ANT4, amplified by the low noise amplifier LNA2, and fed into the RF port of MXR5.

The LO port of MXR5 is sourced by OSC1 through CLPR4, CIRC2, and Delay 2. CIRC2 reduces reflected signals from the mismatch at the LO port of MXR5. DELAY2 serves no purpose in this design; it is simply a carry-over from the range-gated FMCW radar front end.

The IF output of MXR5 is amplified by the wide-band amplifiers AMP8-9 and fed into the Channel A input on the sampling oscilloscope where the time domain range profile is acquired. The vertical output from the sampling oscilloscope is fed into ACH0 on the PCI-6014 data acquisition card where it is sampled at 200 KHz which is more than enough sample rate to digitally acquire the output of this sampling scope.

The horizontal output from the sampling scope is fed into an LM339 comparator circuit from which a digital trigger is derived (Fig. 7.7) and fed to PFI0 on the PCI-6015 data acquisition card. When the horizontal ramp crosses a threshold set by the 10K potentiometer then the output of this circuit goes high thereby triggering the PCI-6014 to acquire samples of the vertical output.

Acquired data is displayed in real time using a Labview GUI and saved on the PC for off-line analysis in MATLAB®.

7.4.1.2 Expected Performance

To determine the maximum range and the range resolution of the X-band UWB impulse radar, specifications (Table 7.2) are substituted into the radar range equation for impulse radar (7.7).

The antenna gains G_{rx} and G_{tx} were estimated based on aperture size and the noise figure is an educated guess based on the apparent age of the surplus microwave amplifier LNA2. $T_p = 500$ ps because the effective bandwidth of the sampling scope was measured to be about 2 GHz. Therefore $B_n = 1/T_p = 2$ GHz. The expected maximum range is 16.8 m and expected range resolution is 7.5 cm. This can be estimated using the MATLAB script [11].

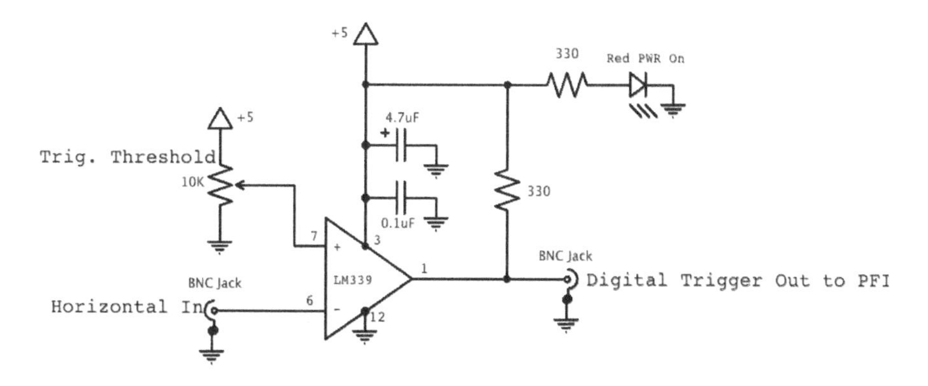

FIGURE 7.7
Comparator circuit that derives a digital trigger for the PCI-6014 from the sampling oscilloscope's horizontal output.

TABLE 7.2
X-band UWB impulse radar specifications.

$P_t =$	$10 \cdot 10^{-3}$ (watts)
$G_{tx} =$	17 dBi antenna gain (estimated)
$G_{rx} =$	17 dBi antenna gain (estimated)
$A_{rx} =$	$G_{rx}\lambda_c^2/(4\pi)$ (m^2)
$\lambda_c =$	c/f_c (m) wavelength of carrier frequency
$f_c =$	10 GHz center frequency of radar
$\rho_{rx} =$	1 because antenna efficiency is accounted for in antenna gain
$\sigma =$	10 (m^2) for automobile at 10 GHz
$L_s =$	6 dB miscellaneous system losses
$\alpha =$	0 attenuation constant of propagation medium
$F_n =$	4 dB receiver noise figure
$B_n =$	2 GHz
$(SNR)_1 =$	13.4 dB
$N =$	1 number of range profiles used in synthesizing an aperture

7.4.1.3 Ranging Example

A demonstration of the X-band impulse radar's ranging capability is shown in a demo video [12] and the results are discussed here.

In the first experiment the radar is placed outdoors (Fig. 7.8) and the author walks down range and back. The resulting RTI plot is shown (Fig. 7.9) where the author is walking outbound between 5 and 10 s, inbound between 10 and 15 s, outbound from 15 to 20 s, and inbound again from 20 to 25 s. At 25 s the author picks up two large copper cylinders (6 and 12 inches in diameter) and walks out to 20 ft away from the radar and begins rotating the

FIGURE 7.8
Target scene for the UWB impulse radar ranging example.

cylinders in a circle starting at 27 s and ending at 42 s. Finally, the author walks back to the radar from 42 to 46 s.

It is clear from this result that the author cannot be detected past 60 ns which corresponds to a maximum range of approximately 10 m. This result agrees with the maximum range of prediction of 16 m for a 10 dBsm target because for every 12 dB decrease in RCS the range prediction is reduced by half. This data is available for the reader to process [13].

RTI data was acquired in a cluttered environment by placing two large cylinders down range from the radar sensor (Fig. 7.10). The author walked down range and back a number of times and data was recorded and RTI

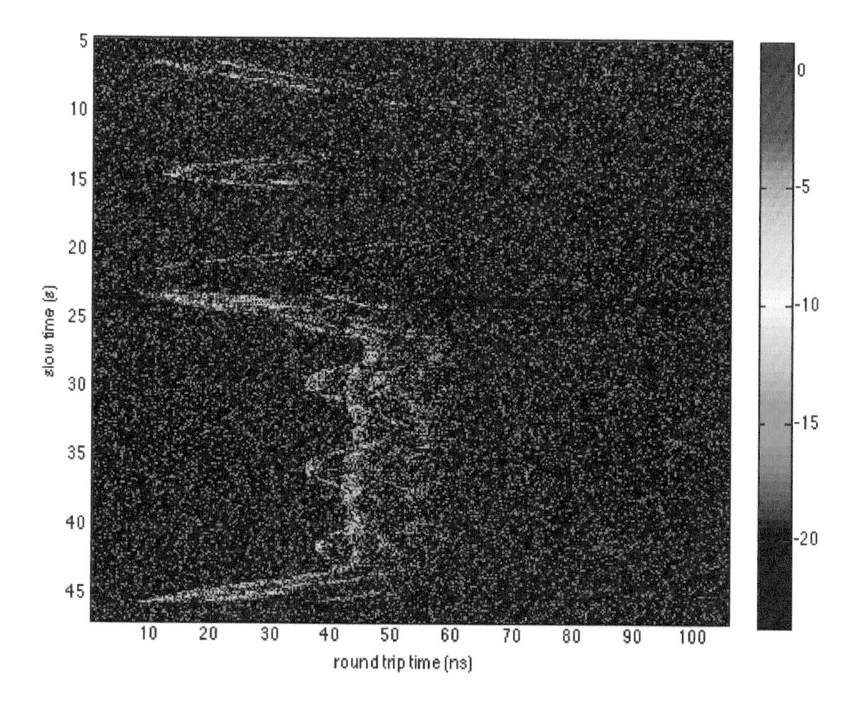

FIGURE 7.9
RTI plot of the author walking away from and back toward the radar, then while holding two large copper cylinders, rotates the cylinders in a circle approximately 20 ft down range.

plotted (Fig. 7.11). The author's location was clearly shown as he traveled away from and back toward the radar. In addition to this, the cylinder targets were also present in the RTI data as bright persistent vertical streaks.

To test this radar's ability to reject clutter, CCD was applied by subtracting the current pulse from the previous pulse and the results were plotted (Fig. 7.12). In this data the persistent returns from the two cylinders are significantly attenuated to the point where they are below the radar's noise floor, demonstrating the radar's ability to coherently subtract out stationary clutter returns.

In summary, maximum range for this radar is in agreement with the maximum range predictions. Additionally, this radar device was shown to be effective at reducing stationary clutter by using two-pulse CCD.

FIGURE 7.10
Target scene for the UWB impulse radar ranging example with two large stationary clutter copper cylinder targets.

7.4.2 X-Band Impulse SAR Imaging System

SAR imaging using an impulse radar will be shown in this section. Details on the SAR implementation are described followed by analysis of its expected performance. A description of data acquisition procedures and processing block diagram will also be shown and measured imagery discussed.

7.4.2.1 Implementation

The X-band impulse radar discussed (Sec. 7.4.1.1) is mounted onto the linear rail and motion control system described previously (Sec. 5.1). The resulting block diagram is shown (Fig. 7.13).

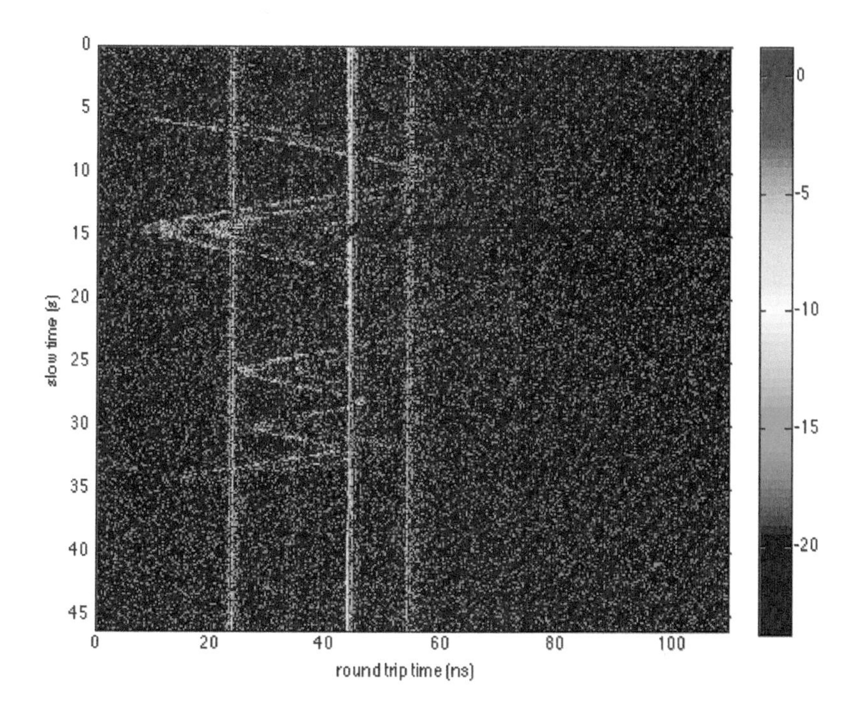

FIGURE 7.11
RTI plot of the author walking away from and back toward the radar with two large stationary clutter targets present (two large copper cylinders).

The X-band front end is mounted onto the linear rail. A stepper motor is connected to the motion controller in the radar CTRL subsystem. This radar is moved in 0.5 inch increments 95 inches down the linear rail and range profiles are recorded at each increment (Fig. 7.14).

7.4.2.2 Expected Performance

To determine the maximum range and range resolution, specifications (Table 7.3) are substituted into the radar range equation for impulse radar (7.7).

The antenna gains G_{rx} and G_{tx} were estimated based on aperture size and the noise figure is an educated guess based on the apparent age of the surplus microwave amplifier LNA2. $T_p = 500$ ps because the effective bandwidth of the sampling scope was measured to be about 2 GHz. Therefore $B_n = 1/T_p = 2$ GHz. The expected maximum range is 62.5 m, expected range resolution is

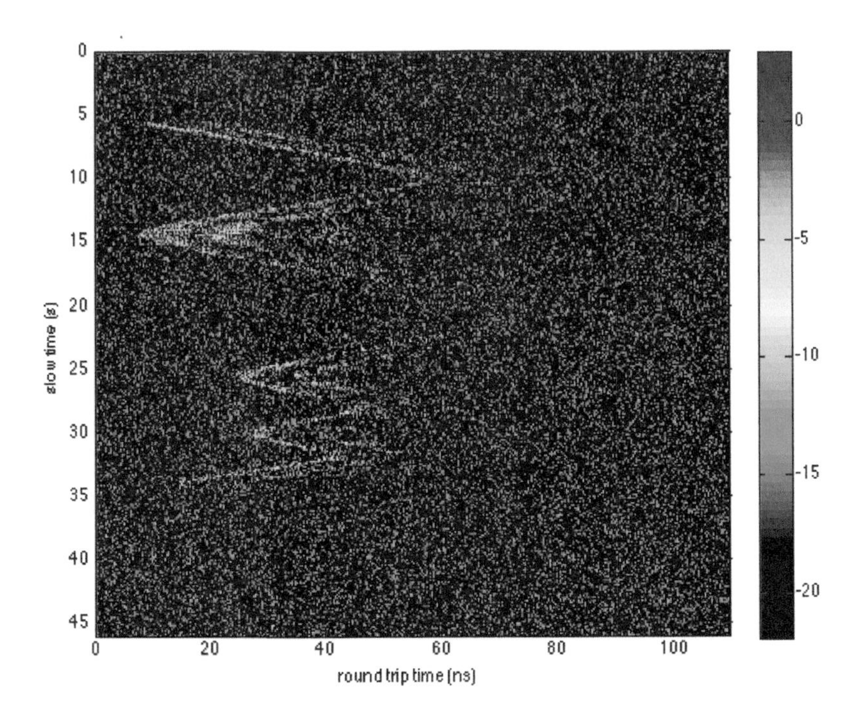

FIGURE 7.12
RTI plot using two-pulse CCD of the author walking away from and back toward the radar with two large stationary clutter targets present (two large copper cylinders).

7.5 cm, and expected cross range resolution is 3.2 cm at 5 m down range from the rail. This can be estimated using the MATLAB script [11].

7.4.2.3 Impulse SAR Data Acquisition and Processing

Image formation for this SAR is less complicated than for previous FMCW SAR imaging systems because calibration and background subtraction were not used. Calibration was a critical step in all previous SAR examples, but attempts to calibrate this system did not work because of insufficient SNR when measuring the calibration target. Additionally, unlike previous SAR examples coherent background subtraction was not used because clutter returns were negligible. To process an image, data is simply acquired then fed directly to the image formation algorithm (Fig. 7.15).

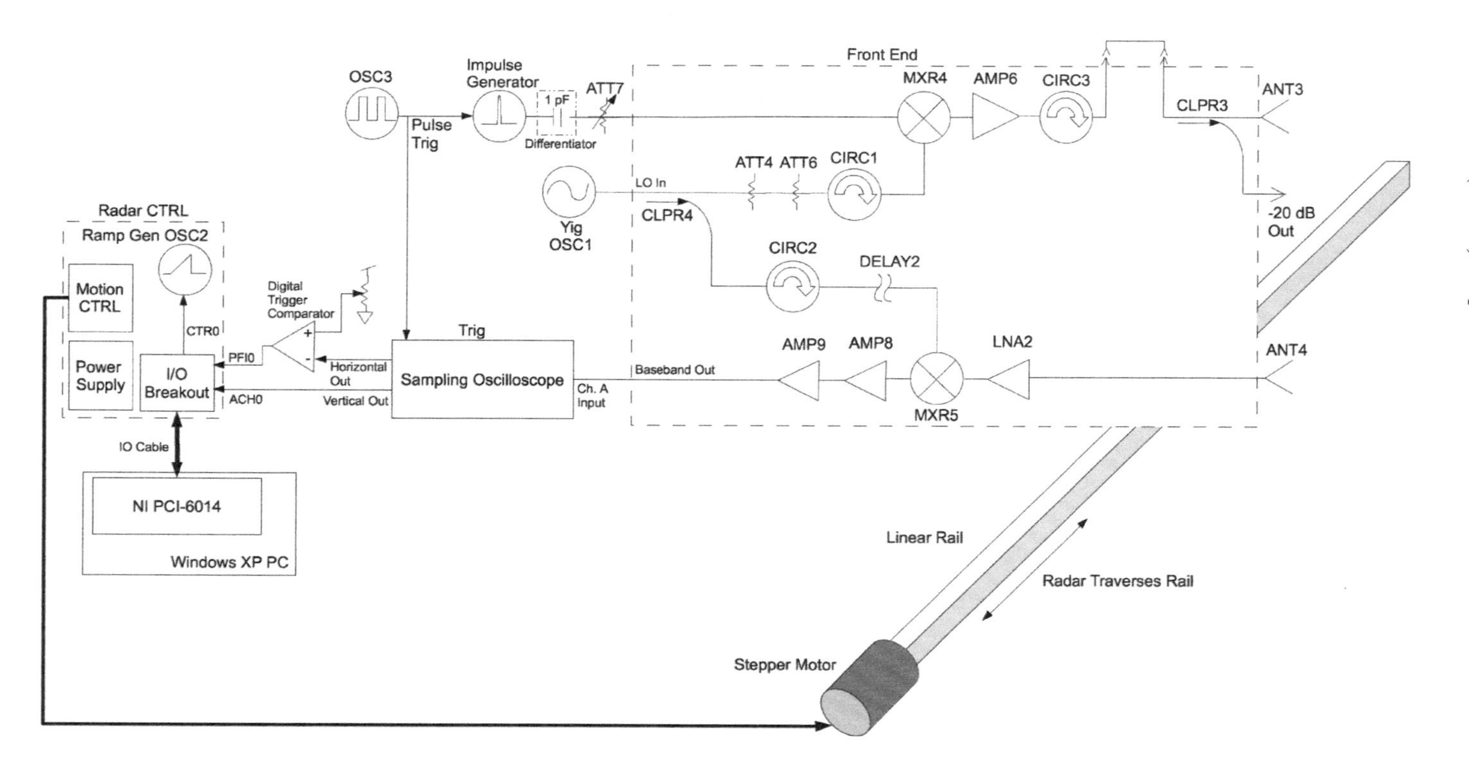

FIGURE 7.13
Block diagram of the X-band impulse SAR imaging system.

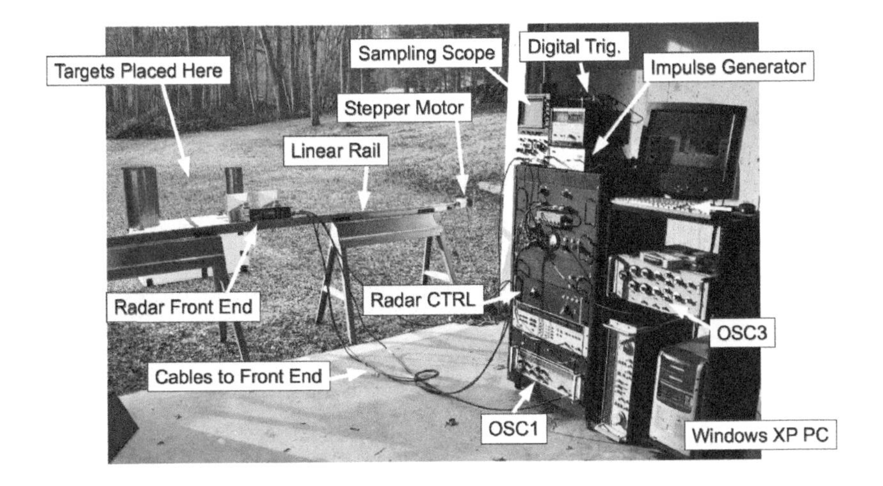

FIGURE 7.14
Target scene and call-out diagram of the X-band UWB impulse SAR imaging system.

TABLE 7.3
X-band UWB impulse SAR imaging system specifications.

$$
\begin{aligned}
P_t &= 10 \cdot 10^{-3} \text{ (watts)} \\
G_{tx} &= 17 \text{ dBi antenna gain (estimated)} \\
G_{rx} &= 17 \text{ dBi antenna gain (estimated)} \\
A_{rx} &= G_{rx}\lambda_c^2/(4\pi) \text{ (m}^2\text{)} \\
\lambda_c &= c/f_c \text{ (m) wavelength of carrier frequency} \\
f_c &= 10 \text{ GHz center frequency of radar} \\
\rho_{rx} &= 1 \text{ because antenna efficiency is accounted for in antenna gain} \\
\sigma &= 10 \text{ (m}^2\text{) for automobile at 10 GHz} \\
L_s &= 6 \text{ dB miscellaneous system losses} \\
\alpha &= 0 \text{ attenuation constant of propagation medium} \\
F_n &= 4 \text{ dB receiver noise figure} \\
B_n &= 2 \text{ GHz} \\
(SNR)_1 &= 13.4 \text{ dB} \\
N &= 190 \text{ number of range profiles used in synthesizing an aperture}
\end{aligned}
$$

7.4.2.4 Imaging Example

This impulse SAR is demonstrated in two imaging scenarios where large cylinders and a group of 11 copper pipes are imaged. A video of this demonstration is shown [14].

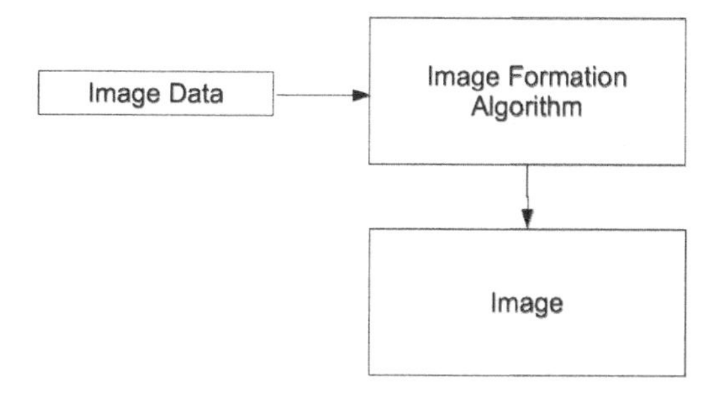

FIGURE 7.15
Signal processing work flow for creating an impulse SAR image.

An image of two cylinders that are 6 and 12 inches in diameter at two different ranges and cross ranges is shown (Fig. 7.16). The location of each cylinder is clearly differentiated in range and cross range. Some clutter is present further down range, likely due to artifacts from the simple differentiator circuit used to generate the impulse waveform.

A group of 6 inch tall 0.5 inch diameter copper pipes in a block S configuration was imaged (Fig. 7.17). Each pipe is present in this image but some are of lower intensity than others.

The −3dB resolution was measured for the pipe at the lower left of the block S. Range resolution was measured to be 8 cm which is close to theoretical best range resolution of 7.5 cm. The cross range resolution was measured to be 4.85 cm which is not very close to the theoretical best of 1.6 cm. This result was likely due to the fact that the target scene was placed close to the rail SAR, limiting antenna coverage of the target scene across all range profiles. Nonetheless, this image shows the SAR's ability to image a slightly more complicated group of point targets.

In summary, this impulse radar imaging system was shown to be effective at imaging relatively large targets and smaller more complicated groupings of point targets. Data and MATLAB processing scripts are available for the reader to process both data sets [15].

7.5 Summary

UWB Impulse radar is a short-pulse variant of traditional radar systems. It is useful for scenarios where range Doppler coupling may be a problem or an application requires a low power density transmit waveform. Applications for

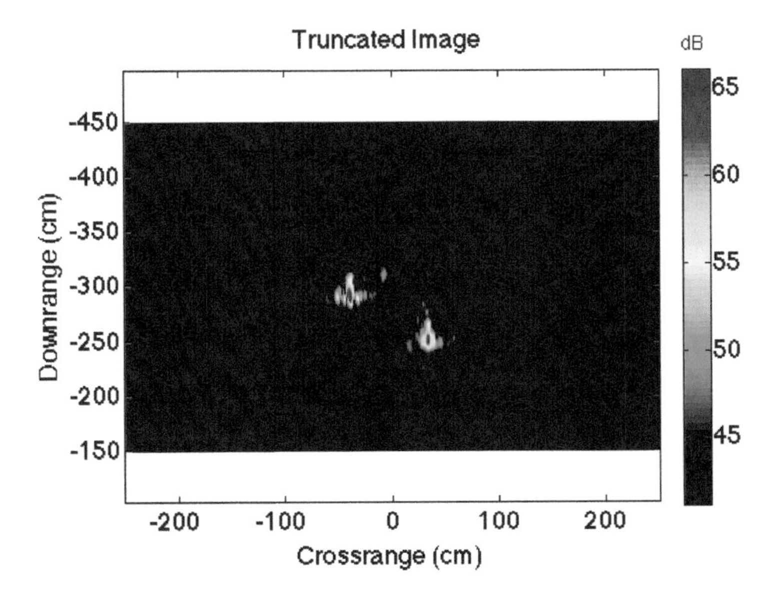

FIGURE 7.16
Impulse SAR imagery of two large cylinders 6 and 12 inches in diameter.

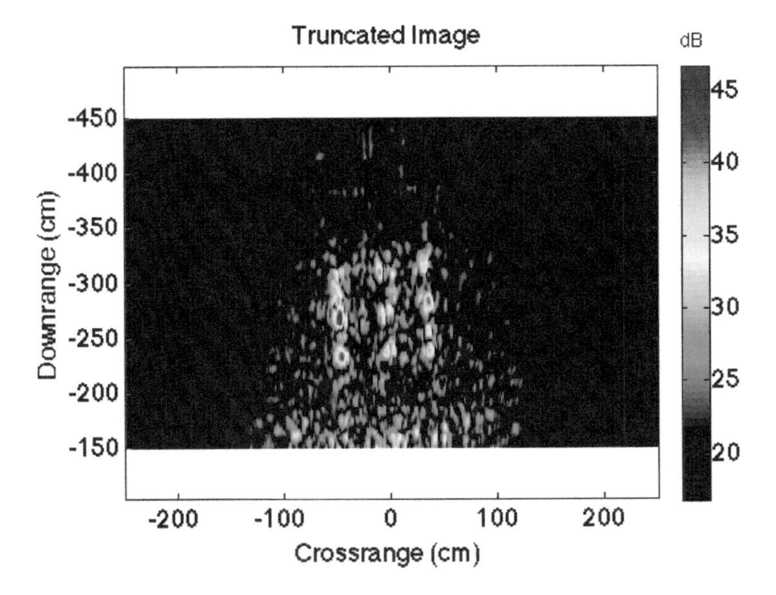

FIGURE 7.17
Impulse SAR imagery of a group of 11 copper pipes 0.5 inch in diameter and 6 inches in length in a block S configuration.

impulse radar include automotive radar, ground penetrating radar, through-wall radar, or for laboratory experimentation. Two examples of impulse radars were shown ranging moving targets with and without clutter and SAR imaging simple and complex target scenes. These results demonstrated impulse radar functionality for use in ranging and beamforming applications.

Bibliography

[1] W. D. Prather, C. E. Baum, R. J. Torres, F. Sabath, D. Nitsch, "Survey of worldwide high-power sideband capabilities," *IEEE Transactions on Electromagnetic Compatibility*, Vol. 46, No. 3, 2004, 335–344.

[2] E. J. Rothwell, W. M. Sun, "Time domain deconvolution of transient radar data," *IEEE Transactions on Antennas and Propagation*, Vol. 38, 1990, 470–475.

[3] E. J. Rothwell, K. M. Chen, D. P. Nyquist, J. E. Ross, "Time-domain imaging of airborne targets using ultra-wideband or short-pulse radar," *IEEE Transactions on Antennas and Propagation*, Vol. 43, 1995, 327–329.

[4] Y. Dai, E. J. Rothwell, "Time-domain imaging of radar targets using algorithms for reconstruction from projections," *IEEE Transactions on Antennas and Propagation*, Vol. 45, 1997, 1227–1235.

[5] M. Schacht, "Radar identification of hidden objects using direct time-domain measurements," M. S. Thesis, Department of Electrical and Computer Engineering, Michigan State University, East Lansing, MI, 1999.

[6] M. Schacht, E. J. Rothwell, "Time-domain imaging of objects within enclosures," *IEEE Transactions on Antennas and Propagation*, Vol. 50, 2002, 895–898.

[7] M. A. Morgan, "Ultra-wideband impulse scattering measurements," *IEEE Transactions on Antennas and Propagation*, Vol. 42, 1994, 840–846.

[8] W.G. Carrara, R.S. Goodman, and R.M. Majewski, *Spotlight Synthetic Aperture Radar Signal Processing Algorithms,* Artech House, Boston MA, 1995, eq. 2.1, p. 27.

[9] *Operating and Service Manual, Oscilloscope 182C,* Hewlett Packard Company, Colorado Springs Division, February 1974.

[10] *Operating and Service Manual, 1 GHz Plug-in (Sampling) 1810A,* Hewlett Packard Company, Colorado Springs Division, August 1971.

[11] radar_range_eq_impulse.m, http://glcharvat.com/shortrange/ultrawideband-uwb-impulse-radar/

[12] UWB X-band Impulse Radar Demo, http://glcharvat.com/shortrange/ultrawideband-uwb-impulse-radar/

[13] Xband_UWB_impulse_demo.zip, http://glcharvat.com/shortrange/ultrawideband-uwb-impulse-radar/

[14] Impulse Synthetic Aperture Radar Demonstration at X Band, http://glcharvat.com/shortrange/ultrawideband-uwb-impulse-radar/

[15] xband_UWB_impulse_rail_SAR_experiment2.zip, http://glcharvat.com/shortrange/ultrawideband-uwb-impulse-radar/

Part II

Applications

8

Police Doppler Radar and Motion Sensors

The earliest example of a small and short-range radar device is the proximity fuse (Fig. 8.1). This fuse is mounted on top of an artillery shell and contains a small CW Doppler radar made up of four tubes operating at approximately 100 MHz. First tube is a free-running 100 MHz oscillator, second and third are amplification stages, and the fourth is a thyratron (basically a high current switch with a threshold). When the shell is fired from a gun a glass capsule breaks open, spilling acid onto the contacts of a lead acid battery. Soon thereafter the tube filaments warm up and the oscillator turns on in free space while the shell is traveling through the air, well outside of the gun. The scattered Doppler signal is coupled off of the free-running oscillator. This Doppler is low-pass-filtered, amplified, and fed to a thyratron tube that actuates when a sufficiently high level of scattered Doppler return is detected. The thyratron blows the smaller explosive within the shell that causes the larger one to explode, thus damaging or destroying whatever was within close proximity to the shell [1]. These devices were developed for air defense so that the shell would explode at just the right time near an aircraft.

Shortly after the war, CW Doppler radar was developed for use in law enforcement of speed limits at a stand-off range. Additionally, radar motion sensors have also been in use for decades opening electric doors, alarm systems, and automatically turning on outdoor lights. These post-war technologies were enabled by a low-cost radar front end module known as a Gunnplexer (Sec. 8.1). This was the first low-cost microwave device enabling widespread use police radar (Sec. 8.2) and Doppler motion sensors (Sec. 8.3).

8.1 The Gunnplexer

The Gunnplexer was developed in the 1960's for use as a low-cost radar or microwave communication device (Fig. 8.2). At a high level, this device consists of an oscillator VCO1, coupler CLPR1, circulator CIRC1, and frequency mixer MXR1 configured as a CW radar (Fig. 8.3a) making it ideal for use in CW radar systems.

The oscillator OSC1 is made up of a Gunn diode placed in a resonant cavity where the dimensions of this cavity determine its approximate center

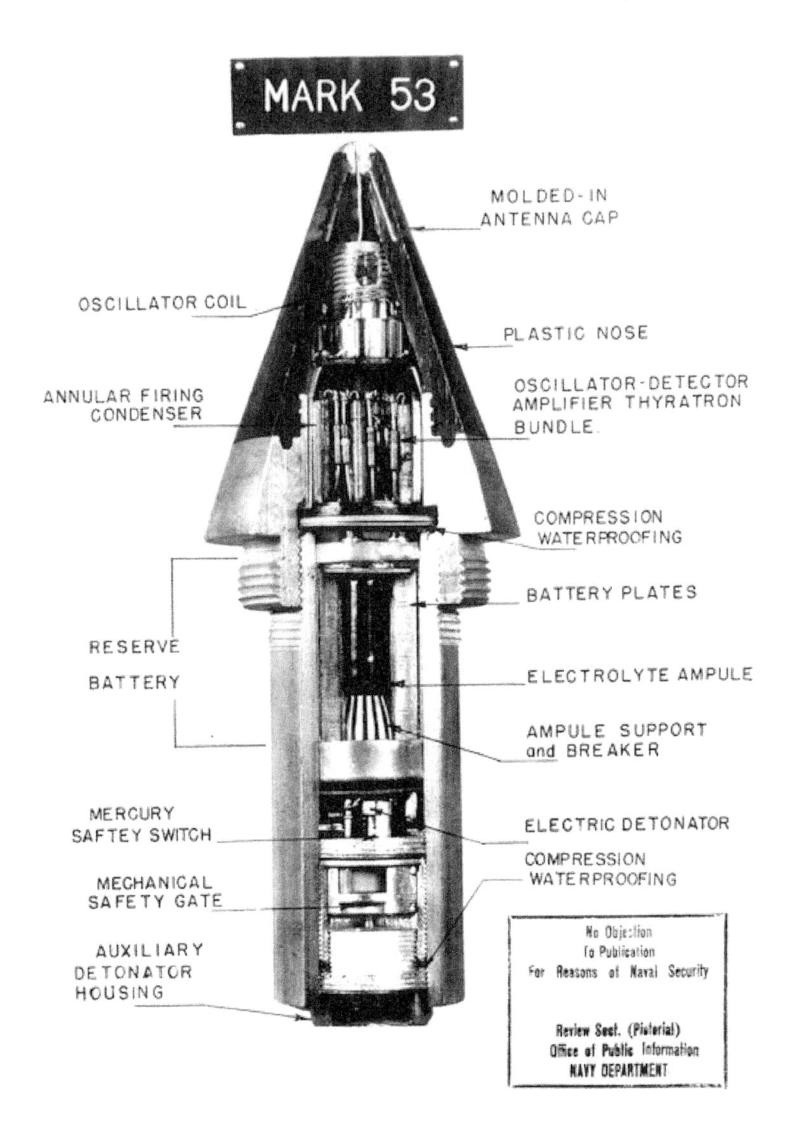

FIGURE 8.1
The proximity fuse, the earliest example of a small and short-range radar device.

frequency (Fig. 8.3b). Optionally, a varactor diode can be placed within this cavity to adjust the frequency of oscillation. Also, adjusting the diode's bias around the voltage at which it oscillates will perturb the operating frequency.

The output of OSC1 is fed through an aperture in its cavity which feeds into a waveguide. This waveguide is fed to the antenna. Within the waveguide

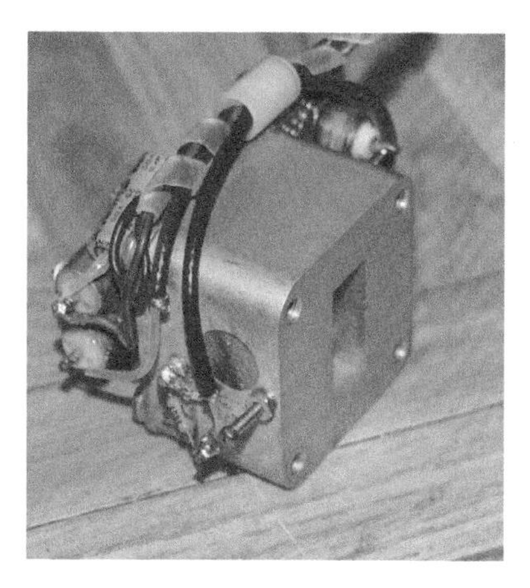

FIGURE 8.2
The Gunnplexer, a low-cost microwave transmit and receive module.

a Schottky mixer diode is placed that couples some power off (CLPR1) of the oscillator, causing it to switch on and off and function as a single-balanced mixer. The frequency mixer MXR1 is made up of this Schottky mixer diode.

A ferromagnetic circulator CIRC1 is also located in the waveguide that forces the transmitter carrier out of the waveguide flange and any reflected power into the frequency mixer MXR1.

This device provides a low-cost CW radar using diodes where the noise figure is the mixer's insertion loss from RF to IF which is typically about 10 dB.

8.2 Police Doppler Radar

Doppler radar is used by law enforcement to measure the velocity of moving vehicles at a stand-off range. Using radar is advantageous over what was previously done to measure vehicle speed, where markers were placed at known distances along a straight road and a stop watch was used to measure the time it would take for a vehicle to pass between the two markers to compute the vehicle's velocity, hence the term 'clocking' your speed.

Police Doppler radar sets are typically referred to as Doppler radar guns, because you point the directional antenna at a moving vehicle to measure its speed. Typical operating frequencies are approximately 10.25 GHz or 24.1 GHz, where 10.25 GHz radar guns are usually of older design.

With little exception, most of these devices are CW Doppler radar sensors that use Gunnplexer front ends (or equivalent). The Doppler frequency is

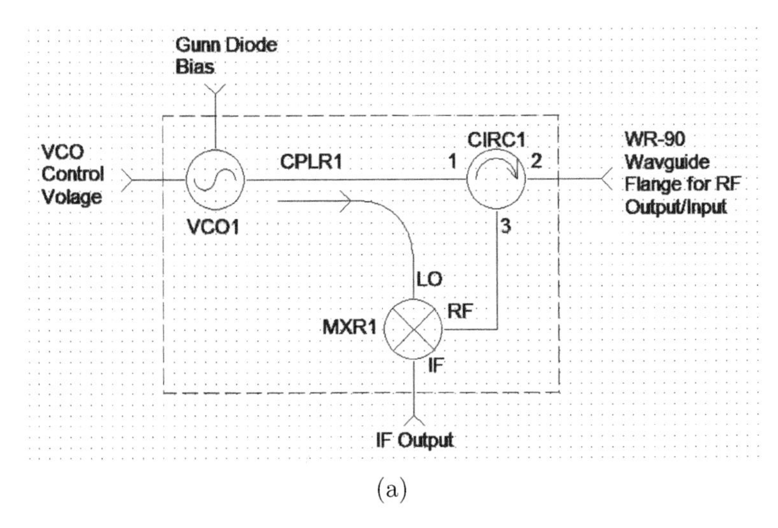

(a)

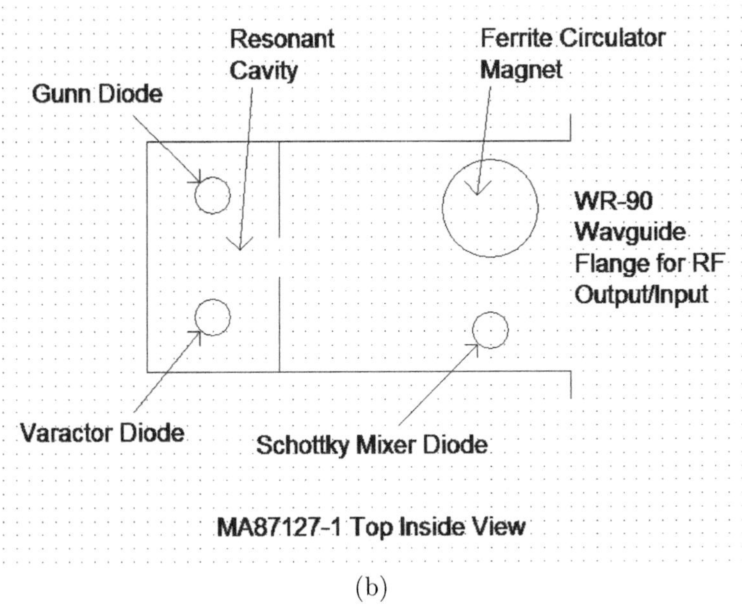

(b)

FIGURE 8.3
Functional block diagram (a) and layout of a Gunnplexer (b).

measured at the IF port of MXR1 which provides the frequency difference between what is transmitted and what is scattered. As described previously (Sec. 2.1), when using a CW architecture, the scattered signal is the difference between the CW carrier and the Doppler-shifted scattered frequency Δf_D

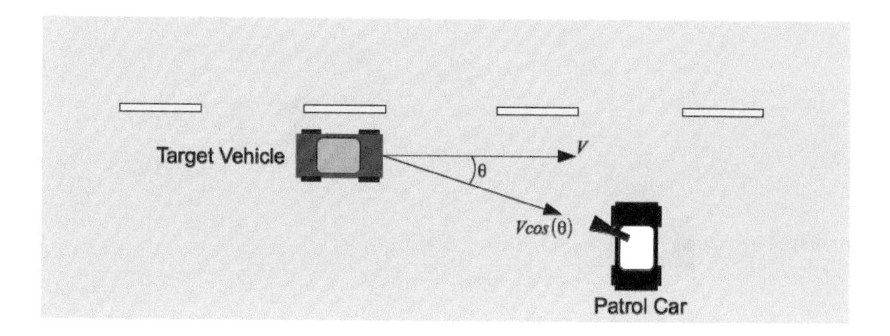

FIGURE 8.4
The cosine effect reduces the measured velocity of a moving vehicle when using police Doppler radar from a parked patrol car.

which is typically an audible frequency in the KHz range. A police radar processes this frequency into velocity by phase locking an oscillator to the scattered Δf_D to provide a stable measurement for a frequency counter [2]. With simple digital circuitry, this count is converted to velocity and displayed.

When deployed in a stationary patrol car, this radar measures the velocity of passing target vehicles for small values of θ, the angle between the forward velocity of the target vehicle and the line between the target vehicle and the radar in the patrol car (Fig. 8.4). When θ is close to 0 the velocity measured by the radar approximates the velocity of the target vehicle $V_{measured} \approxeq V$, but when θ is significant the velocity measured by the radar is

$$V_{measured} = V cos(\theta). \tag{8.1}$$

This error in measurement is in the target vehicle's favor, producing a lower velocity measurement by the patrol car.

Some police radar devices are capable of measuring the velocity of moving vehicles while mounted inside of a moving patrol car. These devices are typically used in highway patrol cars. The radar measures the velocity of the moving patrol car by measuring the ground return Doppler (or guard rails or anything within the field of view) and subtracting this from the measured Doppler of the target vehicle closing with (or traveling away from) the patrol car in the opposite lane.

The two signatures are differentiated from each other because the target vehicle closing in the opposite lane will present a significantly higher (or lower) Doppler return than the persistent ground return. For the case of a target passing the patrol car, this signature would be significantly lower than the persistent ground return.

Measurement errors occur when angles between the target vehicle and its forward velocity θ_T and angle between the patrol car and its forward velocity

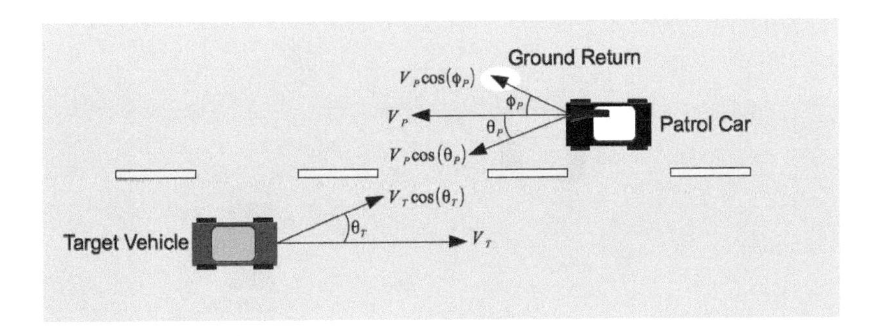

FIGURE 8.5
Police Doppler radar devices are capable of operating in moving mode, where the apparent velocity of the patrol car is subtracted from the apparent velocity of an oncoming car.

θ_P are significant. Additional errors occur when the angle between the radar and its ground return ϕ_P is significant. In these cases, the target vehicle's apparent velocity becomes [3]

$$V_{measured} = V_T cos(\theta_T) + V_P cos(\theta_P) - V_P cos(\phi_P). \quad (8.2)$$

Significant θ_T or θ_P (these values can be different if measuring on a curved road) will result in lower measured velocity $V_{measured}$ than actual target velocity V_T. Significant values of ϕ_P result in higher measured velocity than actual target velocity, although these errors rarely occur in practice because most of the time the road clutter will dominate the patrol car's persistent Doppler spectrum. Additionally, many states follow operational procedures requiring the officer to correlate patrol car's speedometer with measured patrol car's velocity as part of confirming the velocity a target vehicle.

8.2.1 K-Band Police Doppler Radar

A K-band police radar was procured by the author. This unit is designed to be mounted in a moving patrol car, where it measures the velocity of oncoming and outgoing traffic while taking into account the velocity of the patrol car. Specifically, this radar is a Kustom Electronics Trooper built in about June 1988 and operating at $f_c = 24.1$ GHz (Fig. 8.6).

8.2.1.1 Estimated Performance

According to the label on the top of the front end, the radar's center frequency is 24.1 GHz ($= f_c$). After removing the radome from the front end it was realized that the antenna is a horn with a circular aperture with a measured

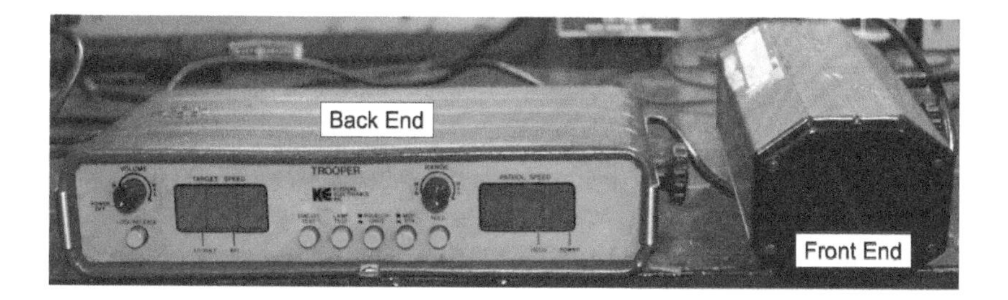

FIGURE 8.6
K-band police Doppler radar system intended to be mounted in a patrol car and capable of measuring the velocities of moving vehicles while the patrol car itself is moving.

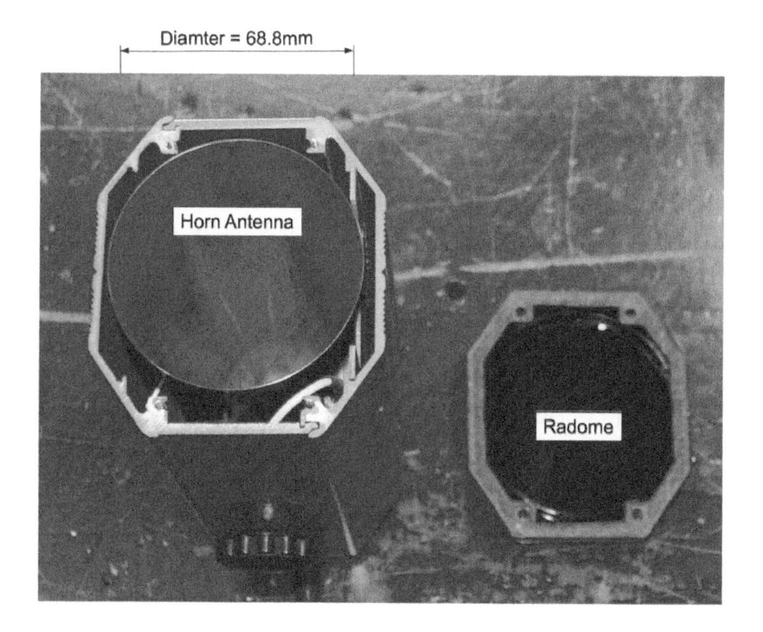

FIGURE 8.7
Horn antenna from the K-band police Doppler radar system.

radius of 34.4 mm (Fig. 8.7) providing an aperture of 3600 mm^2 ($= A_{rx}$). The gain of an antenna with this aperture assuming an efficiency of 0.9 ($= \rho$) is 24 dBi ($= G_{tx}$). Both the transmit and receive antennas are the same for this radar device because there is only one antenna.

TABLE 8.1
K-band police Doppler radar estimated and measured specifications.

$$
\begin{aligned}
P_{ave} &= 10^{-3}\ \text{(watts) assumed transmit power} \\
G_{tx} &= 24\ \text{dBi estimated antenna gain} \\
A_{rx} &= 3600\ \text{mm}^2\ \text{aperture of the antenna} \\
\lambda_c &= c/f_c\ \text{(m) wavelength of carrier frequency} \\
f_c &= 24.1\ \text{GHz center frequency of radar} \\
\rho &= 0.9\ \text{assumed antenna efficiency} \\
\sigma &= 10\ (\text{m}^2)\ \text{for target of interest} \\
L_s &= 6\ \text{dB miscellaneous system losses} \\
\alpha &= 0\ \text{attenuation constant of propagation medium} \\
F_n &= 10\ \text{dB estimated receiver noise figure} \\
B_n &= 80\ \text{KHz assumed system noise bandwidth for} \\
& \quad\ \text{a direct conversion receiver} \\
(SNR)_1 &= 20\ \text{dB assumed SNR requirement}
\end{aligned}
$$

Assumptions were made for several specifications. Transmit power was assumed to be 10 mW ($= P_{ave}$). Given the date of manufacture and after looking inside the back end, this radar does not use digital signal processing methods such as the DFT for Doppler measurement. For this reason its noise power is significantly greater than other Doppler radar devices discussed in this book. The assumed video bandwidth of the phase-locked-loop (PLL) circuitry will be 40 KHz and because this radar front end does not follow an imaging rejection architecture, the effective bandwidth is twice this number 80 KHz ($= B_n$). It is assumed that PLL circuitry is used as part of the Doppler measurement circuitry. For this reason the assumed SNR to achieve stable lock-in is greater than a typical radar system; therefore this SNR will be 20 dB.

The known, measured, and assumed parameters are summarized (Table 8.1). These are substituted into the radar range equation for CW Doppler radar devices (2.6). The maximum range is estimated to be 150 m for a 10 dBsm automobile target using the MATLAB® script [8].

This maximum range would provide adequate performance for measuring the velocity of passing vehicles while patrolling highways and back roads.

8.2.1.2 Experimental Results

The author tested this radar, measuring the velocity of inbound and outbound traffic. The radar was placed in the car on top of the dashboard and the front end was directed out through the windshield. It was observed that the radar detects the Doppler signature of the vehicle's internal fan, so the heating, ventilation and cooling (HVAC) system must be off when using the radar in this configuration.

Cosine errors were not noticeable and it was observed that this radar unit accurately measured the velocity of the author's car. In measuring other

FIGURE 8.8
Example of an X-band police Doppler radar gun.

vehicles, performance was repeatable and radar was reliable. The maximum range seemed to agree with the expected maximum range of 150 m.

It is interesting to note that if basic signal processing in the form of DFT processing were used to calculate the Doppler returns, then maximum range could be significantly increased.

8.2.2 Digital Signal Processing for an Old X-Band Police Doppler Radar Gun

An old X-band police radar gun was modified so that its video output could be digitized and simple signal processing in the form of the DFT applied to measure Doppler signatures of moving targets. It will be shown that when simple DSP is applied to an old radar device its performance and capabilities are dramatically increased.

An old 10.25 GHz police Doppler radar gun is shown (Fig. 8.8). This Doppler radar gun is in effect a CW Doppler radar system (Fig. 8.9 and 8.10), which follows the CW Doppler radar architecture previously discussed in Sec. 8.2.1 except that it uses only one antenna and a circulator to direct the transmitted carrier out the antenna and the scattered energy through to its frequency mixer.

A 10.25 GHz carrier is generated by OSC1 which is a Gunn diode-based cavity oscillator. The output of OSC1 is fed through CIRC1 then out to ANT1. ANT1 radiates toward the target scene. Scattered energy off of the target is collected by ANT1 and fed through the third port of CIRC1 into MXR1, where it is mixed with some power coupled of OSC1. The output product of MXR1 is the Doppler shift between OSC1 and the moving target. This output is amplified by the video amplifier. The output of the video amplifier is usually fed to a frequency counter or PLL circuit as described above, but in

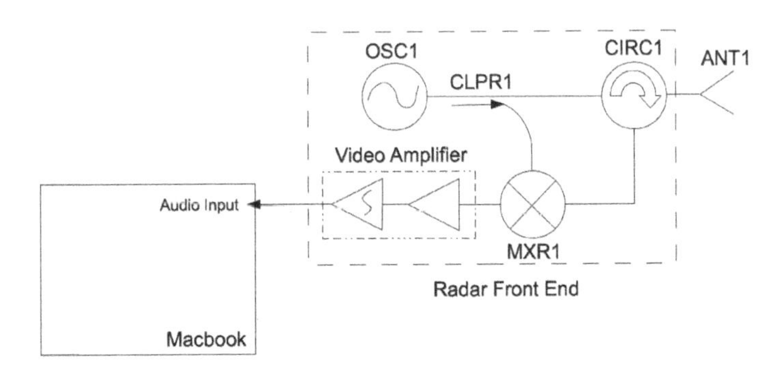

FIGURE 8.9
Block diagram of the X-band police Doppler radar gun.

this example the output will be digitized by the audio input port of a laptop computer and data processed using MATLAB.

8.2.2.1 Expected Performance

A summary of measured, estimated, and assumed specifications is given (Table 8.2). It is important to note that the effective noise bandwidth of this radar device is extremely low because the DFT is applied to 100 ms groups of data providing (for a direct conversion receiver without image rejection) 20 Hz noise bandwidth ($= B_n$). This is significantly lower than the previous K-band police radar and is the primary contributor to its dramatic increase in performance. These parameters are substituted into the radar range equation (2.6), and the maximum range is estimated to be 1500 m for a 10 dBsm automobile target using the MATLAB script [8].

When the DFT is applied, maximum range is significantly greater than the K-band police radar device previously discussed. This would provide more than adequate performance for measuring the velocities of passing vehicles while using low-cost microwave devices in the front end. Alternatively, the transmit power could be reduced proportionally to 150 m range and a low probability of detection police radar design implemented.

8.2.2.2 Working Example

The velocity of approaching vehicles was measured on a stretch of road approximately one quarter mile long with a slight curve in it (Fig. 8.11). Data was acquired in a scenic water front location. Vehicles approaching further down the road (lower apparent signal strength) are approaching at a higher velocity (10 to 15 m/s) and vehicles close to the radar (higher apparent signal

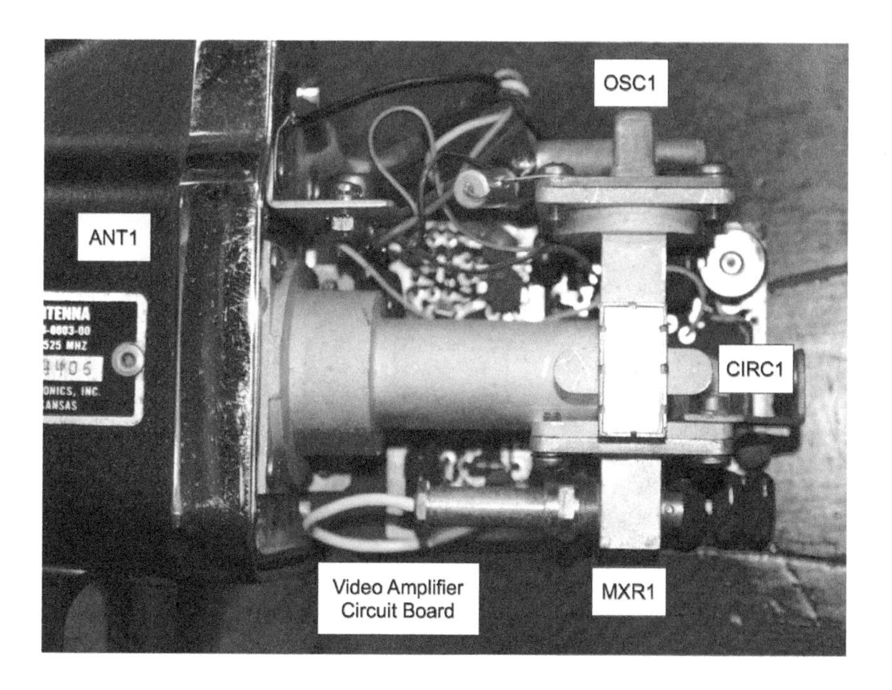

FIGURE 8.10
Call-out diagram of the X-band police Doppler radar gun.

TABLE 8.2
X-band police Doppler radar gun radar range equation parameters.

$$
\begin{aligned}
P_{ave} &= 10^{-3} \text{ (watts)} \\
G_{tx} &= 18 \text{ dBi estimated antenna gain} \\
G_{rx} &= 18 \text{ dBi estimated antenna gain} \\
A_{rx} &= G_{rx}\lambda_c^2/(4\pi) \text{ (m}^2\text{)} \\
\lambda_c &= c/f_c \text{ (m) wavelength of carrier frequency} \\
f_c &= 10 \text{ GHz center frequency of radar} \\
\rho_{rx} &= 1 \text{ because antenna efficiency is accounted for in antenna gain} \\
\sigma &= 10 \text{ (m}^2\text{) for target of interest} \\
L_s &= 6 \text{ dB miscellaneous system losses} \\
\alpha &= 0 \text{ attenuation constant of propagation medium} \\
F_n &= 10 \text{ dB estimated receiver noise figure} \\
B_n &= 2/t_{sample} \text{ system noise bandwidth (Hz) where } t_{sample} = 100 \text{ ms} \\
(SNR)_1 &= 13.4 \text{ dB}
\end{aligned}
$$

strength) are traveling at lower velocities (4 to 7 m/s) to view the scenery. Vehicles close to the radar exhibit a fairly complicated Doppler spectrum,

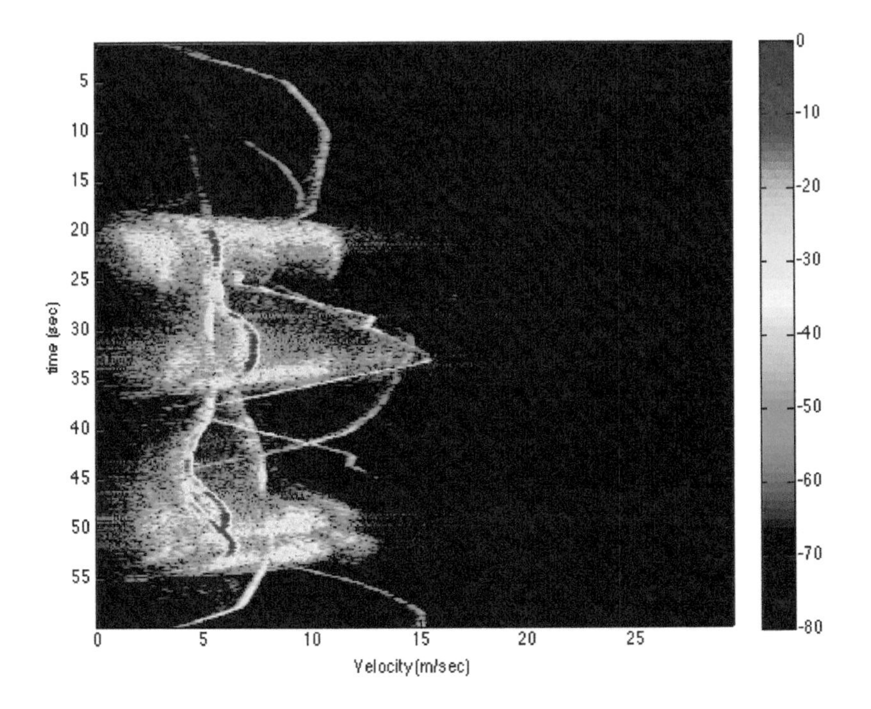

FIGURE 8.11
Vehicles driving down Old Whitfield Street toward the town marina in Guilford Connecticut.

where both the velocity of the vehicle and the apparent velocity of the wheels are shown (at 20 to 35 s and 40 to 55 s). Not only can the velocity of moving targets be shown easily in this plot, but also the velocities of a multitude of targets can be observed simultaneously.

Doppler spectrum of a moving target provides a great deal of information. For example, the Doppler spectrum of the author running is shown (Fig 8.12). The author reaches a maximum speed of about 4 m/s but his arm and leg velocities are much greater and the period of movement can be measured from this data.

Similarly, the Doppler spectrum of a sea gull was measured (Fig. 8.13). Its wing flap period is clearly shown in addition to its velocity. It is very interesting to observe the results of pointing a radar instrument at nature.

A demonstration video is shown [4] and [5] and associated data with MATLAB code for processing is provided [6].

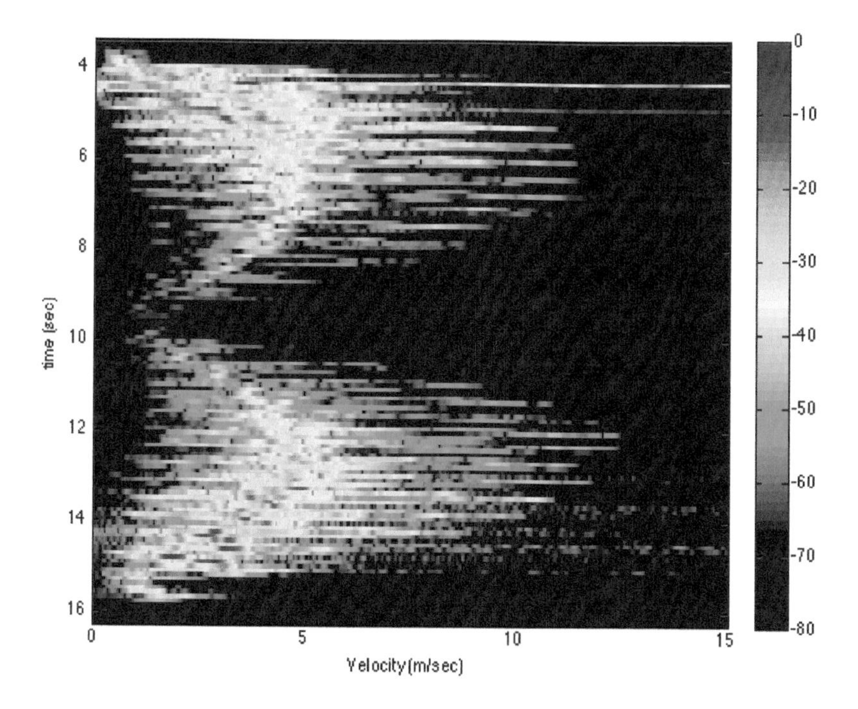

FIGURE 8.12
The author, running as fast as he can toward the police Doppler radar.

8.3 Doppler Motion Sensors

Motion sensors used to turn on outdoor lights or open automatic doors often use CW Doppler as the method of motion detection.

To implement this, the Gunnplexer is typically used as the front end. The IF output of the Gunnplexer would be amplified and fed to an active filter (probably a simple low-pass filter) to select the range of Doppler spectra that the motion sensor should detect (Fig. 8.14). The output of this active filter would then go to a detector followed by a comparator which would trigger only if the amplitude of Doppler exceeded a pre-set threshold. When this threshold is crossed a relay is engaged which turns on a flood light, opens a door, sets off an alarm system, or powers up whatever device to which the radar is connected.

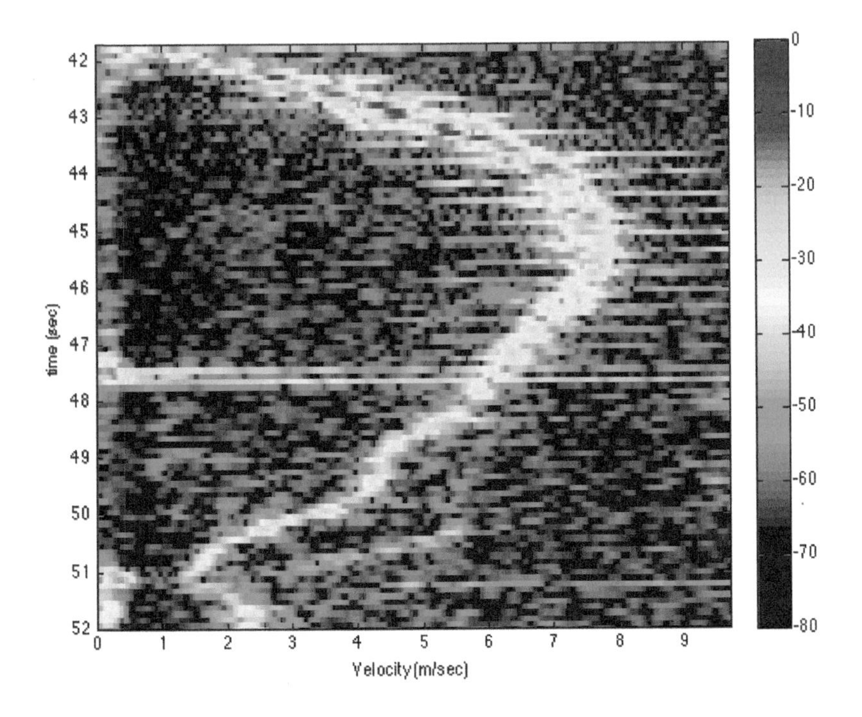

FIGURE 8.13
The Doppler signature of a seagull.

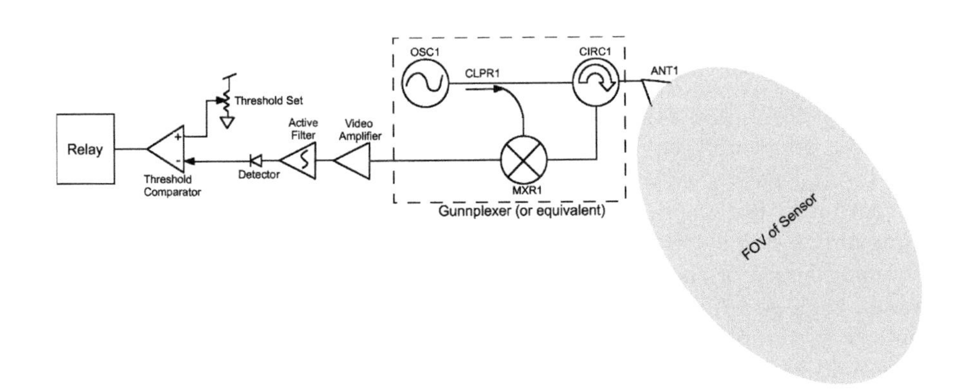

FIGURE 8.14
Block diagram of a typical Doppler motion sensor.

Higher end or lower transmit power motion sensors might use DSP to seek a specific Doppler signature before a detection is made.

These motion sensors are ubiquitous and can be found almost anywhere on both homes, commercial buildings, or at the local home improvement store opening its doors for you.

8.4 Summary

Police radar and Doppler motion sensors were enabled by low-cost diode-based microwave technology. Police Doppler radar devices have been in use for decades in one form or another. Modern devices are capable of measuring the velocity of moving vehicles even when a patrol car is moving. To improve performance, signal processing was applied to an old police radar to great effect, dramatically increasing its maximum range, showing its ability to plot the Doppler signatures of moving vehicles and numerous other targets including people running and sea gulls. Similar to police radar devices, Doppler motion sensors are used to open electronic doors, turn on exterior lighting, and in alarm systems. In summary, simple Doppler radar devices have been in widespread use for decades. The capability of these devices can be greatly increased with basic signal processing.

Bibliography

[1] L. Brown, *A Radar History of World War II*, Carnegie Institution of Washington, Washington DC, 1999.

[2] D. Fisher, "Law enforcement: Shortcomings of radar speed measurement: It's based on sound principles, but present systems have practical limitations and may be misused," *IEEE Spectrum*, Vol. 17, No. 12, 28–31, 1980.

[3] D. Fisher, "Improving on police radar," *IEEE Spectrum*, Vol. 29, No. 7, 38–43, 1992.

[4] hacking a police radar, `http://glcharvat.com/shortrange/police-doppler-radar-and-motion-sensors/`

[5] Doppler radar gun demo connected to audio amplifier demonstration, `http://glcharvat.com/shortrange/police-doppler-radar-and-motion-sensors/`

[6] hacking_a_police_radar.zip, `http://glcharvat.com/shortrange/police-doppler-radar-and-motion-sensors/`

9

Automotive Radar

Shuqing Zeng and James N. Nickolaou, General Motors

At an oval test track in southeast Michigan two vehicles entered. The lead target vehicle launched immediately down the track at 40 mph staying in the right lane. About a minute later the second (host) test vehicle departed at 60 mph. With a quick flip of a switch the host driver took his hand off the wheel. Embedded in the road was a steel cable [1]. The vehicle remained centered in the right lane as the distance closed at 30 feet per second on the target vehicle launched a minute earlier. A red warning light on the dashboard turned on along with the distance readout. At 80 feet the light began to flash and audible increasing pitched alert warned the driver of an impending collision. Finally, just before impact, the brakes were automatically engaged. While no one dared run this vehicle into the back of another vehicle it was demonstrated against several soft targets. The vehicle later went on to demonstrate adaptive cruise control by maintaining a safe and constant distance behind a lead target vehicle even as the target vehicle slowed down.

This was 1959 at General Motors proving ground on a concept Cadillac Cyclone. This was the last "dream car" created during the legendary Harley Earl's tenure as General Motors's vice president of design and the first ever implementation of radar units in vehicles and lane centering prototypes in near-production vehicles (Fig. 9.1). Unfortunately several of the features and the cost of the electronics on the 1959 Cadillac Cyclone would have made it impractical for production and daily use. Today this car sits at the GM Heritage Center in Sterling Heights Michigan.

So, why do we, several decades later, not have radars on every vehicle? The big answer is cost and size. In 1959 this radar vehicle cost the equivalent of hundreds of thousands of dollars in 2010 and barely fit under the fender seen above; and was way out of reach of most folks. The good news is over the years electronic costs, size, and manufacturing have been bringing radar units to the verge of a revolutionary explosion on everyday cars [13].

In this chapter, we consider the basic concepts and technologies that have been or will be deployed in automotive radar design for advanced driver assistance systems (ADAS). We start from Sec. 9.1 with the high-level system requirements and emphasize the challenges of radar design in automotive application domains. Next, in Sec. 9.2 we outline the three key elements of an automotive radar: antenna, analog front end, and radar processor. Then Sec. 9.3 presents several typical waveforms and compares one with each other for their pros and cons.

FIGURE 9.1
1959 Cadillac Cyclone concept equipped with one of the first automotive radar sensors.

We discuss several signal processing methods used by automotive radar for ranging and direction finding in Secs. 9.4 and 9.5, respectively. It is remarkable to note the relation among the linear matched filter, discrete Fourier transform, and radar ambiguity function. We will see that the probability of detection and the accuracy of estimation depends only on signal-to-noise ratio and shape of the ambiguity function.

Section 9.6 describes fusion techniques where data of several modalities are fused to create an accurate surrounding perception for improved decision making such as reduction of rates of false negatives and positives. Section 9.7 presents two automotive applications (i.e., adaptive cruise control and forward collision warning and braking) using a fusion system involving radar. Finally we show in Sec. 9.8 a successful story of using radar technology at an autonomous vehicle competition.

9.1 Challenges in Automotive Domain

Somebody, someplace in the world is having an accident as you read this sentence. Worldwide, millions of collisions happen every year (Fig. 9.2) [2]. That's approximately one person in an accident almost every four seconds.

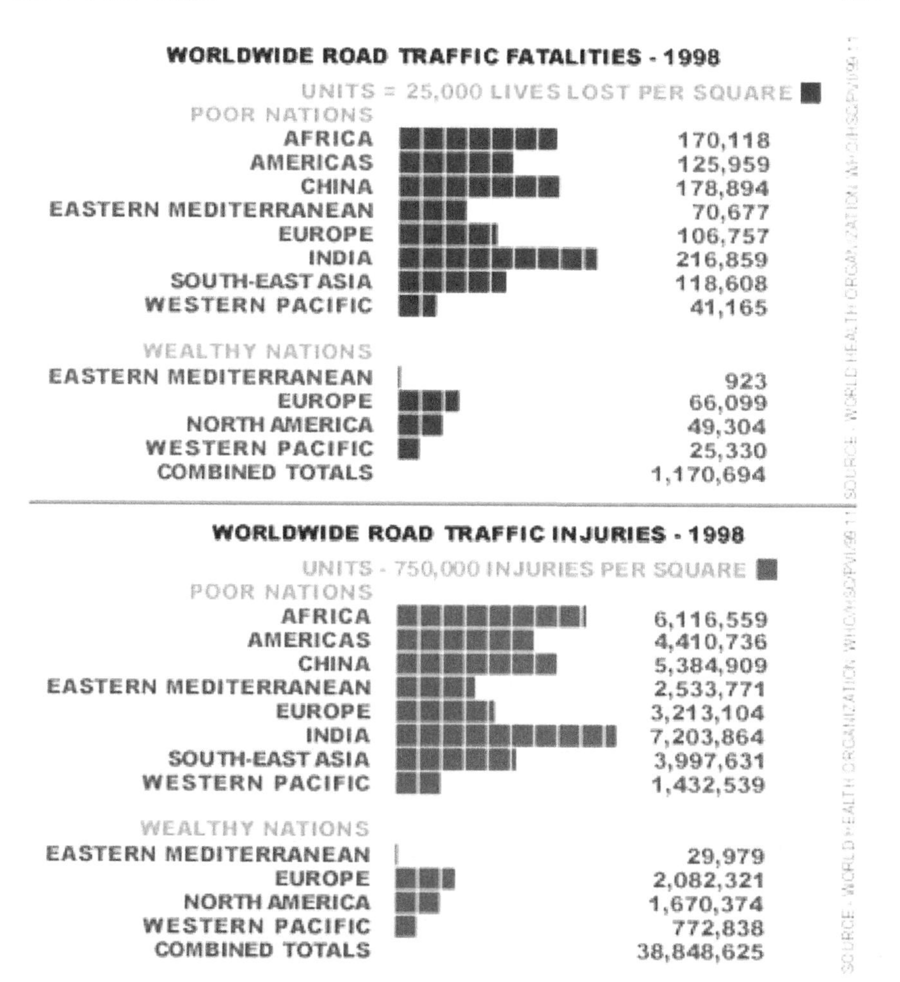

FIGURE 9.2
Estimated global traffic fatalities in 1998.

On top of this, worldwide resources and economic losses caused by traffic accidents are reaching astronomical proportions (tens of billions of dollars per year) for the 38 million traffic injuries. Combined fatalities and injuries account for an incident every 0.7 seconds worldwide resulting in traffic jams daily for millions.

9.1.1 The Automotive Domain Surrounding Sensing

Manufacturers for years were spending millions improving passive safety by adding structural bend points, seat belts, collapsible steering wheels, air bags, etc. as the price, size of electronics and capability further improved for automotive electronic safety. Today all manufacturers now have set their sights on prevention before a collision. Years of analyses of the correlations between collisions and driver behavior, reactions, etc. have shown that a considerable number of accidents can be avoided by recognizing a hazard in sufficient time and making appropriate driving reactions.

Such actions and reactions can be achieved by warning signals to the driver or by automatically taking longitudinal (slow down or speed up) and lateral control (swerve or lane change) of the vehicle. Suitable cost effective sensors are a requirement if the hazardous situation is to be first identified to meet the needs of the functions to achieve an alert, an alert with braking or (ultimately) fully autonomous driving. Along this path a new activity has emerged to identify a suitable human–machine interface.

One such system deployed in several vehicles globally is adaptive cruise control (ACC). Today's ACC systems are mainly based on a long-range radar (LRR) 77 GHz FMCW technology [3]. In a few years as the price points are achieved, LIDAR (light detection and range) units will also be deployed in larger quantities.

Typically FMCW LRR systems allow objects to be detected within a range of 1 to 150 m. At the same time, their distance and speed relative to the host vehicle and with the right number of antennas, and also their angle to the longitudinal axis of the vehicle are determined. Initial Azimuth FOV started at approximately $\pm 7^\circ$ and radars were mounted in the grill. Today FOVs are 120° out to 60 m at the near range in azimuth width and 18° in azimuth from 60 to 150 meters. Most elevation openings on radars are around 4°.

Short range radars (SRRs) are employed today in many automotive active safety applications ranging from parking aid and blind spot detection to more advanced applications such as crash mitigation and collision avoidance. They operate at 24 to 26 GHz and are implemented in either ISM (industrial scientific or medical) or ultrawide band (UWB) frequencies systems. Worldwide radar frequencies are allocated by governments. Because of a lack of worldwide standards across the automotive frequency spectrum, radars have been slow to be installed in many vehicles. In fact, most SRRs are moving toward 76 to 77 Ghz which is globally adopted for vehicle use (Fig. 9.3).

The advantages of radar over other sensing technologies like ultrasound, lidar, and camera (cf. Sec. 9.6.1) come from its unique combination of properties that include the direct measurement of range and range rate information, high tolerance to all weather conditions, and ability to be mounted behind a typical automotive fascia without requiring specific cut-outs or similar accommodations. In more advanced applications where data of several modalities are fused to create a higher level of perception for improved decision making, typically the radar data is relied upon heavily for improving the probability of detection and eliminating false alarms. In addition, radar offers cost benefits over other competitive sensing methods, with the potential for additional major cost reductions as technology progresses.

FIGURE 9.3
Photo of a typical automotive radar sensor.

The basic requirements for automotive SRRs are to be able to detect objects of interest with high probability, high accuracy, and a low false alarm rate. Table 9.1 provides a list of key performance and application attributes that are desired for SRRs. Since historically automotive radar technology has been carried over from the defense domain, today's state-of-the-art short-range radars are still not optimized for achieving the desired specifications while at the same time meeting the aggressive cost constraints that the automotive community demands. Given the increased usage of SRRs in automotive applications and the limitations inherited from existing radar solutions, the need becomes more pressing to develop new radar sensor hardware dedicated specifically to the automotive domain.

While we are talking about SRRs we want to acknowledge briefly the existence of long- and mid-range radar (LRR and MRR), camera or lidar applications as well as fusion of modalities of sensors to improve overall performance and reduce false positives. While these types of devices and functions share some features and technology, they are considerably different in many of their requirements although some of the challenges and difficulties are common. This focus is on SRRs as exemplars in this section, and different sensors and function details are left out.

Several major challenges in SRR integration exist in production automotive systems. In each of the following sections we describe an issue, indicate its significance, and discuss possible solutions. It is not our intention in this section to cover all the technical details related to SRR design and architecture but to illustrate the major gaps that need to be addressed. We leave some technical details to later sections. Needless to say a usable SRR must comply with a full list of other automotive specific requirements (such as ambient

TABLE 9.1
Key desired specifications for SRRs.

Minimum range to perform parking aid	m	0.2 m
Maximum range	m	90 m on vehicle (required for lane change alert)
Range accuracy	m	±0.005 m at the near range and < 1 m at the far range
Velocity range from ... to ...	km/h	144 km/h closing to 54 km/h opening
Opening angle horizontal	°	150°
Power consumption	W	< 5
Sensor size (W × H × D)	mm	< 90 × 70 × 25
Sensor weight	g	< 150
Operating temperature	°C	−40° to +85°
Misalignment detection		Yes
Blockage detection		Yes
Range resolution	m	0.2 m for ranges < 2.5 m, 10% of range for longer ranges
Angle accuracy, horizontal	°	±2° with 15° of boresight and ±5° outside of 15° from boresight
Clustering nearby detections into separate targets		Yes
Working behind fascia		Yes
Performing well under medium levels of rain, snow, and mud accumulation		Yes
Low cost	$	Yes
Start up time	ms	< 500

temperature working range, dimension, weight, power consumption etc.). While each of these requirements is important for a successful product, specifying these details is well outside the scope of book.

9.1.2 Performance Limitations of Today's Automotive SRRs

One of the key performance limitations of today's SRRs is azimuth resolution. Unlike aerial scenarios, the automotive environment is packed with both relevant objects as well as clutter. Angular resolution is therefore a key factor in the radar's ability to correctly detect and classify the perceived scenario. As an example, monopulse (c.f. Sec. 9.5.3) sensors with only two beams (typically referred to as sum and difference channels) spanning the entire field of view have been shown to be problematic in certain road driving situations.

For example, when driving down a narrow street with infrastructure objects symmetrically present on both sides of the street, the objects fall in the same range and range rate bins and thus the wide angle single beam monopulse radar cannot resolve them as separate objects. The result is these targets are inaccurately placed by the monopulse radar sensor directly in the path of the vehicle. In the case where these suboptimal sensors feed an algorithm that is responsible for automatic collision breaking, this would result in unwanted braking events or annoying false alerts.

We will see that the azimuth resolution is determined by beam width of the main lobe of the antenna, which is inversely proportional to the antenna aperture in unit of wavelength (cf. Equation (9.4) in Sec. 9.2.1). With the same antenna size, the effective aperture of a 77 GHz radar is three times larger than that of a 24 GHz radar. To cover a wide field of view with a fine azimuth resolution, electrically scanning technique with multiple beams is needed and this topic is treated in Sec. 9.5.

Another key performance limitation of today's automotive radars is the lack of any information about the relative elevation and height of targets, both clutter and obstacles of interest. To keep cost down, the majority, if not all, of SRRs available in today's automotive domain do not provide any level of information on target elevation and height. One challenge is the need for the SRR to differentiate between lower lying "clutter" objects (such as soda cans, water bottles, manhole covers or other road debris) over which a vehicle could safely drive and taller obstacles of interest (such as vehicles, humans, posts or chain link fences) over which a vehicle could not safely drive, as illustrated in Fig. 9.4. A similar need is to differentiate overhead objects such as bridges and traffic signs which a vehicle can safely drive beneath from the obstacles of interest.

One approach applied to utilize SRRs with no elevation information and resolve this issue is to tilt the SRRs slightly up so that lower height objects start to fall under the sensor's field of view (FOV) sooner than the higher objects. Unfortunately this technique has so far shown limited success. In particular this approach proves to be problematic in scenarios with overhead infrastructure (e.g. parking garages, overhead lights) that reflect the radar

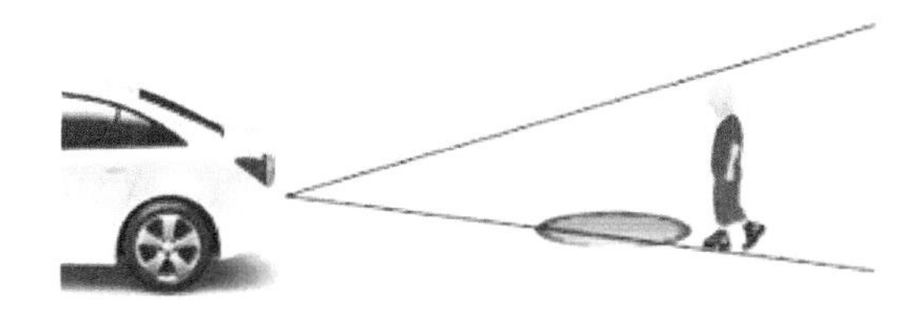

FIGURE 9.4
SRRs need to separate objects based on height.

energy, resulting in detections that fall within the vehicle path. Additionally, different vehicles have different heights and loading scenarios, so designing robust detection algorithms to accommodate all of these cases becomes a significant challenge.

An alternative approach to address the elevation challenge is to use monopulse in elevation with these radar sensors. To investigate this, experimental radar sensors with elevation information have been characterized both in the lab and on actual ground proving vehicles. The collected data suggests that monopulse in elevation could provide the ability to characterize low profile targets on the road surface or targets completely above the vehicle as obstacles not of interest, while still correctly reporting the weighted average height for vertically distributed objects that must be avoided.

Figure 9.5, shows the data collected from a radar sensor with monopulse in elevation mounted on a vehicle driven 5 mph toward a 10 cm corner reflector fixed on the ground. Once the vehicle reaches the target, it is driven at −5 mph back to the starting position. The SRR was mounted at 0.4 m above the ground and the vehicle to target starting distance was 50 m. The top graph is a plot of the reported return amplitude (green) and range (blue) versus time. The middle graph is a plot of reported azimuth angle versus time and the bottom graph is a plot of the reported elevation angle (positive angle toward ground). Looking at the sensor's reported elevation angle of the target, we notice that the reported elevation angle starts initially at zero and then clearly goes up at a range of about 15 m as the angle between the sensor and the target starts to increase. We can also see that the target is only lost at distances closer than 3 m, when the corner reflector begins to move out of the FOV of the

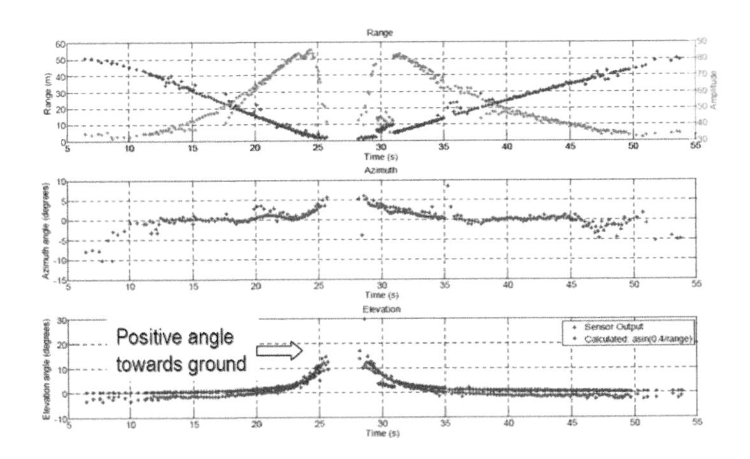

FIGURE 9.5
Radar tracks against road targets from SRR equipped with monopulse in elevation.

sensor. From this data, it can be seen that monopulse in elevation provides a significant indication that the object is located only on the ground and does not represent a danger to the vehicle at 10 m to the object. In contrast, counting on the drop in signal amplitude from the ground target moving out of the radar sensor's FOV only starts to provide information around 5 m to the object. The additional 5 m of separation that elevation monopulse information provides is very significant in collision breaking scenarios and can help minimize unwanted false alarms.

Polarization is another key design consideration for automotive radars. Reference [4] suggests that the best polarization to reduce backscattering from the road surface is achieved when using opposite polarization between the transmitted and received signals (i.e., transmit with a vertical polarization and receive with horizontal polarization or transmit with horizontal polarization and receive with vertical polarization). However, using cross polarization also reduces the return from targets of interest. In addition, the aperture size implication of such dual polarized designs will probably render it less attractive for automotive applications. Following this line, [5] suggests that the preferred polarization is the horizontal over the vertical as the backscattering from road surfaces with horizontal polarization is less, resulting in lower road noise for automotive radar applications. On the other hand, it can be argued that the more backscattering associated with vertical polarization would result in less forward scattering from the ground, and therefore reduce multipath fading effects. While the question of polarization of choice remains open, it will continue to be an important parameter to consider in SRR design.

The next areas of interest are minimum range detection and range resolution. For example, if the radar were to be used as a parking aid sensor, then the SRR minimum range needs to be approximately 20 cm or less. The minimum range and resolution of a radar sensor are determined by its bandwidth per the following equation [4] (cf. Equation (9.10) in Sec. 9.4.3):

$$R_{min} = \Delta r = \frac{c}{2B} \tag{9.1}$$

where R_{min} is the minimum range, Δr is the range resolution, B is radar transmit bandwidth, and c is the speed of light.

To achieve the minimum range and range resolution required for parking aid and similar type applications, large frequency bandwidth is required. For example, to achieve 20 cm resolution, a bandwidth greater than 1 GHz is needed. One such class of radars that can achieve this range resolution is ultra-wideband (UWB) radars. Unfortunately the lack of consistent UWB radar frequency regulations between the major automotive markets (e.g., China, North America, Europe, and Japan) has impeded the development and adoption of UWB automotive radars. Worldwide harmonization of UWB automotive radar regulation is needed to encourage development of these systems, and until that is achieved the effective production of automotive radar systems will be inhibited.

Besides the range resolution, Doppler effect is useful in separation of moving objects from stationary background. In real traffic scenarios, objects or different parts of an object may move at different radial velocities with respect to the radar, thus a high resolution in Doppler measurement is critical in generating an accurate sensing map. According to Sec. 9.4.3 the minimum range rate and resolution are determined by its radar carrier frequency and the dwell time, that is,

$$\Delta v = \frac{c}{2f_c T}$$

where f_c denotes the frequency of millimeter wavelength RF and T is the dwell time interval on the target. Therefore a 77 GHz radar will have three times better range rate resolution than its counterpart of 24 GHz.

It is also worth mentioning that one significant limitation to 24 GHz UWB is the low power limit (EIRP restriction) of −41.3 dBm/MHz, which limits the maximum detection range. This limitation on maximum detection range reduces the ability of 24 GHz UWB radars to be used in applications such as lane change alert where the target vehicles could be as far as 90 m behind the host vehicle.

An ideal sensor would be one that can switch between an UWB beam used for distances less than 20 m and higher EIRP bands (such as 24 GHz ISM) that can be used for detecting objects at distances greater than 30 m. This scheme will also coincide with the required range resolution as the higher accuracy is only needed at short ranges.

Another significant implementation that needs consideration in the design of radars is immunity to mutual interference. As an example, leading vehicles may have SRRs aimed backward, directly illuminating the forward looking SRR of a host vehicle. This typically raises the SNR noise floor and detection performance tends to be reduced. Vendors need to protect against a degradation of performance in such an environment through applying different coding and multiplexing schemes, as well as considering innovative ways of processing signals in both time and frequency domains.

9.1.3 Challenges with Vehicle Integration

The packaging and mounting of radar sensors behind vehicle fascias are significant issues for auto manufacturers, typically requiring considerable resources and significant tradeoffs to be made. Figure 9.3 shows one example of a case where a SRR that is used for blind spot detection cannot be mounted in an otherwise desirable location due to blockage caused by the sheet metal in a particular vehicle.

9.1.4 SRR Packaging Challenges

In some forward-looking applications mounting constraints force the radars to be placed in lower than ideal locations on vehicles, making the radars more susceptible to environment risks, ground clutter false positives, and missed detections. As an example, when mounted very low, radar returns can undershoot the rear of a large leading vehicle (such as a bus) causing serious overestimations of leading vehicle distances. Likewise, curved leading edges of vehicles or creases through the radar sensor aperture can create non-uniform radiation patterns and unwanted loss in signal strength.

The radar vehicle aperture also has implications to the topic of radar calibration. In order to make sure the radar correctly accounts for the parasitic absorption and reflection of the host vehicle body, an advanced calibration process is required. While this can in principle be performed at the factory level, such a solution is difficult in practice due to the additional manufacturing overhead as well as the potential need for re-calibration when small paint or body damages occur. Accommodating vehicle integration with no extra calibration is crucial in successful automotive SRR deployment.

Recent pedestrian regulation also represents a challenge when placing radars at the bumper height of vehicles. This is because the preferred radar mounting location in the vehicle is in the crush zone and must be filled with absorbent materials. Other under fascia structures such as tow bar connections and license plates all create havoc on radar placement through the automotive design process.

While not as significant an issue, one that still is a concern for both manufacturers and consumers is the cost of repair in the case of a collision. SRRs in forward-facing locations, like the bumper, pose a challenge for repair and alignment at dealerships and repair shops.

For cosmetic reasons, another key element for automotive radars is that they must be packaged behind production plastic fascias which could be covered with multiple layers of metallic paints. Data collected suggests that for 77 GHz radar, reflections from and losses through painted fascias are much more significant than at 24 GHz. Also, the experimental data suggests that mounting tolerances for 77 GHz radars are three times tighter than at 24 GHz in order to achieve the required sensor performance when installed behind painted fascias.

Likewise, mounting height (from the ground) and varying distance to fascia (as vehicles typically have curved surfaces) all create challenges in automotive integration. In practice, SRRs (in particular the antenna(s) of the radar sensors) need to be designed for the vehicle styling and structural constraints, as opposed to the other way around. Although not available from the automotive supplier base today, the use of flexible, multiple, and controllable beam antennas could provide significant integration and performance advantages for next-generation SRRs.

9.1.5 Automotive 77 Ghz vs. 24 Ghz Radar Bands

Currently the two primary automotive radar bands are around 22 to 29 GHz (referred to as the 24 GHz band and includes both the 24 GHz ISM band

and U.S. UWB) and 76 to 81 GHz (referred to as the 77 GHz band and includes both the worldwide LRR band and European Union UWB band). Irrespective of the frequency regulation issue discussed later, there are certain attributes that make each band more or less attractive for different automotive applications.

The strengths of 24 GHz relative to 77 GHz include hardware cost and ease of installation. The cost of analog RF components that make up the front ends of the automotive radar sensor tend to increase with frequency. This is because the active device performance drops off with frequency, thus less stringent design tolerances can be considered. In addition, assembly tolerances are relative to the wavelength and therefore assembly requirements and costs of 24 GHz sensors are typically less. As suggested above, the ease of installation comes from the fact that 24 GHz radiation is less affected (less attenuation and reflection) by the painted fascia and the sensor installation criteria are rough one third as stringent.

The strengths of 77 GHz relative to 24 GHz are performance related, both in terms of Doppler discrimination and angular resolution. The Doppler shift introduced by the relative velocity difference between the source and a target is directly related to the carrier frequency. Thus a 77 GHz radar sensor has roughly three times greater Doppler spread than 24 GHz radar sensor. Perhaps more notably, the beam width of an antenna is proportional to its electrical size; thus for the same sized antenna aperture a 77 GHz radar sensor can have three times the angular resolution.

9.1.6 Cost and Long Term Reliability

One thing all people working in the automotive industry agree on is the need for a low-cost and reliable product. Radar technology is still relatively expensive and a tradeoff between cost and performance continues to be a sensitive topic. The development of integrated circuits to perform the RF functions offers the potential to achieve smaller sized sensors with higher reliability at a lower cost.

Silicon semiconductor technologies for microwave and millimeter wave analog components (e.g., SiGe BiCMOS and RF CMOS) have advanced significantly over the past few decades and now dominate commercial wireless application marketplaces that once were covered by discrete GaAs component solutions [6]. The advantages in unit cost, reliability and size afforded by a silicon solution come from the capability to integrate multiple analog functions into a single monolithic integrated circuit. This component integration yields: smaller sized radar sensor packaging, simplified sensor, manufacturing, assembly, and fewer millimeter wave interconnects.

Additionally, the economies of scale and fabrication infrastructure investment by the worldwide silicon semiconductor marketplace results in reduced component costs if the volumes are high enough.

The major impediment to the adoption of silicon semiconductors by automotive radar suppliers is the significantly higher upfront development cost. Development costs are high due to the increased design time needed to ensure that complete circuit RF performance can be achieved with minimal design passes, and the substantial mask cost of the advanced Si technology nodes needed to achieve the millimeter wave bands. These challenges stem from the fact that silicon is fundamentally a less capable semiconductor technology for microwave and millimeter wave analog circuits relative to traditional millimeter semiconductor technologies (e.g., GaAs) due to the lower electron mobility and greater variation with temperature, plus the very tight spacing between components in integrated circuits.

Achieving good noise figure and gain with low cross talk between different channels in the relevant millimeter wave bands is very challenging in silicon technologies. The large upfront investment needed to transform the industry toward silicon technology can only be justified if large volumes will be associated with such a product. This might require commonality between the requirements of different automotive features as well as global harmonized regulations.

Still, while many challenges exist, the trend toward silicon-based technologies for automotive radar sensor front ends is a potential cost barrier breaker.

9.1.7 Regulatory Issues

Different regulations across the globe make it challenging to find a single sensor that can be accepted in different regions around the world [7]–[12]. For applications where minimum range and range resolution are critical, as in the case for parking aid applications where a minimum of 20 cm detection range is required, moving to UWB products becomes necessary. In the US, the 24 GHz UWB band (22 to 29 GHz) is approved for automotive application usage. However in Europe, regulation is pushing the industry to move to the 79 GHz band frequency band (77 to 81 GHz) which has been designated and made available for permanent usage for automotive SRR. On the other hand this band has not been made available in the US. In other major automotive markets, such as China, UWB automotive SRR regulations are still under development.

In addition, there seems to be a potential convergence of telematic technologies and spectrum. For example, the IEEE 802.11ad spectrum could potentially be used in SRRs and with the same device providing communication to and from other devices and vehicles.

These unknowns and different regulatory requirements make it very difficult to develop a single global sensor hardware standard that can meet the regulation of all different export regions and yet provide the design parameters required for the most challenging active safety applications.

9.1.8 Blockage

Certain environmental conditions can result in scenarios where the SRRs are blocked and therefore lead to an increased chance of a target being dropped after being detected or missed all together. Automotive radar sensors need to be able to tolerate normal levels of rain, snow and mud exposure with attenuation factors anywhere between 5 and 15 dB. At the same time, the SRRs need to have a robust way of detecting blockage and reporting such a condition back to the driver so that the driver will know not to rely on the sensor in cases where performance is degraded.

Examples of such cases are driving in heavy rain storms or getting layers of wet mud built up on the fascia directly in front of the SRR.

Blockage hazard is a likely and natural event in automotive domains. The extent to which material build-up on a fascia results in decreased radar sensor performance will vary with composition of soil. Mud shields might be needed for cases where a lot of direct exposure to mud and water splash from road surfaces is expected. When blockage effects occur to the point where there is a serious degradation in performance, a corresponding blockage indication needs to be provided to the driver.

Blockage determination is a very difficult problem, and different systems utilize different blockage detection approaches. One approach assumes a sensor is blocked when the total energy coming back to the receiver drops below a set threshold. Unfortunately this approach reports false blockage alarms in open field-of-view environments. Another approach assumes blockage based on seeing only close objects indicating a reduced detection range. This too has limitations. One interesting option is for an SRR sensor to have a dedicated blockage antenna that points at the ground through the same portion of the fascia with an expected level of return from the road surface. Although this will come at extra cost, detecting and reporting blockage with such a solution is a superior way for resolving the blockage issue

Automotive short-range radar sensor development is far from mature. Significant SRR performance challenges and capability enhancements that need to be addressed to improve overall automotive active safety system performance include:

- Improved azimuth resolution to provide more accurate target extent/separation
- Target elevation information to decrease false alarms caused by roadway and overhead features
- Multimode operation, i.e., the ability to switch between a low power UWB operation mode used for applications requiring high range resolution as in the case with parking aid and a higher power
- Narrowband operation mode used for applications requiring object detection at distances greater than 30 m

- Packaging flexibility to reduce vehicle integration complexities
- Self-calibration to improve vehicle manufacturing efficiencies and reduce re-work difficulties
- Improved blockage determination

In addition to the aforementioned enhanced performance and capabilities, sensor cost reductions are critical for next-generation SRRs. Advances in Si integrated circuit technology offer great promise in this area.

If these challenges are addressed, large improvements in automotive active safety systems effectiveness and driving comfort can be attained.

9.2 Elements of Automotive Radar

Figure 9.6 illustrates a disassembled automotive radar comprising four plates: radome materials, analog front end, RF shielding, and radar processor, listed from the top to bottom. Radome is a waterproof and weatherproof enclosure with minimally attenuation of microwave signals that protects electronic components of the radar. Analog front end includes the antenna and RF circuits, and the RF shielding plate aluminium shields electromagnetic interference for the radar processor plate where the digital radar signal processing is located.

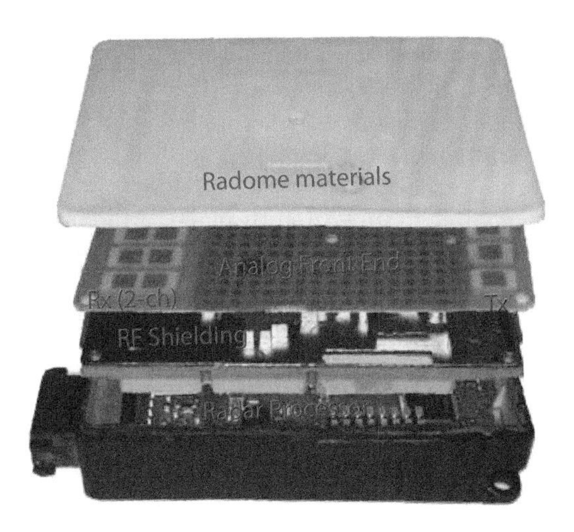

FIGURE 9.6
A typical short range radar (SRR) consists of four plates (Courtesy of Autoliv).

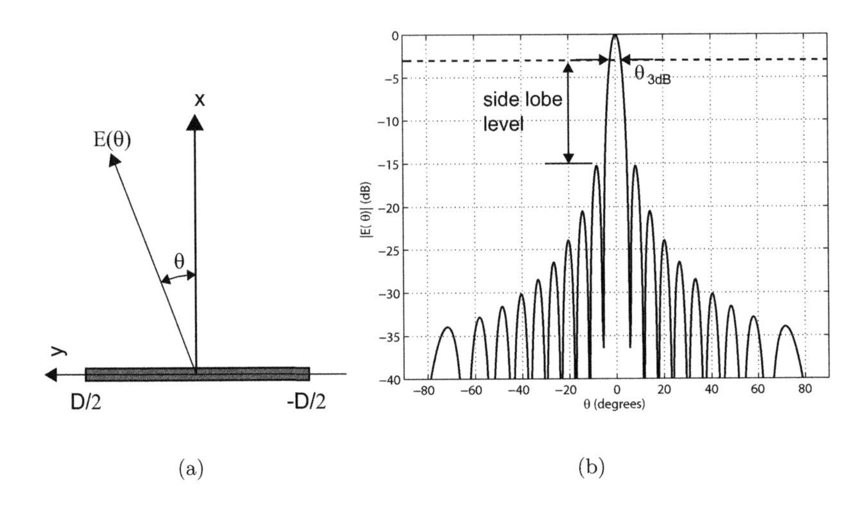

FIGURE 9.7
(a) One-dimension electric field on a rectangular aperture. (b) Radiation pattern as a function of azimuth angle (θ). The 3 dB beam width or the field of view of the antenna ($\theta_{3\text{dB}}$) is illustrated.

9.2.1 Antenna

The antenna plays a major role in determining the sensitivity and angular resolution of the radar. For a perspective of signal processing, the frequently used properties of an antenna are the gain, beam width, and sidelobe levels. Each of these derives from consideration of the antenna power pattern. The power pattern P can be described as the radiation intensity relative to the antenna boresight. Power pattern is related to electric field intensity as $P = E^2$.

In Fig. 9.7(a) one-dimension (azimuth plane of radar) normalized electric field $E(\theta)$ is calculated as a function of the azimuth angle θ on a rectangle aperture, where D is the aperture size and λ is the wavelength of the radar carrier wave.

Let $I(y)$ be the distribution of current across the aperture in azimuth plane. Under the assumption of far field assumption, $E(\theta)$ can be written as [14]

$$E(\theta) = \int_{-\frac{D}{2}}^{\frac{D}{2}} I(y) e^{j\frac{2\pi y}{\lambda}\sin\theta} dy \tag{9.2}$$

If we define $s = \sin\theta$ and $\bar{y} = y/\lambda$, (9.2) becomes

$$\bar{E}(\theta) = \int_{-\frac{D}{2}\lambda}^{\frac{D}{2}\lambda} I(\lambda\bar{y}) e^{j2\pi\bar{y}s} d\bar{y} \tag{9.3}$$

where $\bar{E}(\theta) = E(\sin^{-1}\theta)/\lambda$. This indicates the electric field is the inverse Fourier transform of the current across aperture. This Fourier transform property will allow the use of linear system technique to study the angular resolution problem to achieve spatial separation for multiple targets. We will revisit this for linear antenna array in Sec. 9.5.

A special case of (9.2) occurs when the current is a constant (i.e., $I(y) = I_0$). The normalized voltage pattern $E(\theta)$ is a sinc function, i.e.,

$$E(\theta) = \frac{\sin(\pi\frac{D}{\lambda}\sin\theta)}{\pi\frac{D}{\lambda}\sin\theta}$$

Fig. 9.7(b) shows the voltage pattern with two important properties: beam width and sidelobe level. The angular resolution depends on width of the main lobe, and is represented by the 3 dB beam width ($\theta_{3\text{dB}}$). The smaller the beam width, the finer for radar in angular resolution. $\theta_{3\text{dB}}$ can be found by set $E(\theta) = \frac{1}{\sqrt{2}}$, and by numerical approximation we obtain

$$\theta_{3\text{dB}} = 0.89\frac{\lambda}{D}\text{radians} \tag{9.4}$$

Thus, a smaller beam width needs a larger aperture or a shorter wavelength. From (9.4) one can easily see that for an antenna of the same aperture size a 77 GHz radar sensor can have three times finer angular resolution than that of a 24 GHz counterpart.

The *sidelobe level* in Fig. 9.7(b) represents the level of interference from sidelobes that affects the radar's performance in distinguishing a target from its neighboring clutters. The sidelobe level of 13.2 dB for a uniform pattern as shown in the figure is regarded to be too high for a practical radar system. This can be reduced by use of nonuniform aperture arrangement, referred to as *tapering of shading* technology [14].

9.2.2 Analog Front End

As shown in Fig. 9.6 *the analog front end (AFE)* is a part of an automotive radar circuit, operating at millimeter microwave frequencies (24 GHz or 77 GHz). AFE includes the transmitter antenna arrays (Tx) and receiver antenna arrays (Rx), and is shielded from digital part of the circuit by an aluminum frame.

AFE consists of monolithic microwave integrated circuits (MMICs) that perform functions such as microwave mixing (mixer), power amplifier (PA), low noise amplification (LNA), low-pass filtering (LPF), voltage controlled oscillator (VCO), and local oscillator (LO).

Figure 9.8 shows the signal diagram of an automotive radar. On the analog side, the transmitter is implemented using a VCO to generate the baseband signals controlled by the radar signal waveform from digital-to-analog

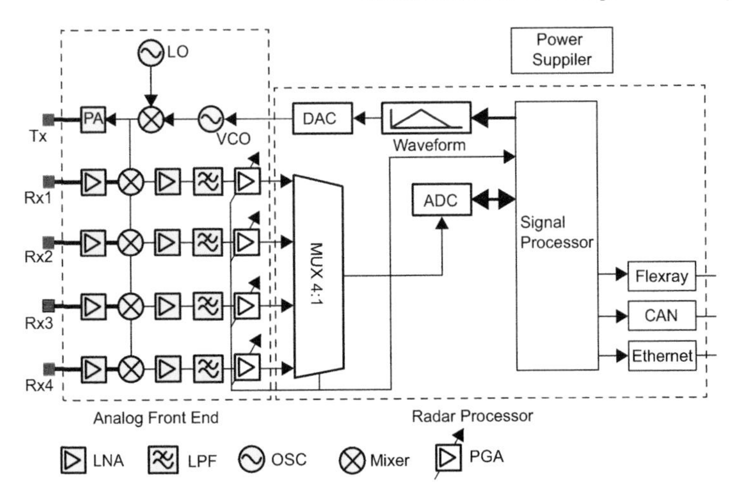

FIGURE 9.8
Signal diagram of an automotive radar.

converter (DAC). The baseband signal is mixed with reference microwave frequency signals from LO. The modulated signals are amplified by PA and fed to Tx.

Figure 9.9 illustrates a typical waveform for frequency modulated continuous wave (FMCW) radar. Consider the case of frequency sweep at 500 MHz in 0.5 millisecond, or 1 MHz in 1 microsecond. Providing the speed of light at $c = 3 \times 10^8$ m/s, the echo signal from the target at distance of 1 m is 7 KHz offset during the ramp intervals. Meanwhile, the echo signal from the object at a 150 m distance will be a 1 MHz offset during the ramp intervals.

In the receiver, four different signal channels in front end are illustrated in Fig. 9.8 and are connected to the elements Rx_1, Rx_2, Rx_3, and Rx_4, respectively. Each channel requires an LNA, followed by an analog mixer. The mixer down-converts the received microwave frequency signal with the ramping transmit signal, outputting a baseband signal signal that contains the difference between the the transit and receive waveforms at any given instant. The ramping is canceled out, as we see fixed frequencies depending upon the range and Doppler shift of the target returns. Again, the high-frequency filtering at microwave frequency can be implemented using etched passive components. The output of the mixer will be at low frequency, up to 1 MHz at maximum range for sensing horizon of 150 m. Therefore, traditional passive components and operational amplifiers can be used to provide anti-aliasing LPF prior to the analog-to-digital converter (ADC).

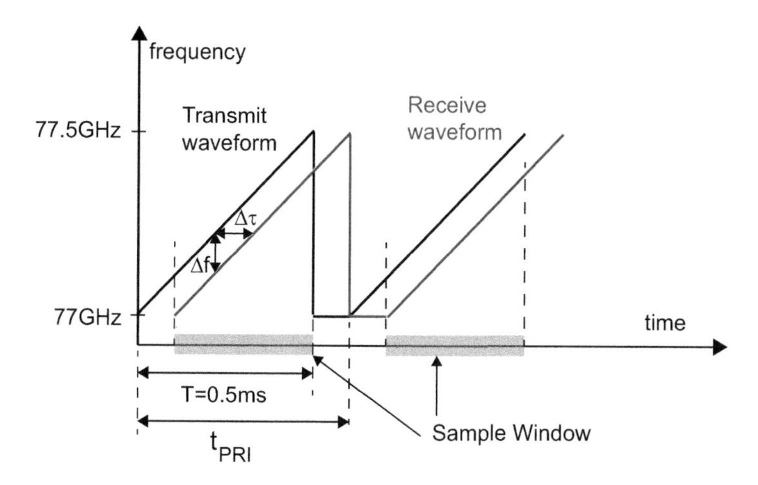

FIGURE 9.9
Waveform for FMCW and sample intervals for radar signal processing.

9.2.3 Radar Processor

We outline how an automotive radar system was built using a digital processor, and the relative strengths and weaknesses of embedded technologies are reviewed for radar signal processing design. Then we present a customizable hybrid design as a cost-, performance- and power-effective tradeoff among them.

In Fig. 9.8 there are two separate signal flows for transmitter and receiver. For transmission process, the *radar processor* calculates radar waveform profile (e.g., chirp ramp or pulse) and outputs to *DAC* (digital-to-analog converter) module such that a voltage signal can be generated to modulate radar radio frequency using MMIC block VOC.

On the other hand, the receiving process first samples and digitizes analog baseband signal, and then it is fed to the *radar processor* for target detection and tracking.

To meet the Nyquist criterion, the sampling frequency of ADC for baseband input must be at a minimum of 2 MHz for the waveform depicted in Fig. 9.9, given the maximal unambiguous range is 150 m. If oversampling scheme is used, for example, a 64 $\times$ sampling frequency of 128 MHz is used, followed by a 64:1 digital delta-sigma filter, then approximately 6 bits of additional resolution can be achieved. This sampling scheme allows a 16 bit ADC to effectively operate in the range of 22 bits, achieving well over 60+ dB of dynamic range. The next step of signal processing is to perform spectral analysis using a fast Fourier transform (FFT). This bit growth can be handled by implementing the FFT in single-precision floating-point processing where the 24 bit precision (23 bits plus sign) can be preserved, and will easily accommodate up to 100+ dB dynamic range in the yielded spectrum. With this high dynamic range, the radar system will not be saturated by strong scatter signals from nearby targets and thus can avoid weak signals from remote targets being blinded by strong nearby reflections.

Finally, tracked targets are output through the automotive network communication infrastructure such as CAN (control area network), Flexray, or automotive Ethernet.

A successful radar processor should pursue optimization in the following perspectives: i) adequate precision in signal representation (e.g., single precision floating point versus fixed point with limited word length); ii) sufficient computational horsepower to meet requirements of realtime signal processing; iii) the time taken to develop the signal processing software; iv) nonrecurring engineering (NRE) costs such as manufacturing, the cost of hardware and software development, and validation.

There are four approaches exist to address these challenges:

- Microcontroller (MCU) is a general-purpose digital processor for computation and control that can be used for a wide variety of applications by developing application software. MCU usually has a floating point coprocessor, and is quite adequate for target detection and tracking software, but is not fast enough to implement computationally intensive operations such as FFT. NRE costs are amortized among all the users of a particular MCU architecture.

- Digital signal processor (DSP) is specific-purpose digital processor that optimizes some basic functions using by many signal-processing algorithms such as multiply-accumulates, single instruction multiple data (SIMD), and very long instruction word (VLIW), at the expense of flexibility. DSP is a good choice for spectrum analysis operations such as FFT and matching filter for waveform signals. The down side is that DSP is usually designed for fixed point operations whose limited range of value for signal representation is problematical for high dynamic range scenarios.

- Field programmable gate array (FPGA) offers the flexibility of implementing customized signal processing in a hardware configuration. FPGA can share NRE costs among a very large population of users, at the expense of high level of transistor redundancy (and therefore high unit costs), limited optimization of clock cycles, and heat dissipation problem. Modern FPGAs come with pre-configured blocks for DSP computation (e.g., DSP slices), which allow designers to design floating point signal processing diagrams without the use of general FPGA fabric. Therefore, it is possible to design floating-point spectral calculation using these DSP slices.

- Application-specific functions (ASICs) are custom-designed for a particular radar system, possibly embedding one or more MCU or DSP cores, with as much as possible of the total system functionality implemented (such as analog front end) on a single die. This optimizes the number of transistors and clock cycles (and therefore unit cost and power consumption), at the expense of development time and NRE cost that are generally an order of magnitude higher than those for MCU, DSP or FPGA.

These four approaches represent different tradeoffs in addressing these challenges. For any particular design the choice is an engineering compromise, and no single approach is perfect. Different approach mixes are often most appropriate at different stages of radar system design.

During prototyping and production ramp-up, a flexible and scalable platform may be preferable in order to reduce development time and cost. A hybrid approach (i.e., known as system on chip (SoC) FPGA) is one preferred choice that incorporates a new breed of low-cost FPGA fabric and powerful multi-core CPU (e.g., ARM Cortex-A9 processors). Automotive-grade versions of these devices are available (e.g., Xilinx Zynq-7000 and Altera Cyclone V) and capable of up to 1 GHz CPU clock rates. SoC FPGAs can provide a solution from radar to light detection and ranging (lidar), infrared and visible cameras, and even to automotive fusion systems. Moreover, both the FPGA hardware and ARM software implementations use floating-point processing, which provides superior performance in radar applications compared to traditional fixed-point implementations in DSP or MCU. These low-cost SoC devices support volume applications, while providing much faster time to market.

When the radar goes into high volume, its functionality can be re-implemented in an ASIC that embeds the CPU core from the standard product, and absorbs the logic from the FPGA, thereby optimizing die size, unit cost, clock cycles and power consumption without the need to rewrite the software. The high NRE costs associated with ASIC development are amortized over the high volume.

9.3 Waveforms for Automotive Radar

Radar baseband signal has variants of waveforms for different system design requirements. The waveforms are commonly modulated with a millimeter wavelength RF (e.g., 24 GHz and 77 GHz) before transmission, and reconstructed from the echoed signals through demodulation. The reconstructed waveforms are correlated using matched filters with the transmitted reference for range and Doppler processing. The spectral analysis is performed on the mixed signals, and the detected spectral peaks are treated as target candidates.

We focus on continuous wave (CW) waveforms due to their low cost compared to classic pulse radars. As an example, for SRR specified in Table 9.1, ultrashort pulse width (6 ns) is needed for a pulse radar to have a range resolution of 0.2 m, which requires expensive components such as ultrafast ADC and high-end DSP chipset.

In this section we discuss several typical radar waveforms that are employed in automotive radar design, including the phenomenon of Doppler shift, which occurs when the target is moving with a nonzero radial velocity relative to the radar. As we will see, linear frequency modulation (LFM) waveforms require low sample rate and, thus, imply low-cost implementation for signal processors. However, the frequency shift of the peak of the spectrum is contributed either by the target's range or by Doppler effect and thus introduces an ambiguity in determining range and range rate. While multiple ramps are needed to address the range–Doppler ambiguity, the association of peaks from different ramps gives rise to new problems.

Alternatively, frequency shift keying (FSK) offers a solution to simultaneously find range and range rate of a target, but it only has Dopper resolution and cannot distinguish targets of the same radial velocity. Although a hybrid waveform of FSK and LFM with multiple ramps gives more power in resolving the range–Doppler ambiguity, "ghost" targets are possibly generated due to incorrect association.

Finally the compression LFM pulse waveform is outlined with the advantage of decoupled range-Doppler processing, but at the cost of fast analog-to-digital converter (ADC) and fast digital signal processing (DSP) systems.

9.3.1 Doppler Shift

Consider the waveform of the baseband signal $s(t)$, $0 \le t \le T$. The modulated signal to be transmitted is $x_T(t) = s(t)\exp(j2\pi f_c t)$ where f_c is the carrier frequency of millimeter wavelength. This signal is received by a receive antenna. Suppose the one-way range from the radar to the target is a function of time as $r(t)$. The received signal x_R is a delayed version of the transmitted signal, ignoring amplitude scaling factors:

$$x_R(t) = x_T\left(t - \frac{2r(t)}{c}\right)$$

Now assume the target is moving toward the radar at a constant radial velocity v. This means $r(t) = r_0 + vt$ where r_0 is the range at $t = 0$. Thus the received signal $x_R(t)$ can be written as

$$\begin{aligned} x_R(t) &= s\left(t - \frac{2(r_0 + vt)}{c}\right)\exp\left(j2\pi f_c\left(t - \frac{2(r_0 + vt)}{c}\right)\right) \\ &\approx s\left(t - \frac{2r_0}{c}\right)\exp\left(-j\frac{4\pi f_c r_0}{c}\right)\exp\left(-j2\pi\frac{2f_c v}{c}t\right)\exp(j2\pi f_c t) \quad (9.5) \end{aligned}$$

In (9.5) the first exponential term is a constant phase factor, while the last is the carrier frequency. The middle exponential term is a sinusoid of frequency $-2f_c v/c$ that is caused by Doppler shift.

In deriving (9.5), we ignore the term $2vt/c$ in the parameters of baseband signal. This represents time compression of the baseband waveform due to the motion. The largest possible value t is the waveform duration T. So the ratio

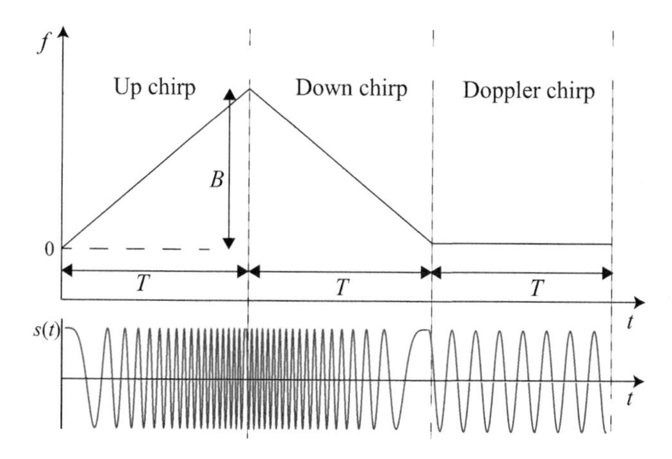

FIGURE 9.10
Multiple LFM waveforms.

of the compression is $(2vT/c)/T = 2v/c$. Thus even for fastest targets in automotive scenarios (e.g., 100 m/s) this compression in waveform is negligible.

An important implication of (9.5) is that the time delay for the one-way propagation due to the motion is ignored. This assumption is often called *stop-and-hop assumption*. In the measurement process in waveform duration $0 \leq t \leq T$, the target remains in r_0 with the radial velocity v, and then hops forward by vT meters before stopping for another cycle of measurement.

9.3.2 Linear Frequency Modulation

Figure 9.10 shows linear frequency modulation (LFM) waveforms with multiple ramps, respectively. The following characteristics exist for this waveform:

- Each ramp duration T can be treated as a coherent processing interval T_{CPI}.
- Coupling of range and Doppler gives rise to an ambiguity in determining the range and range rate of a target.
- Multiple ramps are needed to resolve the ambiguity between range and range rate.
- Possible long dwell time for good Doppler resolution.
- Low sample rate (e.g., 50 KHz) in each ramp duration and, thus, less stressing for signal processor. This implies that low-end DSP and MCU have sufficient computational horsepower for realtime processing.

While we consider the computational model for processing up-ramp waveform, similar derivation can be obtained for other ramp waveforms. The up-ramp LFM waveform is defined by

$$s(t) = \exp(j\phi(t)) = \exp(j\pi\frac{B}{T}t^2), \quad 0 \leq t \leq T \tag{9.6}$$

where $\phi(t)$ is the phase function, T is the chirp duration, and B is the frequency bandwidth of the ramp. The instantaneous frequency of this waveform is the derivative of the phase function:

$$f(t) = \frac{1}{2\pi}\frac{d}{dt}\phi(t) = \frac{B}{T}t$$

Assume the received signal is scattered by a target at range r with radial velocity v. We consider the phase of the down-converted receive signal (beat signal) $\Delta\phi(t)$ from the mixed signal between the transmitted signal $x_T(t)$ and the delayed received signal $x_R(t)$. By plugging (9.6) to (9.5), we have

$$\begin{aligned}\Delta\phi(t) &= \phi(t) - \phi(t - \frac{2r}{c}) \\ &= 2\pi\left(\frac{2f_c r}{c} + \frac{2f_c v}{c}t + \frac{2Br}{Tc}t - \frac{2Br}{Tc}\frac{r}{c}\right) \\ &\approx 2\pi\left(\frac{2f_c r}{c} + \frac{2f_c v}{c}t + \frac{2Br}{Tc}t\right)\end{aligned} \tag{9.7}$$

where we replace r_0 with r because of the *stop-and-hop assumption.* The approximation in the last line from (9.7) follows due to the fact for automotive radar $\frac{r}{c} \ll t$. The last term in the second line of (9.7) is insignificant compared with the first term and can be neglected.

Thus the received beat signal can be modeled as

$$b(t) = a\exp\left[j2\pi\left(\frac{2f_c v}{c} + \frac{2Br}{Tc}\right)t\right] \tag{9.8}$$

where the constant phase term $4\pi f_c r/c$ is absorbed into the amplitude a.

The beat signal (9.8) is sampled with an interval T_A in the chirp duration T. The samples are zero padding before a fast Fourier transform (FFT). We can detect a peak at the resulted frequency f_{IF}, i.e.,

$$f_{\text{IF}} = \frac{2f_c v}{c} + \frac{2Br}{Tc} \tag{9.9}$$

For (9.9), note that the frequency resolution is $\Delta f = F_s/N_Z$ where $F_s = 1/T_A$ is the sampling frequency, and $N_Z = T/T_A$ is the number of FFT data points used. Neglecting the Doppler shift term (the first term) in (9.9), it can be easily seen that the range resolution is determined by the bandwidth B:

$$\Delta r = \frac{Tc}{2BT_A N_Z} = \frac{c}{2B} \tag{9.10}$$

Similarly, the range rate resolution is determined by carrier frequency and chirp-ramp duration

$$\Delta v = \frac{c}{2f_c}\Delta f = \frac{c}{2f_c T} \tag{9.11}$$

If the measured frequencies f_{IF} are multiplied by the sweep time T, we obtain the measured spectral index (FFT bin) $\kappa = T f_{\mathrm{IF}}$, and $\kappa \in [0, N_Z)$ that is the peak index

$$\kappa^{\mathrm{Up}} = \frac{1}{\Delta v}v + \frac{1}{\Delta r}r \tag{9.12}$$

The above equation gives rise to an interesting phenomenon of FMCW in which range and Doppler are coupled. We only have resolution in one dimension, and two targets can be separated only when they are located in FFT bins with different indices κ.

Using a single LFM waveform we cannot simultaneously solve target range and radial velocity [16, 18]. For resolving this measurement ambiguity a second chirp and a third chirp need to be transmitted, as shown in Fig. 9.10. Similarly as in up-ramp, we can detect a spectral peak at spectrum for the target in down and Dopper ramps in Fig. 9.10, respectively,

$$\kappa^{\mathrm{Dn}} = \frac{1}{\Delta v}v - \frac{1}{\Delta r}r \tag{9.13}$$

$$\kappa^{\mathrm{Dp}} = \frac{1}{\Delta v}v \tag{9.14}$$

After target detection for the up-, down-, and Doppler-chirp spectra, the spectral peaks from the different chirp signals are combined to find the target range and radial velocity for a certain target. In a single target situation, it is straightforward that range and range rate can be determined by (9.12) and (9.13). However, in multiple target situations, the association among the peaks from the different chirp sweeps is indeed ambiguous. Without a careful treatment, ghost targets may be generated due to erroneous association among the spectral peaks.

To resolve the ambiguity, the third ramp is introduced. Figure 9.11 illustrates an example of potential ghost targets that may generated by erroneous association. Spectral peak indices κ_1^{Up} and κ_2^{Up} are detected during the up-chirp sweep, and κ_1^{Dn} and κ_2^{Dn} are detected during the down-chirp sweep in the meanwhile. Without the help from the indices (κ_1^{Dp} and κ_2^{Dp}) from the Doppler-chirp sweep, we cannot determine Tg_1 and Tg_2 as the true targets and reject Tg_g as false ones.

9.3.3 Frequency Shift Keying

Figure 9.12 illustrates a typical frequency shift keying (FSK) waveform, where two discrete radio frequencies f_a and f_b are transmitted in an alternated form. The frequency step $f_{\mathrm{step}} = f_b - f_a$ is small and designed based on the maximum

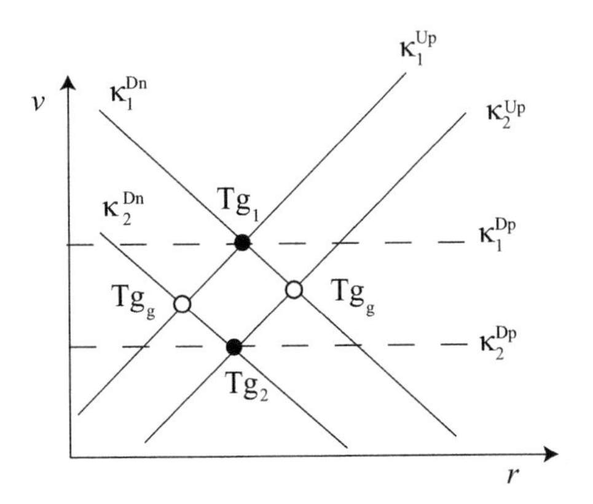

FIGURE 9.11
Ghost targets induced by erroneous association. Tg_1 and Tg_2 are true targets. Tg_g represents potential ghost targets generated from false association.

unambiguous range measurement. For example, a frequency step of 1 MHz is desirable for sensing horizon of 300 m.

In the receiver, the radar echo signal is down converted by mixing with LO output into baseband, and sampled N_Z times inside the coherent processing interval T_{CPI}. The digitized signal is then separately Fourier transformed for the upper and the lower frequencies in T_{CPI}. The targets are detected by an amplitude threshold using constant false alarm rate (CFAR) algorithm (cf. Sec. 9.4.1). Because of small frequency step between f_b and f_a and the alternated pattern, a target will be detected by nearly the same amplitude and at the same FFT bin for the two separated frequency spectra, but with different phase information. Let ϕ_a and ϕ_b be the phase angles of the target

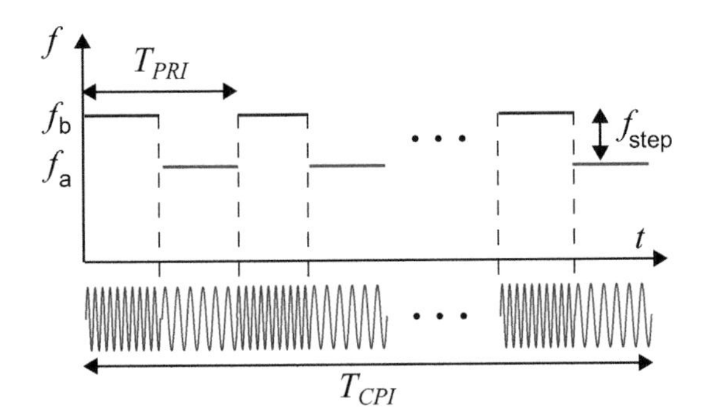

FIGURE 9.12
FSK CW waveform. T_{PRI} and T_{CPI} denote pulse repetition interval and coherent processing interval, respectively.

in the two separated spectra and $\Delta\phi = \phi_b - \phi_a$. From (9.8) with $B = 0$, we can obtain phases of baseband signals as follows:

$$\phi_a = \frac{4\pi r f_a}{c}$$
$$\phi_b = \frac{4\pi r f_b}{c}$$

Thus the range measurement can be computed as

$$r = \frac{c\Delta\phi}{4\pi f_{\text{step}}} \tag{9.15}$$

As in (9.11) for the LFM case, the resolution of range rate for the radar can be written as

$$\Delta v = \frac{c}{2 f_a T_{\text{CPI}}}$$

As an example, we consider a 77 GHz radar with $T_{\text{CPI}} = 40$ ms, and the range rate resolution is about 0.05 m/s. Thus radar targets with a slightly different radial velocity can be separated by the FSK radar waveform. It is important to note that the FSK waveform cannot distinguish targets with the same radial velocity with different ranges, which is one of the major disadvantages for the FSK measurement technique.

In summary, FSK waveform has the following characteristics

- Two discrete radio frequencies with a small difference are transmitted in an alternated form. The target's range is proportional to phase difference between two frequencies.
- Good resolution in radial velocity.
- Targets can be separated only when they have different radial velocities with respect to the radar.

9.3.4 Hybrid Waveform of FSK and LFM

LFM CW waveform cannot simultaneously estimate range and range rate in a single ramp. We need multi-ramp waveform. However, in multi-target scenarios, spectral peaks from multiple ramps have to be associated to find the range and range rate for a certain target.

A hybrid waveform of an FSK and an LFM offers an unambiguous solution for range and range rate of a target [17, 18]. As shown in Fig. 9.13, the transmit waveform is a stepwise frequency modulated signal which consists of two linear frequency modulated up-chirp signals which are transmitted in an intertwined way. The two chirp waveforms have an identical slope and bandwidth and only differ by a small frequency shift f_{step} step as shown in Fig. 9.13.

Combining (9.9) and (9.15), we can solve the target range and range rate

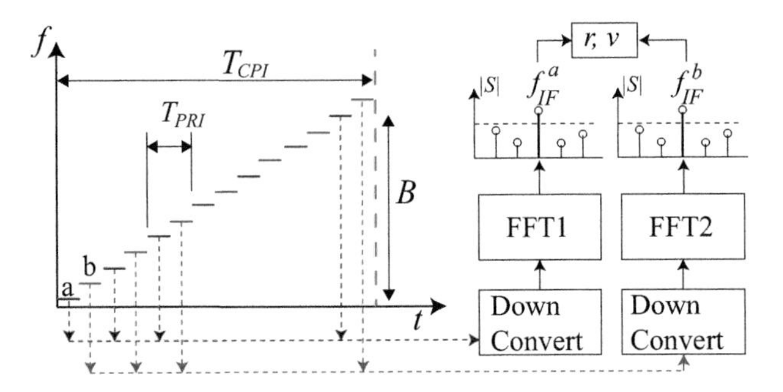

FIGURE 9.13
A waveform design of FSK-LFM radar with a chirp ramp and two FSK frequencies, a and b.

in each chirp range. Similarly as in LFM waveforms we still need multiple ramps to separate targets with different range and range rate and this hybrid waveform provides an additional equation in estimating range and, thus, improves accuracy.

9.3.5 Pulse Compression LFM Waveform

The flip side of the LFM waveform and its variants is the weakness in resolving ambiguity of combining spectral peaks from the different ramps for complex scenarios. Although the third ramp and even the fourth ramp are introduced to resolve the ambiguity, false targets are observed due to erroneous association in clutter scenarios which is typical in urban scenes (e.g., buildings, vehicles, traffic signs, curbs, and bushes). If the designer wants the radar system to have independent separation along both the range and range rate axes, this simple LFM waveform is not adequate because of the coupling characteristic between the range and range.

Figure 9.14 illustrates the waveform that combines *pulse compression* and LFM waveform. This waveform usually has the following characteristics:

- Compression pulse with LFM waveform
- Fast frequency ramp for high distance frequency shift. Duration of pulse is very short such as $T = 20\mu s$. Comparing with the frequency contributed by round trip of flight, the Doppler shift is insignificant in such a short period of time. Thus the target range can be determined instantaneously in every pulse.
- From narrow to ultra-wide bandwidth. For example, B varies from 0.2 GHz to 2.0 GHz for radar at 24 GHz.
- Fast sample rate (e.g., 40 MHz) and signal processing bottleneck can be solved using FPGA.

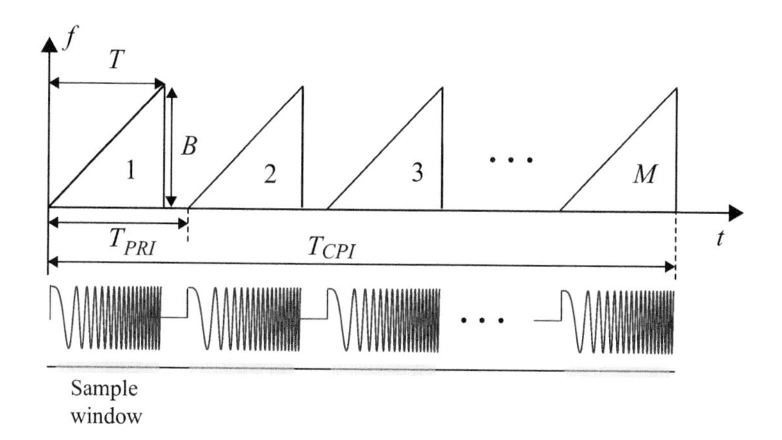

FIGURE 9.14
Waveform of pulse compress LFM.

- Pulse burst waveform. The total duration MT_{PRI} is the coherent processing interval (t_{CPI}), which determines the Doppler resolution $\Delta f = \frac{1}{MT_{\text{PRI}}}$.
- Processing of range and Doppler are decoupled, and we have independent separation along both range and range rate axes.

We define the pulse train LFM waveform as

$$s(t) = \sum_{m=0}^{M-1} s_p(t - mT_{\text{PRI}}) \tag{9.16}$$

where $s_p(t)$ is the single LFM pulse waveform with duration T

$$s_p(t) = \begin{cases} \exp(j\pi \frac{B}{2T} t^2) & 0 \leq t \leq T \\ 0 & \text{Otherwise} \end{cases}$$

T_{PRI} denotes the pulse repetition interval, and M is the number of pulses in the burst. The total duration MT_{PRI} is the coherent processing interval (CPI), also sometimes called a dwell.

Consider a target at range r and radial velocity v. By plugging (9.16) to (9.5), similarly as in Sec. 9.3.2 we can derive the beat signal

$$b(t) = a \exp\left[j2\pi \frac{2Br}{Tc}(t - mT_{\text{PRI}})\right] \exp\left[j2\pi \frac{2f_c v}{c} t\right] \tag{9.17}$$

for $(m-1)T_{\text{PRI}} \leq t < mT_{\text{PRI}}$ and $0 \leq m < M$.

The above beat signal is sampled with an interval T_A in each pulse duration T. Figure 9.15 illustrates the arrangement of sample points in two-dimensional array. Each row denotes a fast-time sequence that contains samples of a pulse.

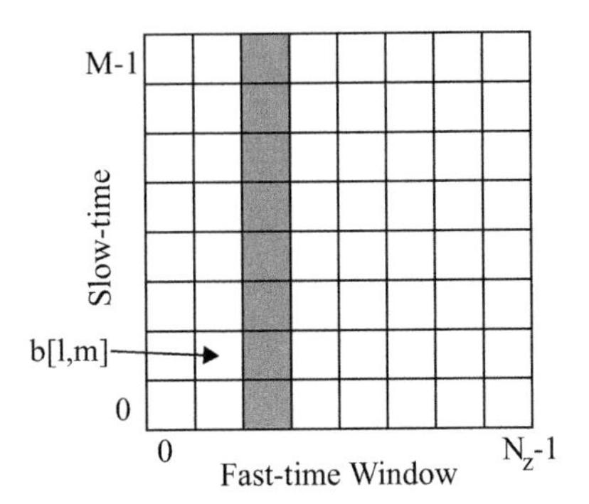

FIGURE 9.15
2D arrangement of data samples for the beat signal.

Slow-time sequence (e.g., the shaded vertical slice) consists of all samples at the same offset from the beginning of different pulse waveforms.

We have the 2D sampled beat signal as

$$b[l,m] = a \exp\left[j2\pi\left(\frac{2f_c v}{c} + \frac{2Br}{Tc}\right) lT_A\right] \exp\left[j2\pi\frac{2f_c v}{c} mT_{\text{PRI}}\right] \tag{9.18}$$

where l and m denote the indices of the fast and slow time, respectively.

Taking fast Fourier transform (FFT) on $b[l,m]$, we have

$$B[p,k] = \frac{1}{\sqrt{N_Z M}} \sum_{l=0}^{N_Z-1} \sum_{m=0}^{M-1} b[l,m] \exp\left[-j2\pi\left(\frac{lp}{N_Z} + \frac{mk}{M}\right)\right] \tag{9.19}$$

A peak can be detected associated at the following indices with the target at the resulted 2D spectrum f_{IF}, i.e.,

$$p = \left(\frac{2f_c v}{c} + \frac{2Br}{Tc}\right) T \tag{9.20}$$

$$k = \frac{2f_c v}{c} MT_{\text{PRI}} \tag{9.21}$$

From (9.20) and (9.21) we can see that the resolution of range and range rate for *pulse compression LFM waveform* is

$$\Delta r = \frac{c}{2B}$$

$$\Delta v = \frac{\lambda}{2MT_{\text{PRI}}} = \frac{\lambda}{2T_{\text{CPI}}}$$

Therefore in this manner, the *pulse compression LFM waveform* achieves the similar resolution of range and range rate as LFM and simultaneously resolves the ambiguity of range and Doppler. The cost, of course, is the time required

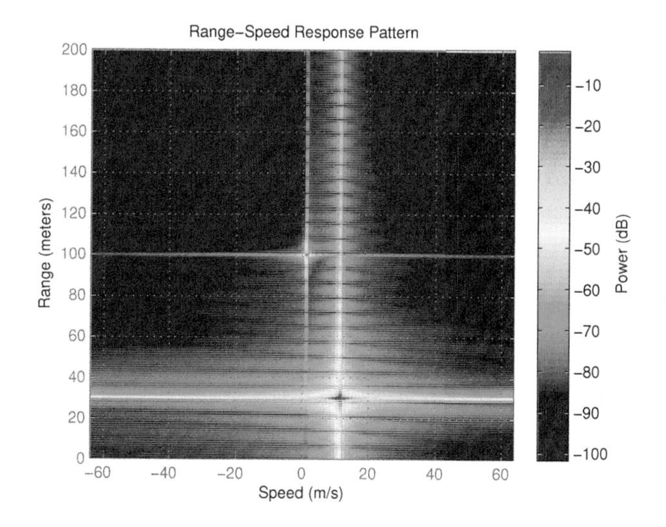

FIGURE 9.16
The range-Doppler response for a simulated 77 GHz FMCW radar.

to transmit and receive M pulses instead of one ramp, and the computational load of processing M rows of samples along slow-time axis.

In Fig. 9.16 we depict the range–Doppler response (i.e., the power of FFT output $B[p, k]$) for a simulated radar with the waveform. Table 9.2 summarizes the radar parameters. There are two targets with relative radial distances of 30 m and 100 m and radial velocities of 11.1 m/s and 1.1 m/s, respectively, which respond to the two peaks in the range–Doppler response.

TABLE 9.2
Parameters of the simulated radar for pulse compression LFM waveform.

System Parameter	Value
Operating frequency (GHz)	77
Maximum target range (m)	200
Range resolution (m)	1
Maximum target speed (km/h)	62.2
Sweep time (microseconds)	7.33
Sweep bandwidth (MHz)	150
Maximum beat frequency (MHz)	27.30
Sample rate (MHz)	150
Range rate resolution (m/s)	1
Number of pulses (M)	256

9.4 Range and Range Rate Estimation

The basic function of a radar is to detect multiple targets in its field of view, and to measure the key attributes of the targets such as range and range rate.

In Sec. 9.3 we discussed several basic representation and properties for a few typical automotive waveforms. In this section we will extend this discussion to the performance characteristics in presence of noise. The emphasis is on the *compression pulse LFM* waveform. The focus on this particular waveform is due to its superiority in range–Doppler resolution and its market penetration in automotive safety systems. The reader is referred to [15] for an extensive treatment of the topic.

The basic idea of matched filtering is correlating a known signal, or template, with an unknown signal to detect the presence of the template in the unknown signal. We will see the matched filter has the following properties: i) optimality in maximizing the signal-to-noise ratio (SNR) of its output, under assumption of Gaussian additive noise model, and ii) equivalence to discrete Fourier transform (DFT).

We will introduce the concept of ambiguity function that determines the resolution and unambiguous scope of range and range rate for the corresponding waveform. Finally, it is shown that the probability of detection and the accuracy of measurement depend only on SNR and the shape of ambiguity function. That is also demonstrated in the derived Cramer-Rao lower bound.

9.4.1 Target Detection

The radar signal processing begins with target detection whose function is to separate an echoed target signal from background signals. This is achieved by comparing the output power of the matched filter with a threshold. If the measured power amplitude exceeds the threshold, a target is detected and the associated signals are further processed to derive the measurements of the range, range rate, and azimuth angle for the target. On the other hand, the signals are regarded as thermal noise or other interference sources (e.g., ground and clutter) if the power amplitude is under the threshold.

The detecting threshold is chosen to comply with the *Neyman-Pearson criterion* where we want to achieve highest detection probability but maintain a constant probability of false alarm for a given SNR. If the statistics of the noise are known a priori, we can derive a threshold to meet the requirement for constant false alarm rate (CFAR). In many cases, the form of noise or signal is modeled as a Rayleigh distribution

$$p(z) = \begin{cases} \frac{z}{\sigma^2} \exp(-\frac{z^2}{2\sigma^2}), & z \geq 0 \\ 0 & z < 0 \end{cases} \tag{9.22}$$

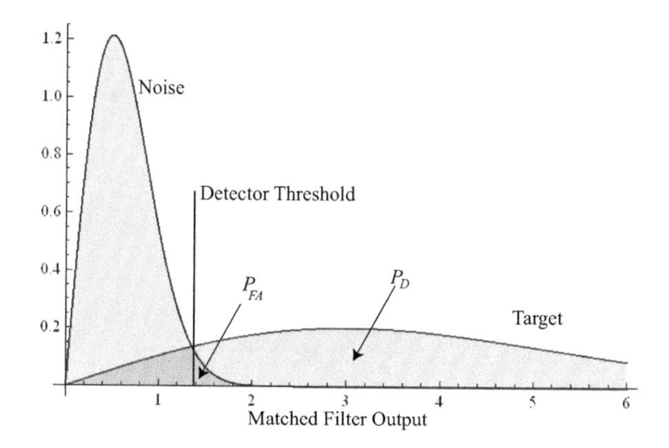

FIGURE 9.17
The Rayleigh distributions of a target and background noise. The detector threshold is set based on false detection rate P_{FA}. P_D is the probability of target detection.

where z is the amplitude of a complex signal.

Note that (9.22) only has one parameter, and this parameter can be estimated from samples from signals $\{z_1, z_2, \cdots, z_N\}$. The maximum likelihood estimate of σ^2 is

$$\sigma^2 = \sum_{i=1}^{N} z_i^2/N \tag{9.23}$$

Figure 9.17 illustrates the distribution of the background noise and a target in amplitude of matched filter output, respectively. Let σ_n^2 and σ_s^2 denote variance of the noise and signal, respectively. The probability of a false alarm is computed by integrating the noise distribution from the threshold T to infinity, or

$$\begin{aligned} P_{\text{FA}} &= \int_T^{+\infty} \frac{z}{\sigma_n^2} \exp(-\frac{z^2}{2\sigma_n^2}) dz \\ &= \exp\left(-\frac{T^2}{2\sigma_n^2}\right) \end{aligned}$$

which gives

$$T = \sqrt{-2\ln(P_{\text{FA}})}\sigma_n \tag{9.24}$$

Similarly the probability of detecting the target is

$$\begin{aligned} P_{\text{D}} &= \exp(-\frac{T^2}{2\sigma_s^2}) \\ &= \exp(\frac{\ln(P_{\text{FA}})}{\sigma_s^2/\sigma_n^2}) \\ &= \sqrt[SNR]{P_{\text{FA}}} \end{aligned} \tag{9.25}$$

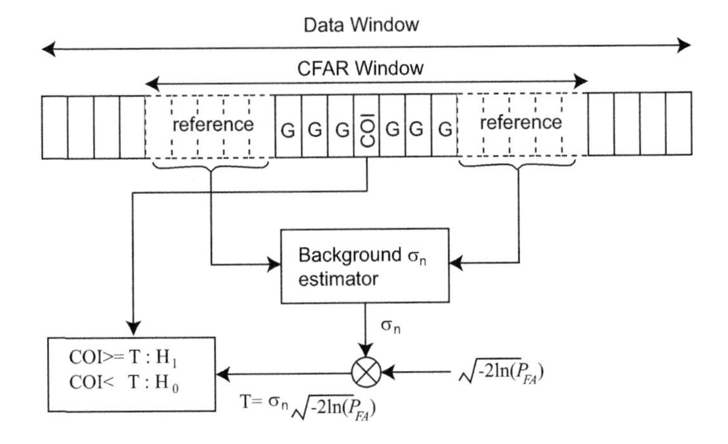

FIGURE 9.18
The CFAR detector using samples to derive the detection threshold. H_0 and H_1 denote the two hypotheses in which target is detected and no target is detected, respectively.

where $SNR = \sigma_s^2/\sigma_n^2$ is the signal-to-noise ratio. It is important to note that the higher the signal-to-noise ratio, the higher rate of detection, given the same tolerable false alarm rate.

We now are ready to outline the CFAR detector. A general CFAR detector is illustrated in Fig. 9.18 with a one-dimension data window where amplitudes of the processed data are stored from a time window within a coherent process interval (e.g., matched-filter outputs or computed FFT spectra). From the figure, we can see that the CFAR window is scanning the whole data. The cell of interest (COI) is located in the center of the CFAR window, and the objective is to determine whether the COI contains a target (H_1 hypothesis) or not (H_0 hypothesis). Several neighboring cells on either sides of COI are defined as gap cells, indicated by Gs. Measurements in the gap cells are not used in estimating the noise statistic due to possibility of containing returns from the target in the cell of interest.

Outside the gap cells there are the reference windows that are used to estimate the statistic of the noise. Let $\{z_1, z_2, \cdots, z_N\}$ denote these cells in reference windows. The background noise variance σ_n^2 can be estimated by using $\sigma_n^2 = \sum_{i=1}^{N} z_i^2/N$.

We compute the threshold $T = \sigma_n\sqrt{-2\ln(P_{\mathrm{FA}})}$. If the measurement z_{COI} at the COI cell is larger than T then a target is detected, and measured SNR for the target is

$$SNR = \sigma_s^2/\sigma_n^2 = z_{\mathrm{COI}}^2/\sigma_n^2$$

Otherwise, no target is detected at the COI cell, and the CFAR window is shifted to the next possible cell for another round of hypothesis testing.

9.4.2 Matched Filter and Ambiguity Function

Target detection in Sec. 9.4.1 needs to decide if a target echo signal is present in the received signal (hypothesis H_1) or target echo is not present (hypothesis H_0). This is achieved by using a matched filter where the received signal is correlated with the template of target echo to detect the target's presence or absence.

Consider the received down-converted signal $x(t)$ to be a noise corrupted signal of the beat signal $b(t)$ (cf. (9.17) for *pulse compression LFM waveform*), i.e.,

$$x(t) = b(t) + n(t)$$

where $n(t)$ denotes the noise term of zero-mean Gaussian distribution with variance σ^2. The discrete version of above equation, sampled at $t = 0, T_A, 2T_A, \cdots$, can be expressed in vector form

$$\mathbf{x} = \mathbf{b} + \mathbf{n}$$

and $\mathbf{n} \sim N(\mathbf{0}, \sigma^2\mathbf{I})$. For the *Neyman-Pearson criterion* we do the optimal test by calculating the test statistic D defined as

$$D = \frac{1}{||\mathbf{b}||^2}|\mathbf{b}^H\mathbf{x}|^2 \tag{9.26}$$

and compare it with a threshold T. If the threshold is exceeded we decide "target presence" (H_1), if not we decide otherwise (H_0).

Two questions need to addressed. The first one is the optimality of (9.26). On the other hand, we note that computing the test statistic D (9.26) requires knowledge of $\mathbf{b}$ and, thus, knowledge of target parameters r and v (cf. (9.8) or (9.17)). When these parameters are unknown, a search is required in a discrete set of hypothetical target parameters, specified on a certain grid. The second question is what granularity is appropriate or how dense and how wide should this grid be.

For the first question we show that the matched filter is the optimal linear filter for maximizing the signal-to-noise ratio (SNR) in the presence of additive stochastic noise. Let us call output of the filter, $\mathbf{y}$, the inner product the filter coefficient, and the observed signal $\mathbf{b}$

$$\mathbf{y} = \mathbf{h}^H\mathbf{x} = \mathbf{h}^H\mathbf{b} + \mathbf{h}^H\mathbf{n}$$

The signal-to-noise ratio (SNR), the objective function, is defined to be the ratio of the power of the output due to the desired signal to the power of the output due to the noise:

$$\frac{|\mathbf{h}^H\mathbf{b}|^2}{E\{|\mathbf{h}^H\mathbf{n}|^2\}} = \frac{|\mathbf{h}^H\mathbf{b}|^2}{\mathbf{h}^H\mathbf{h}\sigma^2} \leq \frac{\mathbf{b}^H\mathbf{b}}{\sigma^2}$$

where $E\{\cdot\}$ is the expectation and the last inequality follows by applying

Cauchy-Schwarz inequality $|\mathbf{h}^H\mathbf{b}|^2 \leq (\mathbf{h}^H\mathbf{h})(\mathbf{b}^H\mathbf{b})$. The upper bound is achieved if and only if $\mathbf{h} = \alpha\mathbf{b}$. Usually we choose the constant $\alpha = \frac{1}{\|\mathbf{b}\|}$. Plugging α back, we have $\mathbf{h} = \frac{\mathbf{b}}{\|\mathbf{b}\|}$, and the power of matched filter output is

$$\|\mathbf{y}\|^2 = \frac{1}{\|\mathbf{b}\|^2}|\mathbf{b}^H\mathbf{x}|^2$$

Comparing with (9.26), we see that the test statistic D is actually the power of matched filter output.

To address the second question, we define normalized target range $\tilde{r}$ and target range rate $\tilde{v}$ (cf. (9.18)), as

$$\tilde{r} = \left(\frac{2f_c v}{c} + \frac{2Br}{Tc}\right)T \tag{9.27}$$

$$\tilde{v} = \frac{2f_c v}{c}MT_{\text{PRI}} \tag{9.28}$$

where r and v are hypothetic target range and range rate, respectively. The useful signal template corresponding to these parameters is

$$\mathbf{b}(\tilde{r},\tilde{v}) = \exp\left[j2\pi\left(\frac{\tilde{r}}{N_Z}l + \frac{\tilde{v}}{M}m\right)\right]$$

Let $\mathbf{x}$ denote the digitized received signal. We write the output of the matched filter as

$$\begin{aligned}\mathbf{y} &= \frac{\mathbf{b}^H(\tilde{r},\tilde{v})\mathbf{x}}{\|\mathbf{b}(\tilde{r},\tilde{v})\|} \\ &= \frac{\sum_{l=0}^{N_Z-1}\sum_{m=0}^{M-1} x[l,m]\exp\left[-j2\pi\left(\frac{\tilde{r}}{N_Z}l + \frac{\tilde{v}}{M}m\right)\right]}{\sqrt{N_Z M}}\end{aligned} \tag{9.29}$$

where $x[l,m]$ denotes the data points at cell indices $[l,m]$.

The important implication of (9.29) is that matched filtering is equivalent to discrete Fourier transform (DFT). This can be easily seen by comparing with (9.19). $\mathbf{y}$ is the discrete Fourier transform (DFT) spectrum of the signal $\mathbf{x}$ with p and k replaced by $\tilde{r}$ and $\tilde{v}$, respectively.

Using this notation and assuming that the target is present, we have $\mathbf{x} = \mathbf{b}(\tilde{r}^+,\tilde{v}^+) + \mathbf{n}$ where $\tilde{r}^+$ and $\tilde{v}^+$ are the true normalized range and range rate, respectively. The detection statistic (9.26) defined for potential normalize target parameters $(\tilde{r},\tilde{v})$ is

$$D(\tilde{r},\tilde{v}) = \frac{1}{\|\mathbf{b}(\tilde{r},\tilde{v})\|^2}|\mathbf{b}^H(\tilde{r},\tilde{v})\mathbf{b}(\tilde{r}^+,\tilde{v}^+) + \mathbf{b}^H(\tilde{r},\tilde{v})\mathbf{n}|^2 \tag{9.30}$$

Let

$$\hat{A}(\tilde{r}-\tilde{r}^{+},\tilde{v}-\tilde{v}^{+}) = \frac{\mathbf{b}^{H}(\tilde{r},\tilde{v})\mathbf{b}(\tilde{r}^{+},\tilde{v}^{+})}{\|\mathbf{b}(\tilde{r},\tilde{v})\|\|\mathbf{b}(\tilde{r}^{+},\tilde{v}^{+})\|}$$

$$= \frac{\sum_{l=0}^{N_Z-1}\sum_{m=0}^{M-1} b_{lm}(\tilde{r}^{+},\tilde{v}^{+})\exp\left[-j2\pi\left(\frac{\tilde{r}}{N_Z}l+\frac{\tilde{v}}{M}m\right)\right]}{N_Z M}$$

where $b_{lm}(\tilde{r}^{+},\tilde{v}^{+})$ is the sampled received signal at indices $[l,m]$. We can define the ambiguity function as the power of $\hat{A}(\tilde{r}-\tilde{r}^{+},\tilde{v}-\tilde{v}^{+})$

$$A(\tilde{r}-\tilde{r}^{+},\tilde{v}-\tilde{v}^{+}) = |\hat{A}(\tilde{r}-\tilde{r}^{+},\tilde{v}-\tilde{v}^{+})|^2 \tag{9.31}$$

We can show that the test statistic D in (9.30) can be expressed in terms of the ambiguity function

$$D(\tilde{r},\tilde{v}) = \left|\sqrt{\mathrm{SNR}}\cdot\hat{A}(\tilde{r}-\tilde{r}^{+},\tilde{v}-\tilde{v}^{+}) + n_1\right|^2 \tag{9.32}$$

where $\mathrm{SNR} = \frac{\|\mathbf{x}\|^2}{\sigma^2}$ is the signal-to-noise ratio of the received signal $\mathbf{x}$ and n_1 is complex normal distribution, i.e., $n_1 \sim CN(0,1)$.

From (9.32) we see that the probability of detection depends only on SNR, shape of the ambiguity function, and how close the hypothetic parameters $(\tilde{r},\tilde{v})$ are to the unknown true parameters $(\tilde{r}^{+},\tilde{v}^{+})$. The width of the main lobe of the ambiguity function indicates the granularity of the hypothetic parameter grid (resolution) and the period of this function specifies the size of the grid (unambiguous region).

9.4.3 Estimation Accuracy

Noting the property of shift invariant for ambiguity function, we may assume for simplicity that $\tilde{r}^{+}=0$ and $\tilde{v}^{+}=0$. Since $b_{lm}(0,0)=1$, the ambiguity function for the waveform of *compression pulse LFM* can be decoupled into a product of two functions depending on range and the function depending on range rate:

$$A(\tilde{r},\tilde{v}) = A_r(\tilde{r})A_v(\tilde{v})$$

$$= \frac{\left|\sum_{l=0}^{N_Z-1}\exp\left[-j2\pi\frac{\tilde{r}}{N_Z}l\right]\right|^2}{N_Z}\frac{\left|\sum_{m=0}^{M-1}\exp\left[-j2\pi\frac{\tilde{v}}{M}m\right]\right|^2}{M}$$

Thus target range characteristic is independent of target motion characteristic. We can call $A_r(\tilde{r})$ range ambiguity function and $A_v(\tilde{v})$ range rate ambiguity function.

Note the equality $\sum_{n=0}^{N}\exp(-j\omega n) = \frac{\sin(\frac{\omega N}{2})}{\sin(\frac{\omega}{2})}\exp\left[-j\frac{N-1}{2}\omega\right]$. We have

$$A_r(\tilde{r}) = \left|\frac{\sin(\pi\tilde{r})}{\sqrt{N_Z}\sin(\pi\frac{\tilde{r}}{N_Z})}\right|^2$$

The zero-to-zero width of the main lobe of this function is equal to 2. Defining the range resolution $\Delta\tilde{r}$ as half of this value, we obtain $\Delta\tilde{r} = 1$. Plugging this to (9.27) and assuming the Doppler term in (9.27) is negligible, we have

$$\Delta r = \frac{c}{2B}$$

The granularity of the range grid on which the detection tests are performed should not be smaller than 1. Furthermore, the period of $A_r(\tilde{r})$ is N_Z and, thus, the maximal unambiguous range for a target is up to N_Z, that is $\tilde{r} \in [0, N_Z)$.

Notice that $A_v(\tilde{v})$ is in the same form as $A_r(\tilde{r})$. Thus similar derivation can be obtained for the ambiguity function of range rate:

$$\Delta\tilde{v} = 1$$

$$\Delta v = \frac{\lambda}{2MT_{\text{PRI}}}$$

The maximal unambiguous range for normalized range rate $\tilde{v} \in [0, M)$. Correspondingly the maximal unambiguous range rate for a target is $v \in [-\frac{\lambda}{4MT_{\text{PRI}}}, \frac{\lambda}{4MT_{\text{PRI}}})$.

Now we consider the accuracy of estimation of a target parameter γ (γ can be range r and range rate v). It depends on the resolution that is determined by ambiguity function, signal-to-noise ratio, implementation imperfections, propagation effects (multipath), interference, and so forth. The accuracy is measured by the mean-square error (MSE) of the estimator:

$$\text{MSE}\{\hat{\gamma}\} = E\{|\hat{\gamma} - \gamma|^2\}$$

where $\hat{\gamma}$ and γ are the estimate and the true parameter. Note that

$$\begin{aligned} \text{MSE}\{\hat{\gamma}\} &= E\{|(\hat{\gamma} - E\{\hat{\gamma}\}) - (\gamma - E\{\hat{\gamma}\})|^2\} \\ &= \text{VAR}\{\hat{\gamma}\} + |\gamma - E\{\hat{\gamma}\}|^2 \end{aligned}$$

For unbiased cases where $E\{\hat{\gamma}\} = \gamma$, $\text{MSE}\{\hat{\gamma}\} = \text{VAR}\{\hat{\gamma}\}$.

The Cramer-Rao bound (CRB) [20] is a useful tool to assess the accuracy of parameter estimation algorithm since it provides a lower bound on the accuracy of any unbiased estimator. For any unbiased estimator $\hat{\gamma}$,

$$\text{MSE}\{\gamma\} = \text{VAR}\{\gamma\} \geq \text{CRB}$$

According to [19, 21], the Cramer-Rao bounds for the normalized range $\tilde{r}$ and the normalized range rate $\tilde{v}$ are

$$\text{CRB}(\tilde{r}) = \frac{3}{2(2\pi)^2 \cdot \text{SNR}}$$

$$\text{CRB}(\tilde{v}) = \frac{3}{2(2\pi)^2 \cdot \text{SNR}}$$

Therefore, the corresponding Cramer-Rao bounds for target range and range rate are

$$\text{CRB}(r) = \frac{3}{2(2\pi)^2 \cdot \text{SNR}} (\Delta r)^2$$
$$\text{CRB}(v) = \frac{3}{2(2\pi)^2 \cdot \text{SNR}} (\Delta v)^2$$

9.5 Direction Finding

Unlike aerospace applications, the automotive scenarios involve both relevant objects and clutter background. Therefore angular resolution is critical for the radar to correctly perceive the surrounding environment. Although we have shown in Sec. 9.4 that the *compression pulse LFM* waveforms have the ability in separating objects in different range and range rate bins, this separation is not sufficient. For an example, in a narrow street with infrastructure objects symmetrically present on both sides of the street, the objects fall in the same range and range rate bins and, thus, the radar cannot resolve them as separate objects. In this section, we will discuss that additional resolution in direction of arrival can be achieved by 3D matched filtering.

Two basic techniques are discussed for direction finding using linear array antenna: beamforming and monopulse. Because of cost considerations or physical constraints such as limited space to place antenna arrays, an automotive radar usually has a small number of elements and a limited effective aperture size. The challenge is how to design an antenna subsystem to deal with the case of large beam width and hence relatively coarse angular resolution.

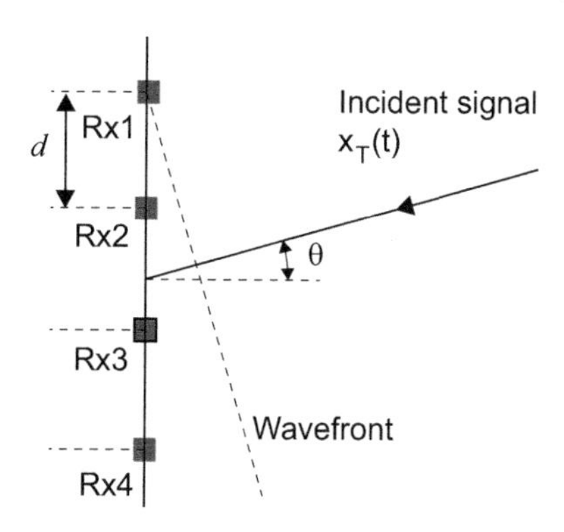

FIGURE 9.19
Four-element uniform spaced linear array.

9.5.1 Linear Array Antenna

Let us consider an antenna transmitting a signal $x_t(t) = s(t)\exp(j2\pi f_c t)$, where $s(t)$ is the baseband signal, and f_c is the carrier frequency of millimeter wavelength. The signal is scattered by a target and arrives at an array of antennas as illustrated in Fig. 9.19. The received signals are delayed versions of $x_T(t)$:

$$\mathbf{x}_R(t) = \begin{bmatrix} s(t-\tau_1)\exp(j2\pi f_c(t-\tau_1)) \\ s(t-\tau_2)\exp(j2\pi f_c(t-\tau_2)) \\ \vdots \\ s(t-\tau_K)\exp(j2\pi f_c(t-\tau_K)) \end{bmatrix}$$

where τ_k is the propagation delay determined by the direction of arrival (DOA) relative to the boresight of the radar. In case of linear array, the delay can be expressed as

$$\tau_k = \tau_0 - \frac{d_k}{c}\sin\theta$$

where τ_0 is the propagation delay measured at a reference point on the array, d_k is the distance of the element from the reference point, and θ is the incident angle.

Without loss of generality we can assume $\tau_0 = 0$ because the contribution of τ_0 is a phase offset $\exp(-j2\pi\tau)$ which is common to the elements of the received signal vector. This further requires that the derived direction-finding algorithm has to be invariant to a common additive phase.

After down-converting $\mathbf{x}_R(t)$ to baseband, we obtain

$$\mathbf{x}(t) = \begin{bmatrix} s(t-\tau_1)\exp(j2\pi f_c\tau_1) \\ s(t-\tau_2)\exp(j2\pi f_c\tau_2) \\ \vdots \\ s(t-\tau_K)\exp(j2\pi f_c\tau_K) \end{bmatrix} \tag{9.33}$$

Let D denote the aperture of the array. Then D/c is the time for the signal to propagate over the whole antenna. Thus $\tau_k < D/c$. If the baseband signal bandwidth is B and $B \ll c/D$, we can approximate $s(t-\tau_k)$ using $s(t)$. Note that the condition $B \ll c/D$ implies that the waveform is narrow band. For example, considering a 77 GHz radar with antenna aperture 3 cm, we can design $s(t)$ with a bandwidth of 500 MHz since this is much less than $c/D = 10$ GHz.

It follows that under the narrow-band assumption, (9.33) can be approximated by

$$\mathbf{x}(t) = s(t)\begin{bmatrix} \exp(j2\pi f_c\tau_1) \\ \exp(j2\pi f_c\tau_2) \\ \vdots \\ \exp(j2\pi f_c\tau_K) \end{bmatrix} = s(t)\mathbf{a}(\theta) \tag{9.34}$$

where

$$\mathbf{a}(\theta) = \begin{bmatrix} \exp(j2\pi d_1 \sin\theta/\lambda) \\ \exp(j2\pi d_2 \sin\theta/\lambda) \\ \vdots \\ \exp(j2\pi d_K \sin\theta/\lambda) \end{bmatrix}$$

is regarded as the response function of the antenna in variation of incident angle.

For a uniformed spaced linear array (ULA) if we choose the element 1 as the reference point, the response function can be expressed as

$$\mathbf{a}(\theta) = \begin{bmatrix} 1 \\ \exp(j2\pi d \sin\theta/\lambda) \\ \vdots \\ \exp(j2\pi K d \sin\theta/\lambda) \end{bmatrix} \tag{9.35}$$

9.5.2 Digital Beamforming

The basic concept of direction finding is by adding the antenna outputs so that the signal from a given direction can be coherently accumulated. Let the incident angle for the target be θ_0. We compute $|\mathbf{a}(\theta)^H \mathbf{a}(\theta_0)|$. If the guessed direction θ aligns with the true incident angle θ_0, we obtain the maximum of the combined $|\mathbf{a}(\theta)^H \mathbf{a}(\theta_0)|$. On the other hand, if $\theta \neq \theta_0$, the received signals are not coherently added and, thus, it is smaller. This technique is called beamforming.

The basic function of beamformer is a matched filter that uses a weight vector $\mathbf{h}(\theta)$ to linearly add the array inputs, and we are interested in the power of the combined output:

$$D(\theta) = |\mathbf{h}(\theta)^H \mathbf{x}|^2 \tag{9.36}$$

where $\mathbf{x}$ is the received baseband signal at time t. For direction-finding purpose, we compute $D(\theta)$ over a range of directions θ. The direction where $D(\theta)$ reaches it largest value is the estimated incident angle. Figure 9.20 illustrates the power of the combined output power $D(\theta)$ from an eight-element ULA. Without loss of generality, the target is at the direction of the boresight of the sensor or $\theta_0 = 0$. As one can see, $D(\theta)$ reaches its maximum at the true direction. Without noise this beamforming technique provides a perfect estimate of direction, but in presence of noise the peak location is shifted and an estimation error may be introduced.

Using similar derivation of the matched filter for range and range rate in Sec. 9.4.2, we have $\mathbf{h} = \frac{\mathbf{a}(\theta)}{\|\mathbf{a}(\theta)\|}$ for direction finding.

Now we show the equivalence between the matched filter and DFT computation. We define the normalized angle $\tilde{\theta}$ as

$$\tilde{\theta} = \frac{d \sin\theta}{\lambda} K \tag{9.37}$$

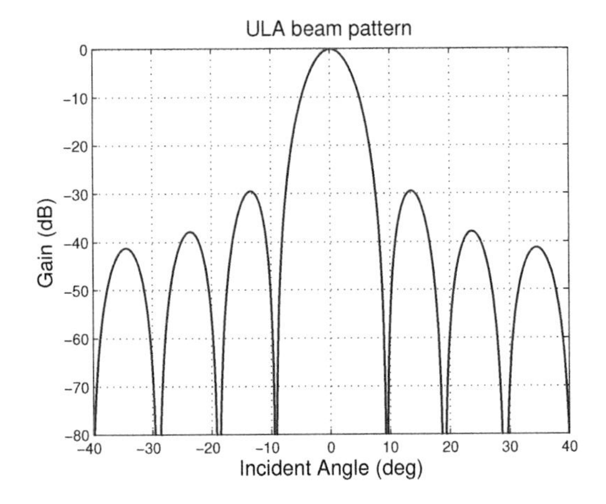

FIGURE 9.20
Beam pattern for eight-element uniform spaced linear array at 77 GHz frequency with the space between two adjacent elements $d = 0.003$ m.

The ULA antenna response function (cf. (9.35)) can be expressed as

$$\mathbf{a}_k(\tilde{\theta}) = \exp\left[j2\pi\frac{\tilde{\theta}}{K}k\right], \quad 0 \leq k < K$$

Let $x[k]$ denote the discrete received signal sampled at the k-th antenna. We have the output of the matched filter as

$$\begin{aligned} \mathbf{p} &= \frac{\mathbf{a}^H}{\|\mathbf{a}\|}\mathbf{x} \\ &= \frac{1}{\sqrt{K}}\sum_{k=0}^{K-1} x[k]\exp\left[-j2\pi\frac{\tilde{\theta}}{K}k\right] \end{aligned} \tag{9.38}$$

One sees that (9.38) is the DFT computation along Rx antenna channels.

In order to analyze the performance of the beamformer in the presence of noise, we need to study the ambiguity function for direction of arrival. Similarly as in Sec. 9.4.3, the following decoupled ambiguity function is defined:

$$\begin{aligned} A_\theta(\tilde{\theta}) &= \frac{|\mathbf{a}(\tilde{\theta})^H|^2}{K} \\ &= \frac{1}{K}\sum_{k=0}^{K-1}\left|\exp\left[-j2\pi\frac{\tilde{\theta}}{K}k\right]\right|^2 \\ &= \left|\frac{\sin(\pi\tilde{\theta})}{\sqrt{K}\sin(\pi\frac{\tilde{\theta}}{K})}\right|^2 \end{aligned} \tag{9.39}$$

The angular resolution is the minimum angular separation that can be achieved when two targets are at the same range and range rate. This important characteristic of a radar is determined by antenna beam width usually represented by the -3dB angle $\theta_{3\text{dB}}$. This angle also is defined by the length between two half-power points of the main lobe in (9.39). Using numerical approximation, we have

$$\theta_{3\text{dB}} = 0.89\frac{\lambda}{Kd}\text{radians} \tag{9.40}$$

Note that this result is consistent with the one determined by the antenna's aperture size (cf. (9.4)) where for ULA antenna the aperture is $D \approx Kd$.

Moreover, with the known ambiguity function we can derive the Cramer-Rao lower bound (CRB) for the unbiased estimator of direction. According to [21], the CRB of arrival azimuth angle θ for a K-element linear array is

$$\text{CRB} = \frac{\lambda^2}{8\pi^2 \cdot \text{SNR} \cdot \cos^2\theta \bar{d^2}}$$

where $\bar{d^2} = \sum_{k=1}^{K-1} d_k^2$ and d_k is the distance of the k-th element from the phase center of the array. For the ULA case,

$$d_k = k\frac{D}{K-1} - \frac{D}{2}$$

where D denotes the array aperture.

9.5.3 Monopulse

Monopulse is a commonly used direction finding method for automotive side-looking radar with wide field of view. For a comprehensive treatment of the subject, the reader is referred to [22]. This method involves computing the difference of the power outputs in slightly two different angles. Let

$$b(\theta) = \frac{D(\theta + \Delta/2) - D(\theta - \Delta/2)}{\Delta}$$

where $D(\theta)$ is the beam pattern defined in (9.36).

Figure 9.21 shows the response curve of $b(\theta)$. Note that $b(\theta)$ is nearly linear over a significant range including the main lobe in Fig. 9.20. As we can see, $b(\theta) = 0$ means the target in the boresight of the antenna. The output is positive if the target is to the right of the direction in which the difference beam is pointed and negative if it is to the left.

Without noise, using the monopulse method we can obtain the perfect estimate of direction. In the presence of noise and practical imperfections, calibration using reference targets can be used to find the mapping from the amplitude to direction of arrival for the signals.

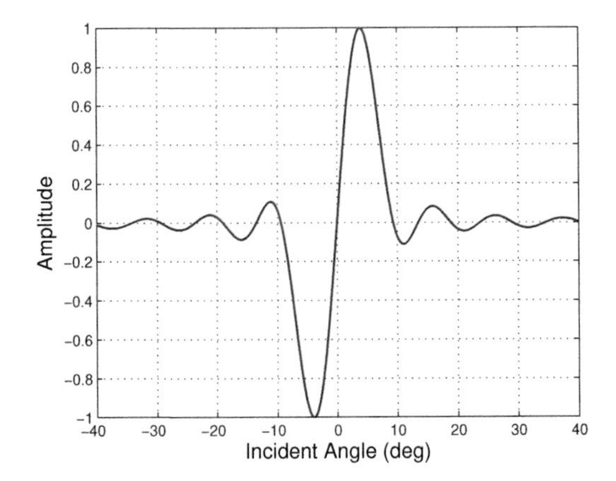

FIGURE 9.21
Response $b(\theta)$ of a monopulse system for eight-element uniform spaced linear array at 77 GHz frequency with the space between two adjacent elements $d = 0.003$ m.

9.5.4 Simultaneous Processing for Range, Doppler, and Angle

We consider the radar system with a design of compression pulse LFM waveform and ULA antenna. Figure 9.22 illustrates the arrangement of sample points in three-dimensional array. One extra dimension is added to the 2D arrangement depicted in Fig. 9.15 for independent samples from different Rx antenna channels.

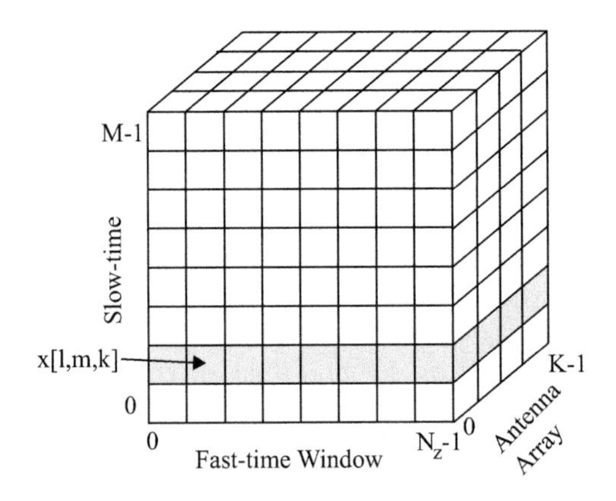

FIGURE 9.22
3D arrangement of data samples.

We define the 3D measurement as $\mathbf{x}$ and $x[l, m, k]$ denotes a cell at indices of fast-time l, slow-time m, and ULA element k. The signal template $\mathbf{h}$ for a target at normalized parameters $(\tilde{r}, \tilde{v}, \tilde{\theta})$

$$h_{lmn}(\tilde{r}, \tilde{v}, \tilde{\theta}) = \frac{1}{\sqrt{N_Z MK}} \exp\left[j2\pi \frac{\tilde{r}}{N_Z} l\right] \exp\left[j2\pi \frac{\tilde{v}}{M} m\right] \cdot \exp\left[j2\pi \frac{\tilde{\theta}}{K} k\right] \tag{9.41}$$

The matched filter output is equal to inner product of $\mathbf{x}$ and $\mathbf{h}$, i.e.,

$$\begin{aligned} \mathbf{y}(\tilde{r}, \tilde{v}, \tilde{\theta}) &= \mathbf{h}^H \mathbf{x} \\ &= \frac{1}{\sqrt{N_Z MK}} \sum_{l=0}^{N_Z} \sum_{m=0}^{M-1} \sum_{k=0}^{K-1} x[l, m, n] \exp\left[-j2\pi \frac{\tilde{r}}{N_Z} l\right] \\ &\cdot \exp\left[-j2\pi \frac{\tilde{v}}{M} m\right] \exp\left[-j2\pi \frac{\tilde{\theta}}{K} k\right] \end{aligned} \tag{9.42}$$

One can easily see that the inner product $\mathbf{y}$ in (9.42) is actually the DFT for the 3D signals $\mathbf{x}$. Thus, an extended target detection algorithm CFAR (cf. Sec. 9.4.1) for 3D grid can be applied to $|\mathbf{y}|^2$ to test the presence of a target in the cell of $(\tilde{r}, \tilde{v}, \tilde{\theta})$ for $0 \leq \tilde{r} < N_Z$, $0 \leq \tilde{v} < M$, and $0 \leq \tilde{\theta} < K$.

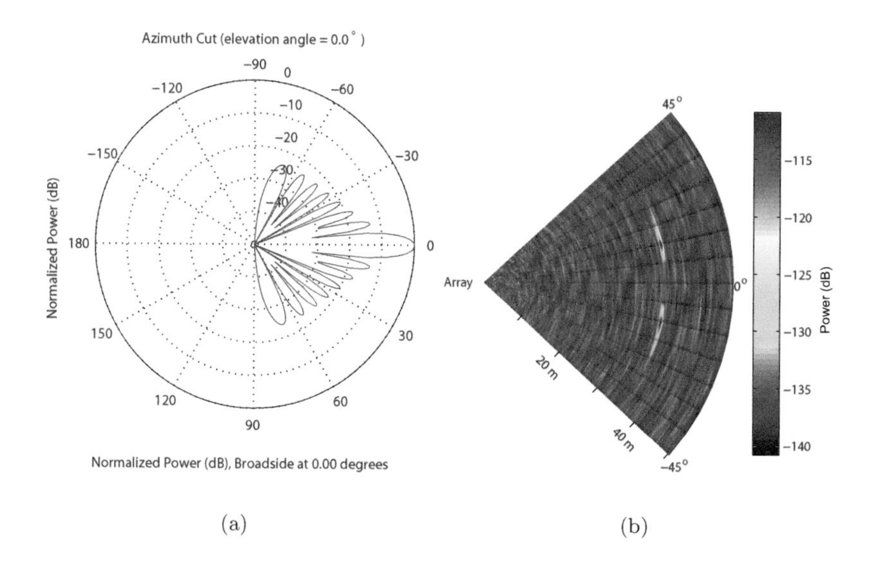

FIGURE 9.23
a) Angular beam pattern for 16-element ULA in polar coordinate. b) Range-angle responses of two targets of the same range and range rate.

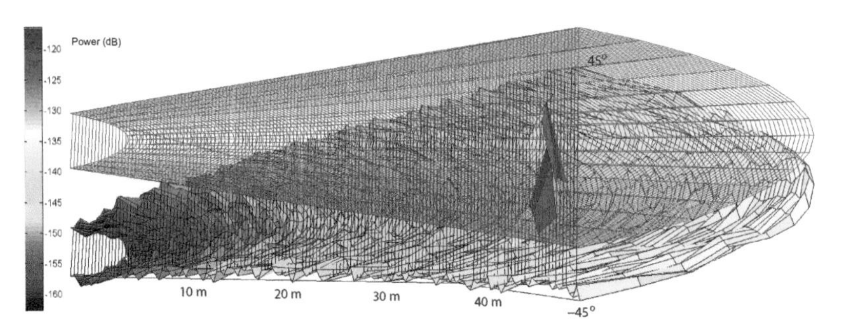

FIGURE 9.24
Target detection. The top red plane denotes the threshold. The bottom surface shows the response (dB) of the two targets in 2D grid of range and angle.

Now we simulate a radar with 16-element ULA using the waveform of the parameters outlined in Table 9.2. The space between two adjacent antenna elements is the half wavelength, and the beam pattern of the antenna is depicted in Fig. 9.23. The beam width $\theta_{3\mathrm{dB}} = 6.37°$ (cf. (9.40)). Assume that there are two point targets with distance, radian velocity and angle of (35 m, 0 m/s, 11°) and (35 m, 0 m/s, −11°), respectively. Since the two targets are in same range and range rate bin, they are generally inseparable in a single receiving channel. However, with 16 receiving channels, the echo signals from the targets can be separated. Consider the digitized signals in the shaded slice $x[l, 127, n]$ in Fig. 9.22(b) for $0 \leq l < N_z$ and $0 \leq k < K$. Figure 9.23 shows the power of the response after the 2D FFT transform. Clearly we can visualize the two peaks corresponding to the two targets, respectively. To better represent the data, we only plot range up to 50 m.

In Fig. 9.24 we can visualize the process of target detection. There are two peaks visible above the detection threshold, corresponding to the two targets we defined earlier. We can find the locations of these peaks and estimate the range and angle of each target.

9.6 Fusion of Multiple Sensors

Combining results of multiple sensors can provide more reliable and accurate information than using a single sensor [33, 28]; this allows either improved performance from existing sensors or the same performance from smaller and less expensive sensors. Consequently, there has been an increased interest in fusing data from multiple sensors to determine the location, moving direction, relative velocity and acceleration of critical objects around the vehicle for advanced driver assistance systems (ADAS). For example, [29] shows that the adaptive cruise control (ACC) is enhanced by fusing a 77 GHz radar and an infrared lidar. Good system performance is obtained in [23, 30] for an autonomous vehicle application by using radar, stereo vision and lidar. A pre-crash system based on a lidar and two short range 24 GHz radars is presented in [32]. A collision mitigation system using lidar and stereo vision is recommended in [27].

The field of view (FOV) of forward-looking ACC long-range radar (LRR) is limited to 15°. For ADAS systems beyond ACC, LRR alone might not be sufficient [29]. To improve ACC stop-and-go, two extra sensor types were added: a forward object fusion system with a camera with object detection capability and multiple low cost 24 GHz short range radars (SRR). The 45° FOV of vision system and 60° FOV of SRR cover a large area left uncovered by the LRR radar.

The measurements of these low-cost sensors are noisy, have errors, and are only a partial view of the world. In addition, we have to evaluate the reliability of sensor data. Reliability represents how much confidence we have in the data reported by the sensor. All of these aspects contribute to an increase in the uncertainty in the fusion system. To address the aforementioned issues, in this section we discuss a fusion system in the following perspectives:

- The architecture is designed in a way so that new generations of sensors can be integrated. In this architecture, we can have fusion in different configurations. We can have fusion from one sensor (time series), redundant sensors, or multiple heterogeneous sensors.

- Use of a weighted-least-squares based method to fuse data from different sources. Different sensor configurations are evaluated to find the optimal setting.

- Use of a sensor registration method to align sensors online. The sensors must be aligned and stay aligned in order to work properly. Without proper alignment, information from one sensor cannot be effectively associated with that from a different sensor. For example, inadequate data association causes overstated confidence in a fused target and unnecessary plural fused tracks that correspond to a single object.

- Time synchronization. A generic hardware mechanism is introduced to estimate and to handle the different latencies of measurement between asynchronous sources.

9.6.1 Automotive Sensor Technology

While the focus is on automotive radars we outline object detection technologies used by ADAS applications. These sensors may potentially be used by a sensor fusion system. Three other technologies of object detection will be discussed: ultrasonic, lidar, and electric-optical or camera-based. Table 9.3 compares the properties for the three technologies with radar based on analysis of echoed waves, showing where each fits into the spectrum.

TABLE 9.3
Key attributes for automotive sensing technologies.

Attributes	Wavelength	Frequency	Illumination
Ultrasonic	4 to 9mm	35 to 80KHz	Active
Radar	3 to 15mm	24 to 79GHz	Active
Lidar	400 to 1500nm	200 to 750THz	Active
Camera	380 to 750nm	400 to 7,500THz	Passive

We first summarize the pros and cons of each technology for automotive applications in Table 9.4, and then outline ultrasonic, lidar, and camera technologies in contrast to radar. The characteristic and value are highlighted in design of an ADAS fusion system for each technology.

9.6.1.1 Ultrasonic

Unlike using electromagnetic waves by radar, the measurement process of ultrasonic sensor is through correlation between the transmitted sound waves and their received echoes. Ultrasonic sensors are typically mounted in the bumper(s) to assist the drivers during parking maneuvers. Figure 9.25 shows an automatic parallel parking system using ultrasonic sensors. Here is how the parking system works: 1) The driver of subject vehicle (i.e., the bottom vehicle with ultrasonic sensors) pulls alongside a possible parking spot and activates the ultrasonic sensors to measure the potential feasible parallel parking space by scanning (i.e., self motion of subject vehicle) the gap between the parked vehicles. 2) The system prompts the driver to accept the system assistance, then the steering system takes over and steers the car into parking space automatically. The driver still needs to control the transmission and press the throttle or brake pedal to drive the car. 3) The system prompts the proximity to the rear object using audible and/or visual cues.

TABLE 9.4
Pros and cons of sensor technologies.

Technologies	Pros	Cons
Ultrasonic	Low cost Small package Large aftermarket	May blocked by dirt Affected by air temperature, pressure, wind Limited to 5 m No angular measurement No range resolution
Radar	Good distance and velocity accuracy Easy object grouping capability Sensor can be mounted behind plastic fascia Insensitive to dust, dirt, inclement weather Fair resolution Up to 200 m detection range	EM radiation regulatory restriction Limited ability in classification Limited azimuth resolution
Lidar	High resolution and accuracy of range Wide FOV and high azimuth resolution	Velocity is calculated not measured Sensitive to inclement weather Limited range detection Complicate signal processing
Camera	Wide FOV with high accuracy in azimuth Ability in target recognition Lane sensing, traffic sign detection Pedestrian detection Middle range	Blockage by dirt; sensitive to inclement lighting Computational complexity in image processing Less robustness in velocity estimation High missing detection rate Limited range detection Less accurate in range estimation

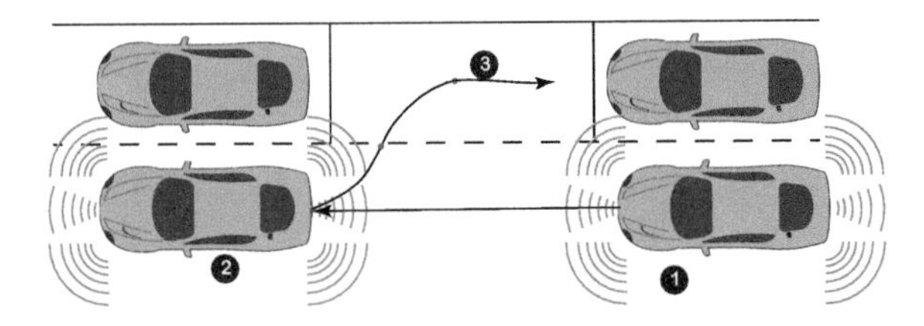

FIGURE 9.25
Automatic parking system using ultrasonic sensors.

Because of its low cost and ability to provide the presence and distance of a target within its field of view in low speed scenarios, ultrasonic is quite successful in mid- and low-end car markets. Usually the maximal detection range is about 5 m but there is no resolution in range measurement and azimuth angle measurement. Incapability of classifying target further limits applying ultrasonic to complicated ADAS systems. However, ultrasonic can serve as an auxiliary sensor in a fusion system to fill in possible gaps not covered by radars or other major sensors.

There are two main methods for ultrasonic to measure distance: time of flight and phase difference. In time-of-flight method the sensor measures the presence of the distance of a target by sending out a pulsed ultrasound wave and then measuring the flight time of the reflected echo wave, in similar way as employed in radar.

For phase difference method, a continuous-wave ultrasonic sensor outputs waves of a certain frequency and uses the Doppler principle to detect the presence and speed of moving objects. The range is measured via computing the phase difference between the transmitted and received signals. Although this method has an integer ambiguity in measuring a distance more than a wavelength, this limitation can be overcome by employing phase-shift keying technology (PSK) similar to radar (cf. Sec. 9.3.3) by creating a second signal at a similar but different frequency, and using the relationship of phase differences detected for both frequencies to resolve the integer ambiguity.

For the future, the single beam ultrasonic can be extended to multi-beam by digital beamforming technology similar to radar by use of an array of ceramic piezoelectric elements. The elements can be arranged and placed in a variety of housings: aluminum, Teflon, PVC, stainless steel, and so on. The unit can be completely sealed for waterproofing.

9.6.1.2 Lidar

Lidar is the acronym for the technology of light detection and ranging. There are two types of lidars: Doppler lidar and pulse lidar. Doppler lidar can measure the velocity directly through Doppler shifts but due to its cost and complexity, the system has not been integrated in automotive applications. An automotive lidar system is pulse lidar that measures time of flight through

FIGURE 9.26
The scan points captured in an automotive scenario by a 3D lidar. The top-right and bottom-right images are acquired by forward-view and rear-view cameras, respectively.

transmitting a train of light wave pulses, in the same concept as a pulse radar system. Velocity is not directly measured but can be estimated by use of tracking algorithms.

Figure 9.26 shows a snapshot of data from a 3D lidar sensor. The red points are point-cloud representations of background obstacles. The magenta points enclosed by green boxes are detected vehicles. The arrow on top of a box indicates the magnitude and direction of the velocity of the vehicle, which is usually estimated by tracking algorithms.

Although automotive lidar systems are not coherent, interference reduction can be achieved through the use of polarization and random allocation of the phase of pulse train. Polarization will not eliminate false detections due to direct illuminations (or false returns from indirect illumination) that originate from vehicles in the same lane. However, the phases of light pulse train would be uniformly distributed at random, therefore reducing the probability of interference from different sources. The power of laser emissions is restricted, and a classification of eye safety is required such that one can look into a lidar transmitter and incur no damage to the eyes.

9.6.1.3 Camera

The technology is originally designed for human vision applications where the output of the camera is consumed by a driver that determines whether there is an imminent threat or not.

With advances in computer vision, a camera-based vision system, to a greater extent, outputs useful environmental attributes and can serve as a major component in a sensor fusion system. Although the vision technology is susceptible to dirt, inclement weather and inadequate lighting, it has the ability to detect targets like a car, truck, bicycle, and/or pedestrian. The disadvantage is that this technology requires image processing that involves large amounts of data to be computed in real time. However, with new technology such as the floating point gate array (FPGA) processor, graphic processing unit (GPU), digital signal processor (DSP), and application-specific integrated circuit (ASIC), sufficient computational horsepower is available for embedded implementation at low cost. On the other hand, advancement in complementary metal oxide semiconductor (CMOS) sensors introduces high dynamic range cameras that are not as susceptible to light and can also work during the nighttime.

The objective of the technology is not only to represent the scene with high resolution image, but also to robustly detect an object of interest (e.g., vehicle, pedestrian, or bicycle) in a variant of situations of different lightings, backgrounds, and so forth. The detection algorithms need to be validated in real traffic, a process that often takes years. Thus, vision based systems for ADAS applications are not as mature as their radar-based counterparts.

A stereo vision camera can provide a rough distance image by computing the disparity map of environment through its binocular lenses and triangulation ranging process. Due to limited baseline between the binocular lenses, the range accuracy degenerates proportionally to quadratic of the range. Figure 9.27 depicts an example of disparity image estimated by a stereo vision system. The blueness and redness are used to indicate the close and far obstacles, respectively.

Monocular vision cameras can estimate the range of object by calculating the number of pixels between the vanish point and object's contact point on the ground, assuming the ground is flat. The target velocity is estimated by

(a) (b)

FIGURE 9.27
Stereo disparity map for an automotive scenario. (a) The image from a camera. (b) The stereo disparity map.

FIGURE 9.28
Monocular camera system with object and lane detection.

tracking range measurement of consecutive time frames. Figure 9.28 shows a snapshot of the scene with vehicles enclosed by rectangles. The red box indicates the vehicle in the same lane as the subject vehicle while the green boxes show other vehicles. The detected lane markings are shown in green dashed lines, which can enable some useful applications such as lane departure warning (LDW).

Generally, radar has a superior performance in distance measurement and is insensitive to a variety weather conditions. Morever, the Doppler effect can be easily employed to measure radial speed of target. For an UWB SRR radar the resolution of range can easily reach 15 cm, which is not easily achieved by cameras. However, camera provides texture pattern of an object for potential classification capability. Naturally we can see that fusion of sensors with complementary performance characteristics is a good strategy to provide a robust sensing solutions for ADAS systems. For example, we may use radar for distance or velocity and camera for object classification.

9.6.2 Fusion Algorithm

Hall [25] describes three basic approaches to data fusion, depending on where in the processing flow fusion is performed: centralized fusion (feature level), distributed fusion (object level) and hybrid.

9.6.2.1 Architecture Aspect

The centralized architecture transmits features (unprocessed data) from several sensors to a central fusion process that performs the functions of data registration and association, followed by correlation, tracking, and target classification. At the opposite extreme, the distributed fusion architecture allows each sensor perform a maximum amount of processing to generate object data,

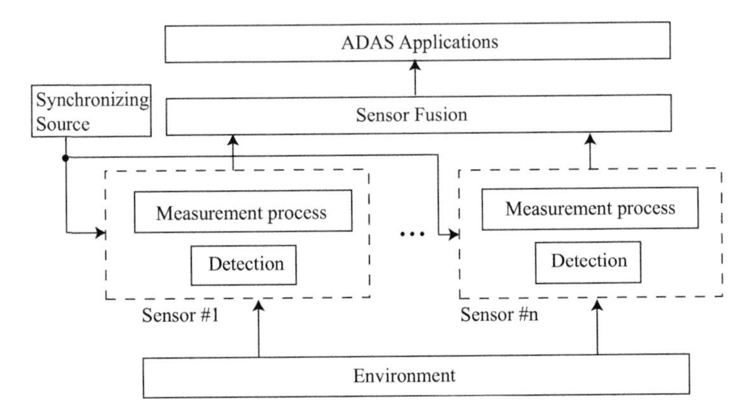

FIGURE 9.29
Abstract block diagram for range sensors.

which in turn is transmitted to the fusion layer that fuses the data. The third architecture is a hybrid combination of the former approaches.

Although better performance is reported to run fusion at the feature level than at the object level [24, 26], this chapter utilizes the distributed architecture because of lack of raw feature data (e.g., unprocessed measurement data) and limited communication bandwidth in the vehicle architecture.

Figure 9.29 shows the fusion architecture of the proposed system. We abstract the sensors to have a detection process and a measurement process [37]. The detection module separates signals echoed by object from background and determines the presence of an object in the scanning window. The measurement process employs a physical process to determine and map the location, direction or speed (i.e., Doppler effect). The fusion module integrates the object maps along with the uncertainty metrics from individual sensors and reports an improved and more consistent object map to ADAS applications. Moreover, the fusion layer hides implementation details of the underlying sensors.

Figure 9.30 outlines the block flow of the fusion system; the details of those blocks will be presented later. The multi-step process assumes that observations are processed sequentially, and begins with the acquisition of the observations from the individual sensors. There are two independent threads.

First, *sensor transformation* transforms the object maps from individual sensors into a unified object map in the vehicle frame based on the estimated pose and the latency of each sensor. *Data association* compares the unified object map against known entities represented by a *fused track list*. The observations may represent the observed position of an entity (e.g., range, azimuth and range rate), identity information, and parameters that can be related to identify confidence level, tracking maturity, and geometric information of the entity. As illustrated in Fig. 9.30, the *data association* block systematically compares observation against the known fused tracks, and determines

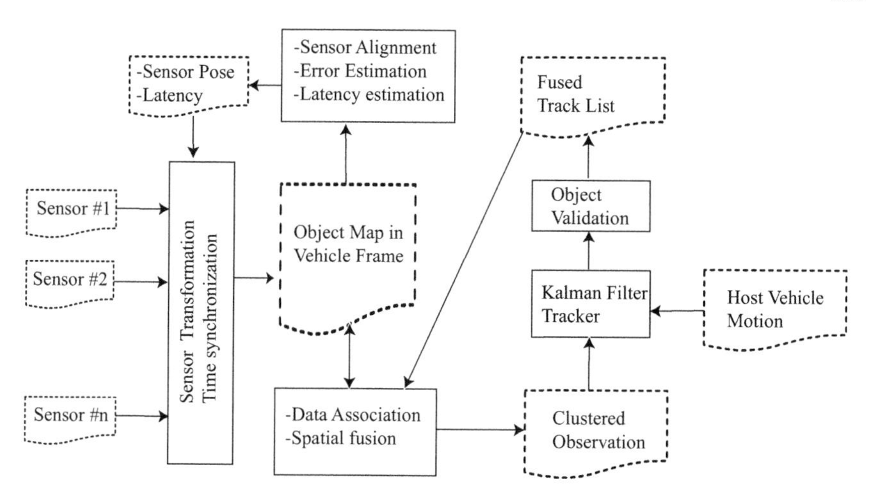

FIGURE 9.30
Flow diagram of the fusion system.

whether or not these observation-track are related. The block *spatial fusion* groups the observations that are associated with the same fused track and outputs *clustered observation*. The *Kalman filter tracker* updates the fused tracks by using *clustered observation* and the vehicle's motion.

In the second thread, starting from the *data association* block, we retrieve the candidate pairs from the observation-track pairs from a particular sensor and then select the pairs with good matching scores to estimate the position and pose of the sensor. The *latency estimation* block uses the synchronizing clock (see Fig. 9.29) as the time reference to find out the latency in each measurement cycle.

9.6.2.2 Error Model of the Sensor

We consider a sensing system with multiple sensors on the host vehicle. Let's first define the system of reference as shown in Fig. 9.31. The sensor k is mounted at the pose $\mathbf{m} = (x_0, y_0, \theta_0)$ with respect to the vehicle frame, where θ_0 denotes the orientation of the sensor's boresight. A measurement of an object is a three dimensional vector $\mathbf{o} = (r, \theta, v_r)$, where r and θ are the range and azimuth angle measurements in the sensor frame, respectively; v_r denotes the range-rate along the azimuth axis. With random error in measurement, the observation in the vehicle frame determined from $\mathbf{o}$ becomes a probability distribution whose extent can be characterized by the sensor's error variances. The error variances $(\sigma_r^2, \sigma_\theta^2, \sigma_{v_r}^2)$ found in the sensor's specification determine the accuracy of the sensor measurement. For radar sensor, these error variances can be determined from the shape of ambiguity function and received signal's SNR. Therefore, the covariance matrix $\mathbf{\Gamma}_o$ of a measurement in the sensor

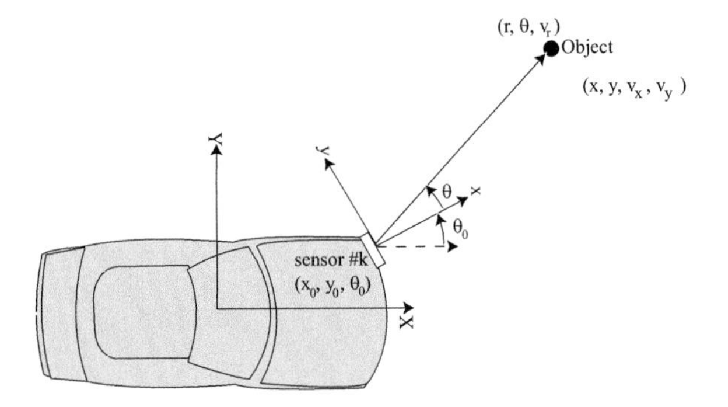

FIGURE 9.31
Reference system.

frame can be represented by

$$\mathbf{\Gamma}_o = \begin{pmatrix} \sigma_r^2 & 0 & 0 & 0 \\ 0 & \sigma_\theta^2 & 0 & 0 \\ 0 & 0 & \sigma_{v_r}^2 & 0 \\ 0 & 0 & 0 & \sigma_{v_t}^2 \end{pmatrix} \tag{9.43}$$

where, σ_{v_t} is an extremely large number or infinity, corresponding to the unobservable tangent velocity v_t. By using a covariance matrix also including the component of tangent velocity v_t, we can treat the sensors with complementary performance characteristics and different orientation in a unified way. Although our analysis is applicable to both polar or Cartesian representation of sensor measurement, we only focus on the polar representation in this section.

We investigate the error in determining the 2D positions (in the vehicle frame) of a set of points visible by a sensor. These points denoted by $\mathbf{y}$ correspond to a set of sensor measured points $\mathbf{o}$.[1] Assume the estimated pose of the sensor to be $\mathbf{m}$. The estimated 2D position of points $\mathbf{y}$ in the above reference is then a function $\mathbf{y} = g(\mathbf{o}, \mathbf{m})$. From Fig. 9.31, it is not difficult to derive the equations to calculate the components of $\mathbf{y}$ as

$$\begin{aligned} x &= x_0 + r\cos(\theta + \theta_0) \\ y &= y_0 + r\sin(\theta + \theta_0) \\ v_x &= v_r\cos(\theta + \theta_0) - v_t\sin(\theta + \theta_0) \\ v_y &= v_r\sin(\theta + \theta_0) + v_t\cos(\theta + \theta_0) \end{aligned} \tag{9.44}$$

[1] Note that $\mathbf{o}$ is now redefined to be $\mathbf{o} = (r, \theta, v_r, v_t)$.

and conversely $\mathbf{o} = g^{-1}(\mathbf{y}, \mathbf{m})$, that is

$$\begin{aligned}
r &= \sqrt{(x - x_0)^2 + (y - y_0)^2} \\
\theta &= \tan^{-1}\left(\frac{y - y_0}{x - x_0}\right) - \theta_0 \\
v_r &= v_x \cos(\theta + \theta_0) + v_y \sin(\theta + \theta_0) \\
v_t &= -v_x \sin(\theta + \theta_0) + v_y \cos(\theta + \theta_0)
\end{aligned} \tag{9.45}$$

To set up notation for later optimization, we write the Eqs. (9.44) and (9.45) in a generic form:

$$\mathbf{g}(\mathbf{o}, \mathbf{y}) = 0 \tag{9.46}$$

where sensor pose $\mathbf{m}$ is known. We write small perturbations of $\mathbf{o}$ by $\delta\mathbf{o}$ and of $\mathbf{y}$ by $\delta\mathbf{y}$. Then the errors are related each other as expressed by the following equation:

$$\frac{\partial \mathbf{g}}{\partial \mathbf{o}}\delta\mathbf{o} + \frac{\partial \mathbf{g}}{\partial \mathbf{y}}\delta\mathbf{y} = 0 \tag{9.47}$$

where

$$\frac{\partial \mathbf{g}}{\partial \mathbf{o}} = \begin{pmatrix} b & -ra & 0 & 0 \\ a & rb & 0 & 0 \\ 0 & -v_r a - v_t b & b & -a \\ 0 & v_r b - v_t a & a & b \end{pmatrix} \tag{9.48}$$

$$\frac{\partial \mathbf{g}}{\partial \mathbf{y}} = \begin{pmatrix} \frac{x - x_0}{r} & \frac{y - y_0}{r} & 0 & 0 \\ -\frac{y - y_0}{r^2} & \frac{x - x_0}{r^2} & 0 & 0 \\ 0 & 0 & b & 0 \\ 0 & 0 & 0 & a \end{pmatrix} \tag{9.49}$$

$$\begin{aligned}
a &= \sin(\theta + \theta_0) \\
b &= \cos(\theta + \theta_0)
\end{aligned}$$

Finally, as an illustrative example, Fig. 9.32 shows the covariance matrices of the observations reported by two sensors with poor azimuth accuracies (e.g., $\pm 10°$). The true object is located at 10 m and 0° azimuth, and the sensors are mounted 2 m apart. The ellipses denote the uncertain measure of the observations.

9.6.2.3 Data Association

Data association answers the question: given N observations, $\mathbf{y}_i$ from one or more sensors, how do we determine which observations belong together, representing observations of the same entity? That is, we seek to determine whether observation $\mathbf{y}_{Ai}$ (the i-th observation from sensor A) results from the same entity as observation $\mathbf{y}_{Bj}$ (the j-th observation from sensor A).

Five steps are illustrated in Fig. 9.33 to determine the association between observation and track. These steps are listed as 1) retrieve candidate entities from the track database, 2) update candidate entities to the observation time

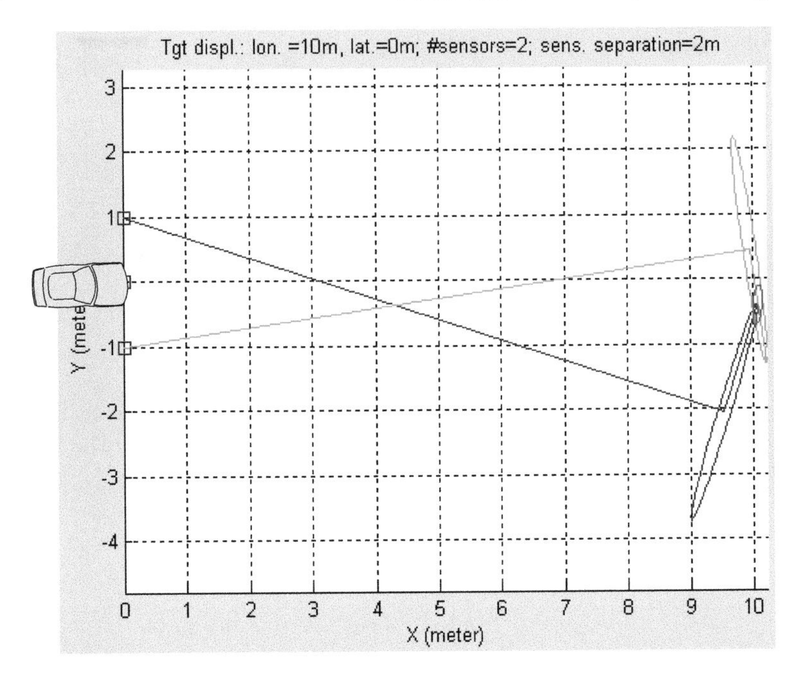

FIGURE 9.32
The covariance matrices of the two observations given by two sensors with poor azimuth accuracy (e.g., $\pm 10°$).

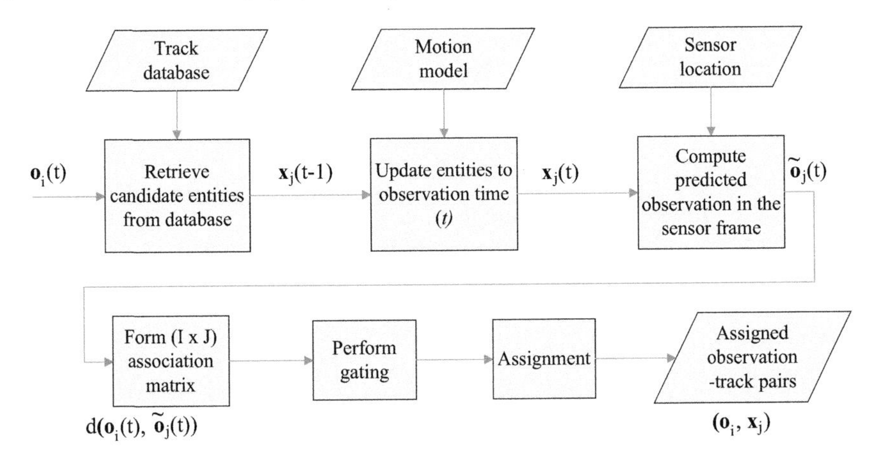

FIGURE 9.33
The data association process.

t, 3) compute predicted observations for each candidate track, 4) compute association matrix, 5) perform gating to eliminate unlikely observation-track pairs, and 6) implement assignment logic.

Given an observation $\mathbf{o}_i$, the first step involves querying candidate entities from the track database containing the previously determined state vector ($\mathbf{x}_j$) of the tracks, which denotes the current estimate of the entities' position, velocity, and classification. Such queries retrieve all entities from the database within specific geometric boundaries (e.g., the sensor's field of view), or having specified attributes (e.g., stationary versus moving entities).

For dynamic situations the second step required for association is to update the state vectors of the tracks to the observation time t. Thus, for each previous determined track, $\mathbf{x}_j(t-1)$, the predicted value of $\mathbf{x}_j(t)$ is computed by

$$\mathbf{x}_j(t) = \Phi(t, t-1)\mathbf{x}_j(t-1) + \mathbf{k} \tag{9.50}$$

where $\Phi(t, t-1)$ represents a transformation that updates the state vector from time $t-1$ to time t, and the vector $\mathbf{k}$ denotes unknown noise modelled as $\mathbf{k} \sim N(0, \mathbf{Q})$.

The third step is required in association unless our sensors are able to observe directly an entity's state. This step utilizes an observation equation to predict observation $\tilde{\mathbf{o}}_j(t)$, which results from entity $\mathbf{x}_j$. That is, $\tilde{\mathbf{o}}_j(t) = g^{-1}(\mathbf{x}_j) + \mathbf{n}_j$ where $\mathbf{n}_j \sim N(0, \mathbf{R}_j)$ denotes a white Gaussian random process.

The fourth association step shown in Fig. 9.33 is to calculate an association matrix. The (i,j)th component of the matrix is the similarity measure that compares the closeness of observation $\mathbf{o}_i(t)$ and the predicted observation $\tilde{\mathbf{o}}_j(t)$. Many measures of similarity are proposed in literature. In this report the Mahalanobis distance [25] is used, that is

$$\mathbf{d}(\mathbf{o}_i, \tilde{\mathbf{o}}_j) = (\mathbf{o}_i - \tilde{\mathbf{o}}_j)^T(\mathbf{P}_i + \mathbf{P}_j)^{-1}(\mathbf{o}_i - \tilde{\mathbf{o}}_j) \tag{9.51}$$

where $\mathbf{P}_i$ and $\mathbf{P}_j$ denote the covariance matrices of the given observation $\mathbf{o}_i$ and the predicted quantity $\tilde{\mathbf{o}}_j$, respectively.

The fifth step is termed gating where the number of possible combinations of observation-track pairs is reduced. In other words, the gating performs a screen to eliminate the unlikely pairs, via heuristic knowledge or statistical hypothesis testing. A straightforward approach is to use the association matrix. All pairs not satisfying $\mathbf{d}(\mathbf{o}_i, \tilde{\mathbf{o}}_j) < T$ will be removed, where T denotes a threshold.

The final step in the association process is the actual assignment of observation to tracks. This assignment step is the invocation of decision logic to declare the association that relates an observation to a candidate track. A hard decision approach is employed in our proposed system; that is an observation is assumed to belong to a single target. In this report, the assignment logic assigns the observation to the nearest adjacent track (namely, the nearest neighbor approach), i.e., $j = \arg\min_j \mathbf{d}(\mathbf{o}_i, \tilde{\mathbf{o}}_j)$. An alternative soft-decision approach is probabilistic data association (JPDA) [31]. In JPDA the observations within the gating window are assigned to a track weighted a posteriori. Hence, a single observation may be assigned as belonging to multiple tracks in a dense target environment. Although JPDA tends to result in track convergence for closely spaced targets [25], we use the nearest neighbor approach due to limited computational resources in the proposed system.

9.6.2.4 Optimization

Having established the association that relates a state vector to predicted observations, a key issue will be addressed in this subsection: how to determine a value of a state vector $\mathbf{x}(t)$ that best fits the observed data.

To illustrate the formulation and processing flow for the optimization process, we consider the weighted-least-squares method to group related observations into a clustered observation $\mathbf{y}$ in the vehicle frame.

One or more sensors observe an object, reporting multiple observations related to the target position $\mathbf{x}$. The (unknown) fused observation in the vehicle frame is represented by a vector $\mathbf{y}$, determined by a time invariant observation equation $\mathbf{g}(\mathbf{o}, \mathbf{y}) = 0$. With actual observation $\mathbf{o}^*$ and estimate $\mathbf{y}^*$, the first order approximation of $\mathbf{g}(\mathbf{o}, \mathbf{y})$ can be written as

$$\mathbf{g}(\mathbf{y}^*, \mathbf{o}^*) + \left.\frac{\partial \mathbf{g}}{\partial \mathbf{y}}\right|_{(\mathbf{y}^*, \mathbf{o}^*)} (\mathbf{y} - \mathbf{y}^*) + \left.\frac{\partial \mathbf{g}}{\partial \mathbf{o}}\right|_{(\mathbf{y}^*, \mathbf{o}^*)} (\mathbf{o} - \mathbf{o}^*) \approx 0 \tag{9.52}$$

Writing

$$\begin{aligned} \mathbf{A} &= \left.\frac{\partial \mathbf{g}}{\partial \mathbf{y}}\right|_{(\mathbf{y}^*, \mathbf{o}^*)} \\ \mathbf{B} &= \left.\frac{\partial \mathbf{g}}{\partial \mathbf{o}}\right|_{(\mathbf{y}^*, \mathbf{o}^*)} \end{aligned}$$

$\mathbf{l} = -\mathbf{g}(\mathbf{y}^*, \mathbf{o}^*)$ and $\varepsilon = -\mathbf{B}(\mathbf{o} - \mathbf{o}^*)$, (9.52) becomes a linearized form as

$$\mathbf{A}(\mathbf{y} - \mathbf{y}^*) = \mathbf{l} + \varepsilon \tag{9.53}$$

The residue $\mathbf{o} - \mathbf{o}^*$ gives the difference between the noise-free observation $\mathbf{o}$ and actual observation $\mathbf{o}^*$. Hence quantity $\mathbf{o} - \mathbf{o}^*$ can be treated as observation noise. Letting $\mathbf{\Gamma}_o$ denote the observation noise, the covariance matrix ($\mathbf{\Gamma}_\varepsilon$) of the residue ε in (9.53) becomes

$$\mathbf{\Gamma}_\varepsilon = \mathbf{B}\mathbf{\Gamma}_o\mathbf{B}^T \tag{9.54}$$

Now, we assume a total of K independent observations, $\{\mathbf{o}_k | k = 1, ..., K\}$, related to the fused quantity $\mathbf{y}$. Hence (9.53) can be extended to be

$$\begin{pmatrix} \mathbf{A}_1 \\ \mathbf{A}_2 \\ \cdots \\ \mathbf{A}_K \end{pmatrix} (\mathbf{y} - \mathbf{y}^*) = \begin{pmatrix} \mathbf{l}_1 \\ \mathbf{l}_2 \\ \cdots \\ \mathbf{l}_K \end{pmatrix} + \begin{pmatrix} \varepsilon_1 \\ \varepsilon_2 \\ \cdots \\ \varepsilon_K \end{pmatrix} \tag{9.55}$$

By the Gauss-Markov theorem [34], obtaining the linear minimum variance estimate of $\mathbf{y} - \mathbf{y}^*$ in (9.55) yields

$$\hat{\mathbf{y}} = \mathbf{y}^* + \left(\sum_{k=1}^{K} \mathbf{A}_k^T \mathbf{\Gamma}_{\varepsilon k}^{-1} \mathbf{A}_k \right)^{-1} \sum_{k=1}^{K} \mathbf{A}_k^T \mathbf{\Gamma}_{\varepsilon k}^{-1} \mathbf{l}_k \tag{9.56}$$

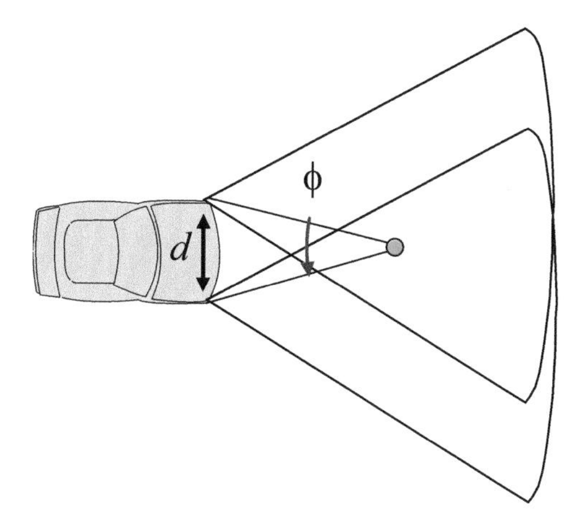

FIGURE 9.34
A simulated vehicle with two sensors.

To further verify the analysis presented above and validate the proposed algorithms, simulations have been conducted under different conditions. As shown in Fig. 9.34, two sensors with poor azimuth accuracy (e.g., $\pm 10^\circ$) are placed in the front bumper 2 m apart.

In the first simulated scenario, both the host and target vehicles are stationary. The target is located at 10 m and 0° azimuth. Hence, the initial state vector $\mathbf{x}(0) \sim N(\mu_0, P_0)$ with $\mu_0 = (10, 0, 0, 0, 0)$ and $P_0 = \text{diag}[100, 100, (1\pi/180)^2, (5\pi/180)^2, 25]$. Assume the motion noise parameters are $\sigma_v^2 = 1$ and $\sigma_\omega^2 = (1\pi/180)^2$ (Cf. (9.57)). The result is shown in Fig. 9.35. Observe that the error in lateral displacement measurement is reduced by about 50%.

It is interesting to observe that the estimated error of lateral displacement is a function of angle ϕ shown in Fig. 9.34. The plot (a) in Fig. 9.36 illustrates the basic idea. We assume that estimation uncertainty can be approximated by a Gaussian distribution, illustrated by an uncertainty ellipsoid in the state space. In the plot, the solid ellipses indicate measurement uncertainties, and the dash ellipses represent the fused estimates described in Sec. 9.6.2.4. Although in both cases, the uncertain areas are reduced, the fused uncertainty in Case (I) along the long principal axis of the ellipse is still big. This is directly caused by the sensor's poor accuracy in measuring azimuth angle, and angle ϕ between the sensors is small. A big improvement is observed in Case (II) when the orientations of the sensors are perpendicular. We can observe that the uncertainty is reduced tremendously along both axes of the fused ellipse. The plot (b) in Fig. 9.36 confirms the above discussion by showing that the standard deviation of the fused estimate decreases as ϕ approaches $\pi/2$.

In the third experiment, we illustrate how the uncertainty in state space reduces with increasing number of sensors under the condition of perfect data

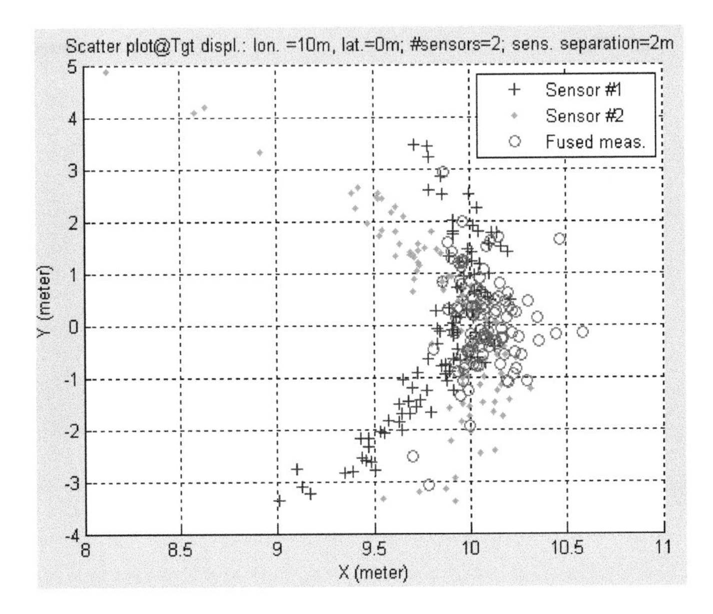

FIGURE 9.35
Scatter plot of raw sensor measurements and fused tracks. The crosses and dots denote the measurements of sensor 1 and sensor 2, respectively. The circles represent the fused tracks.

association; n sensors are placed evenly in the front bumper of the simulated vehicle. Figure 9.37 shows that the standard deviation for the lateral displacement estimate monotonically decreases with increasing number of sensors.

9.6.2.5 Dynamic Models

We assume the target executes a maneuver under constant speed along a circular path. This type of motion is common in ground vehicle traffic.

Consider a two-vehicle scenario, as shown in Fig. 9.38. The host vehicle follows a target vehicle in a lane with constant curvature κ. As described previously, the measurement $\mathbf{y}$ (in vehicle frame) includes x_o, y_o, v_{xo}, and v_{yo}. The target vehicle dynamic state is represented by $\mathbf{x} = (x, y, \psi, \omega, v)$, where the quantities x, y, and ψ denote the pose of the target, and ω and v denote the target's kinematic state.

The kinematic state of the host, modelled as a bicycle model, is represented by yaw rate (ω_H), longitudinal speed (v_{xH}) and lateral speed (v_{yH}). Let ΔT denote the sampling interval from previous cycle $\mathbf{x}$ to new cycle $\mathbf{x}'$. Hence, we can write the dynamic equation $\mathbf{x}' = \mathbf{f}(\mathbf{x})$ of the target state by

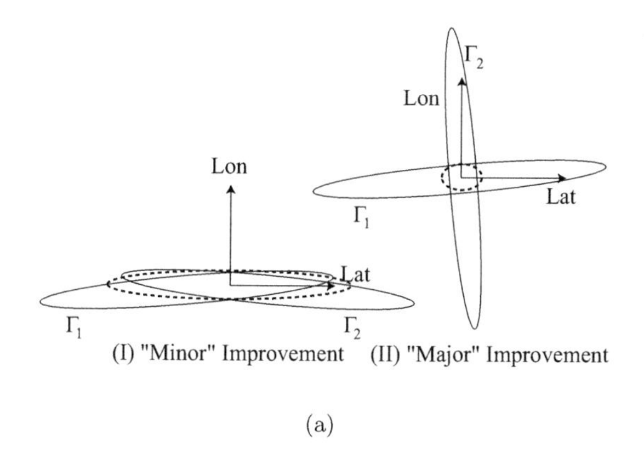

(a)

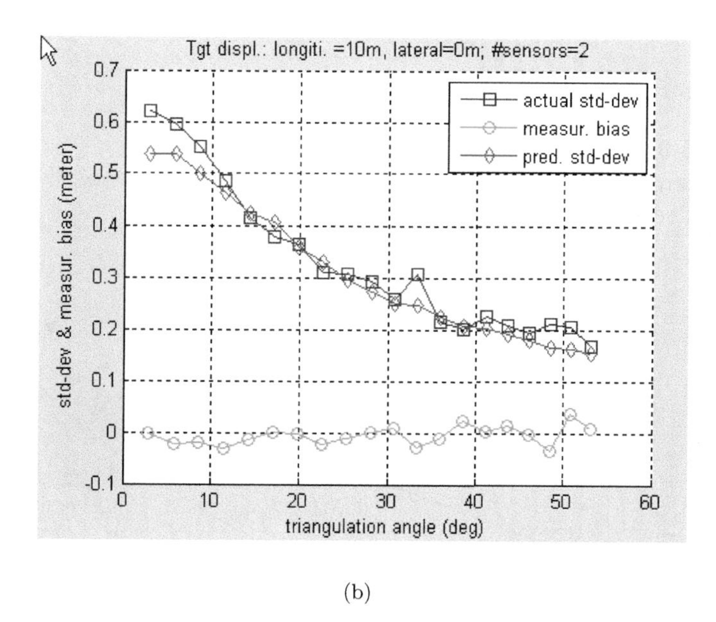

(b)

FIGURE 9.36
(a) Accuracy improvement varies with different sensor configurations. (b) The standard deviation curve of lateral displacement measured with increasing angle ϕ.

$$\begin{aligned} x' &= x + (v\cos\psi + y\omega_H - v_{xH})\Delta T + k_1 \\ y' &= y + (v\sin\psi - x\omega_H - v_{yH})\Delta T + k_2 \\ \psi' &= \psi + (\omega - \omega_H)\Delta T + k_3 \\ \omega' &= \omega + k_4 \\ v' &= v + k_5 \end{aligned} \quad (9.57)$$

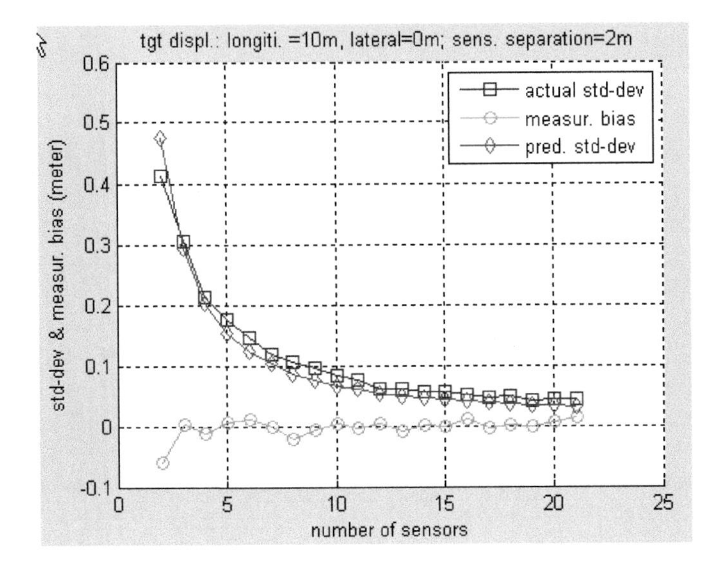

FIGURE 9.37
The standard deviation curve of lateral displacement measured with increasing number of sensors.

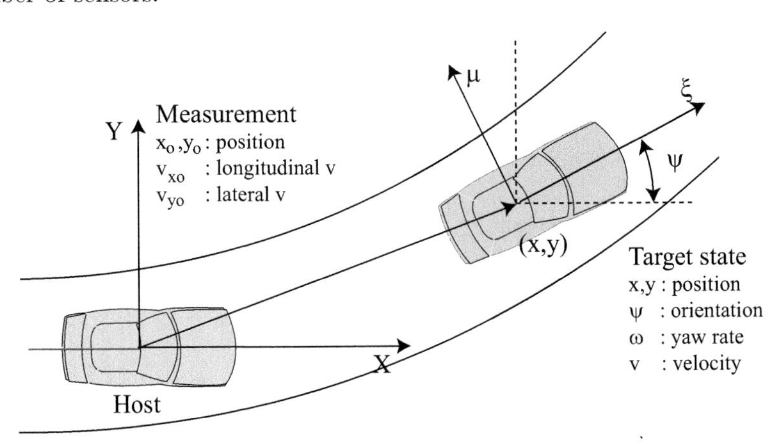

FIGURE 9.38
Coordinated turn model.

and the observation function $\mathbf{y} = \mathbf{h}(\mathbf{x})$ by

$$\begin{aligned} x_o &= x \\ y_o &= y \\ v_{xo} &= v \cos \psi + y \omega_H - v_{xH} \end{aligned} \tag{9.58}$$

Additionally, the Jacobians of Eqs. (9.57) and (9.58) are

$$\mathbf{F} = \begin{pmatrix} 1 & \omega_H \Delta T & -v\Delta T \sin\phi & 0 & \Delta T \cos\phi \\ -\omega_H \Delta T & 1 & v\Delta T \cos\phi & 0 & \Delta T \sin\phi \\ 0 & 0 & 1 & \Delta T & 0 \\ 0 & 0 & 0 & 1 & 0 \\ 0 & 0 & 0 & 0 & 1 \end{pmatrix} \tag{9.59}$$

and

$$\mathbf{H} = \begin{pmatrix} 1 & 0 & 0 & 0 & 0 \\ 0 & 1 & 0 & 0 & 0 \\ 0 & \omega_H & -v\sin\phi & 0 & \cos\phi \end{pmatrix} \tag{9.60}$$

respectively.

After establishing the observation equations that relate a state vector to predicted observations, and also the motion equations for the dynamic system, we can then write the tracking algorithm, a version of extended Kalman filter (EKF), as

Measurement update:

$$\mathbf{K}(t) = \mathbf{P}(t, t-1)\mathbf{H}^T(\mathbf{HP}(t, t-1)\mathbf{H}^T + \mathbf{C})^{-1} \tag{9.61}$$

$$\mathbf{x}(t) = \mathbf{x}(t, t-1) + \mathbf{K}(t)(\mathbf{y}(t) - \mathbf{h}(\mathbf{x}(t, t-1))) \tag{9.62}$$

$$\mathbf{P}(t) = \mathbf{P}(t, t-1) - \mathbf{K}(t)\mathbf{FP}(t, t-1) \tag{9.63}$$

Time update:

$$\mathbf{x}(t, t-1) = \mathbf{f}(\mathbf{x}(t-1)) \tag{9.64}$$

$$\mathbf{P}(t, t-1) = \mathbf{FP}(t-1)\mathbf{F}^T + \mathbf{Q} \tag{9.65}$$

9.6.2.6 Algorithm Summary

Inputs to the algorithm include: 1) an initial estimate of the state vectors $\{\mathbf{x}_j(0)\}$ and initial state uncertainties $\{\mathbf{P}_j(0)\}$ at an epoch $t = 0$ and 2) I_t observations, $\{\mathbf{o}_i(t)|i = 1, 2, ..., I_t\}$, at time t with associated uncertainties $\mathbf{\Gamma}_{oi}(t)$.

At each time t, the system performs a series of calculations:

1. Retrieve the observations, $\{\mathbf{o}_i(t)|i = 1, 2, ..., I_t\}$, measured at time t, , and its associated observational uncertainty, $\mathbf{\Gamma}_{oi}(t)$.
2. For each previous determined track in $\{\mathbf{x}_j(t-1)|j = 1, 2, ..., J_t\}$, calculate the predicted quantities $\tilde{\mathbf{x}}_j(t)$ by (9.57) and $\tilde{\mathbf{o}}_j(t)$ by (9.45).
3. Utilizing the data association algorithm shown in Fig. 9.33, find the associated observation-track pairs $\mathcal{P} = \{(\mathbf{o}_i, \mathbf{x}_j)\}$.
4. For each unmatched observation in set $\{\mathbf{o}_{i_k}| \nexists \mathbf{x} : (\mathbf{o}_{i_k}, \mathbf{x}) \in \mathcal{P}\}$, create a new track $\mathbf{x}_{j_k}$.

5. Remove the unmatched tracks in set $\{\mathbf{x}_{i_k} | \nexists \mathbf{o} : (\mathbf{o}, \mathbf{x}_{j_k}) \in \mathcal{P}\}$ from the track database.

6. For each track $\mathbf{x}_j$, retrieve the associated observations $\{\mathbf{o}_{i_k} \mid (\mathbf{o}_{i_k}, \mathbf{x}_j) \in \mathcal{P}\}$ from the observation-track pairs, and perform the following steps:

   - Letting $\mathbf{y}_j^*$ be the predicted state vector $\tilde{\mathbf{x}}_j(t)$ and $\mathbf{o}_j^*$ be predicted sensor measurement $\tilde{\mathbf{o}}_j(t)$, linearize the observation 9.52 at the point $(\mathbf{o}_j^*, \mathbf{y}_j^*)$.
   - Compute the fused observation $\hat{\mathbf{y}}_j$ in the vehicle frame by (9.56).
   - Apply Eqs. (9.61)-(9.65) to sequentially estimate the state vector $\mathbf{x}_j$.

9.6.3 Online Automatic Registration

In order for the data from different sensors to be successfully combined to produce a consistent object map, the sensor data need to be correctly registered. That is, the relative locations of the sensors and the relationship between their coordinate systems and the vehicle's frame need to be determined. Failing to correctly account for registration errors may result in a mismatch between the compiled object map and the ground truth. Examples would be overstated confidence in a fused target and unnecessary multiplicity of tracks in the tracking database, such as multiple tracks that correspond to a single target.

Therefore, each individual sensor has to be aligned with an accuracy comparable to its intrinsic resolution (e.g., the azimuth accuracy typically is of the order of 0.1 degree). Such a precise mounting is vulnerable to drift during the vehicle's life and expensive if we maintain it manually. It is desirable to have sensors automatically aligned using tracked objects as references.

We present a method to automatically perform an online fine alignment of multiple sensors. The basic idea is illustrated in Fig. 9.39.

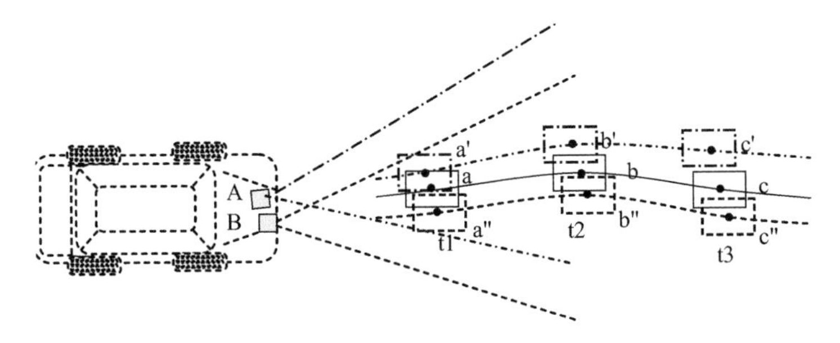

FIGURE 9.39
Schematic illustration of the method to correct sensor positioning.

In Fig. 9.39, A and B denote two sensors mounted at the front of a vehicle. A single target moves away from the vehicle and t1, t2, and t3 denote three consecutive time frames. The dash-dot, solid and dash rectangles represent, respectively, the locations of the target measured by sensor A, fusion processor, and sensor B. The fused track's trajectory is given by objects a, b, and c. Using a large number of associated object pairs, such as $\{(a, a'), (b, b'), (c, c')\}$ and $\{(a, a''), (b, b''), (c, c''))\}$, we can compute, respectively, the positions of sensors A and B by minimizing the residues (i.e., least square method). Here a', b', and c' denote the object map measured by sensor A and a'', b'', and c'' are the object maps observed by sensor B.

Consider a sensor mounted at an unknown position with an unknown orientation. Up to three geometrical parameters, two for location (x_0, y_0), one for bearing alignment (θ_0), can be computed for each sensor on a basis of object trajectories.

Assume that a set of associated observation-track pairs $\{(\mathbf{o}_i, \mathbf{x}_i) \mid i = 1, ..., N\}$ is given. From (9.44), the sensor measurement $\mathbf{o}_i$ is a function $\mathbf{o}(\mathbf{m}, \mathbf{x}_i)$ that can be rewritten as (omit the velocities)

$$\begin{aligned} r_i &= \sqrt{(x_i - x_0)^2 + (y_i - y_0)^2} \\ \theta_i &= \tan^{-1}\left(\frac{y_i - y_0}{x_i - x_0}\right) - \theta_0 \end{aligned} \tag{9.66}$$

with the sensor pose $\mathbf{m} = (x_0, y_0, \theta_0)$.

Initially $\mathbf{m}$ is approximated by a nominal value $\mathbf{m}^*$ determined by a factory (manual) calibration process. Then the corresponding expected sensor measurements ${\mathbf{o}^*}_i = (\tilde{r}_i, \tilde{\theta}_i)$ can be calculated by

$$\begin{aligned} \tilde{r}_i &= \sqrt{(x_i - \tilde{x}_0)^2 + (y_i - \tilde{y}_0)^2} \\ \tilde{\theta}_i &= \tan^{-1}\left(\frac{y_i - \tilde{y}_0}{x_i - \tilde{x}_0}\right) - \theta_0 \end{aligned} \tag{9.67}$$

Using approximates, the corrections $\delta\mathbf{m} = (\delta x_0, \delta y_0, \delta\theta_0)$ for the unknowns $\mathbf{m}$ can be expressed by

$$\mathbf{m} = \tilde{\mathbf{m}} + \delta\mathbf{m} \tag{9.68}$$

where the corrections $\delta\mathbf{m}$ are new unknowns. This means that the original unknowns have been split into a known part (represented by the approximate values $\tilde{\mathbf{m}}$) and an unknown part (represented by the corrections $\delta\mathbf{m}$). The advantage of this split is that the function $o(\mathbf{m}, \mathbf{x}_i)$ is replaced by an equivalent function $o(\tilde{\mathbf{m}} + \delta\mathbf{m}, \mathbf{x}_i)$ which can now be expanded into a Taylor series with respect to the approximate point. This leads to

$$\begin{aligned} o(\mathbf{m}, \mathbf{x}_i) &= \mathbf{o}(\tilde{\mathbf{m}} + \delta\mathbf{m}, \mathbf{x}_i) \\ &= \tilde{\mathbf{o}}_i + \frac{\partial \mathbf{o}_i}{\partial \mathbf{m}} \delta\mathbf{m} \end{aligned} \tag{9.69}$$

where the Jacobian

$$\frac{\partial \mathbf{o}_i}{\partial \mathbf{m}} = \begin{pmatrix} -\frac{x_i - \tilde{x}_O}{r_i} & -\frac{y_i - \tilde{y}_O}{r_i} & 0 \\ \frac{y_i - \tilde{y}_O}{r_i^2} & -\frac{x_i - \tilde{x}_O}{r_i^2} & -1 \end{pmatrix} \tag{9.70}$$

Considering the random noise in the measurement process, the term on the left side of (9.69) will be corrupted with an additive noise

$$\mathbf{o}_i = \mathbf{o}(\mathbf{m}, \mathbf{x}_i) + \mathbf{n}_i \tag{9.71}$$

where, $\mathbf{n}_i$ denotes the white Gaussian random process, i.e., $\mathbf{n}_i \sim N(0, \mathbf{\Gamma}_i)$.

Leaving the terms containing unknowns on the right side, the equation above is rewritten in the component form as

$$-\begin{pmatrix} -\frac{x_i - \tilde{x}_O}{r_i} & -\frac{y_i - \tilde{y}_O}{r_i} & 0 \\ \frac{y_i - \tilde{y}_O}{r_i^2} & -\frac{x_i - \tilde{x}_O}{r_i^2} & -1 \end{pmatrix} \begin{pmatrix} \delta x_O \\ \delta y_O \\ \delta \theta_O \end{pmatrix} = \tilde{\mathbf{o}}_i - \mathbf{o}_i + \mathbf{n}_i \tag{9.72}$$

or in shorthand notation

$$\mathbf{A}_i \delta \mathbf{m} = \mathbf{l}_i + \mathbf{n}_i \tag{9.73}$$

where

$$\mathbf{A}_i = -\begin{pmatrix} -\frac{x_i - \tilde{x}_O}{r_i} & -\frac{y_i - \tilde{y}_O}{r_i} & 0 \\ \frac{y_i - \tilde{y}_O}{r_i^2} & -\frac{x_i - \tilde{x}_O}{r_i^2} & -1 \end{pmatrix} \tag{9.74}$$

and

$$\mathbf{l}_i = \tilde{\mathbf{o}}_i - \mathbf{o}_i \tag{9.75}$$

The linear system in (9.72) comprises three unknowns and two equations. Consequently, two or more control points are needed to solve the problem.

As in Sec. 9.6.2.4, the solution of the correction $\delta\mathbf{m}$ that best fits the known associated observation-track pairs is:

$$\begin{aligned} \delta \mathbf{m} &= \arg\min_{\delta \mathbf{m}} \sum_{i=1}^{N} (\mathbf{A}_i \delta \mathbf{m} - \mathbf{l}_i)^T \Gamma_i^{-1} (\mathbf{A}_i \delta \mathbf{m} - \mathbf{l}_i) \\ &= (\sum_{i=1}^{N} \mathbf{A}_i^T \Gamma_i^{-1} \mathbf{A}_i)^\dagger (\sum_{i=1}^{N} \mathbf{A}_i^T \Gamma_i^{-1} \mathbf{l}_i) \end{aligned} \tag{9.76}$$

where, the covariance matrix $\mathbf{\Gamma}_i = \text{diag}[\sigma_{r_i}^2, \sigma_{\theta_i}^2]$ and the symbol † denotes the pseudo-inverse operator.

We close this section by summarizing the calibration algorithm as below, denoted as Algorithm 1.

Remark: Algorithm 1 uses the position of a fused object as the ground truth. We move each sensor's coordinate system in a way such that position discrepancies between fused objects and sensed objects are minimized. The presented registration algorithm hence will help to create a consistent fused object map. However, in the proposed method there is no mechanism

Algorithm 1 Auto-calibration algorithm

1: **INPUT:** N pairs of the control points and measurements $\{(o_i, x_i)|i = 1, ..., N\}$; the maximum iteration L; the measurement noise model Γ_i. The initial factory default sensor pose $\mathbf{m}^* = (x_0^*, y_0^*, \theta_0^*)$
2: **OUTPUT:** The estimated pose of the sensor $\mathbf{m} = (x_0, y_0, \theta_0)$
3: $t = 0$
4: $\mathbf{m} = \mathbf{m}^*$
5: $\mathbf{A} = 0, \mathbf{l} = 0$
6: **loop**
7:     **for** iter= 1 to L **do**
8:         **for** $i = 1$ to N **do**
9:             Compute $\mathbf{A}_i$ and $\mathbf{l}_i$ by Eqs. (9.74) and (9.75).
10:             $\mathbf{A} = \mathbf{A} + \mathbf{A}_i^T \mathbf{\Gamma}_i^{-1} \mathbf{A}_i$
11:             $\mathbf{l} = \mathbf{l} + \mathbf{A}_i^T \mathbf{\Gamma}_i^{-1} \mathbf{l}_i$
12:         **end for**
13:         $\delta \mathbf{m} = \mathbf{A}^\dagger \mathbf{l}$
14:         $\mathbf{m} = \mathbf{m} + \delta \mathbf{m}$
15:     **end for**
16:     $t = t + 1$
17: **end loop**

to correct the sensor alignment when the fused object map is wrong, which is possible even though the fusion and proposed registration algorithms have tremendously reduced the possibility of such an occurrence. An example would be that the angles of all the mounted sensors drift to the left and the positions of fused targets are drifted to the right.

We hence need a calibration process for the fused tracks. An approach to deal with this issue could be using the fused lane information from map, vision, and yaw rate. Assuming that the leading vehicle is within lane and the road geometry is known, we can use the fused track's lateral offset to correct the angular alignment. However, this is beyond the scope of this chapter.

9.7 Case Studies of ADAS Fusion System

In order to address ADAS challenges, we show a fusion system in Fig. 9.40 with the sensor suite of long-range radar (LRR), short range radar (SRR), and vision system (VIS). The field of view (FOV) of the 77 GHz LRR radar is limited to 15°. This long-range radar alone might not be sufficient [29]. To improve performance two extra sensor types were added: a 45° FOV vision system and four 24 GHz UWB short-range radars (SRR) with 60° FOV, which cover a large area uncovered by the LRR.

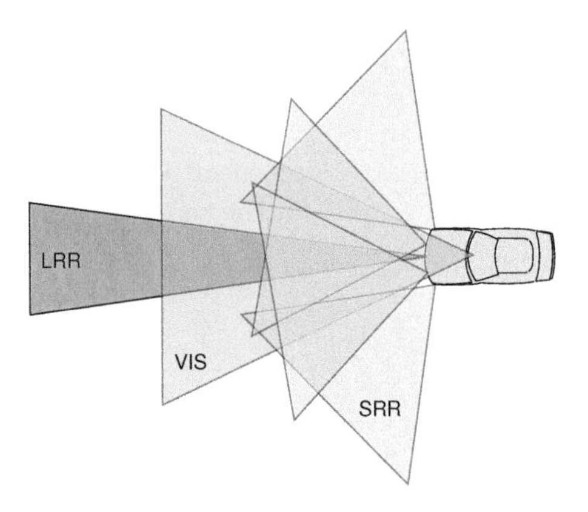

FIGURE 9.40
A sensor suite for ADAS fusion system.

9.7.1 Adaptive Cruise Control

Adaptive cruise control (ACC) is one of the first advanced driver assistance systems (ADAS) launched for luxury models by car makers. Figure 9.41 shows that the subject vehicle (SV) maintains a set speed with an open road, but if there is slower traffic ahead, ACC reduces the vehicle's speed automatically in order to keep a pre-set following distance. The first generation used long-range radar (LRR) and was only operational above about 55 km/h. Speed control was accomplished solely via engine control. Second generation systems are sometimes labeled stop-and-go or full-speed-range ACC (FSRACC) because they can bring the car to a full stop by automatically applying brakes if the vehicle in front stops. The LRR sensor must be supplemented by ultrawide band (UWB) short-range radar or a camera to the cross validate front imminent crashing vehicle. The short-range radars will be used for additional features such as parking assistance or pedestrian protection.

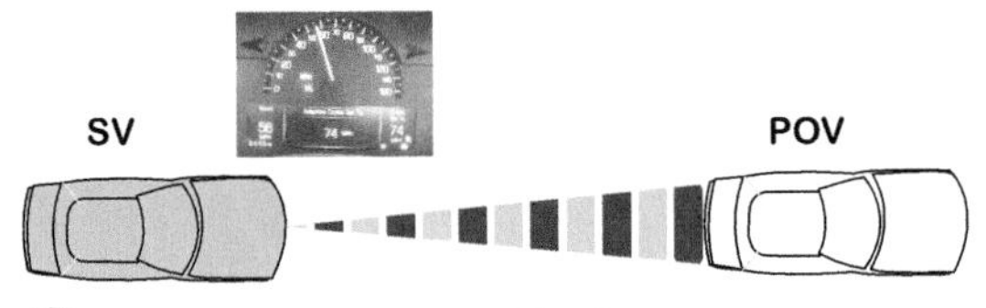

FIGURE 9.41
Adaptive cruise control.

One of the challenges for FSRACC is reducing the rate of false activation. This occurs when a road-side stationary obstacle (e.g., traffic sign, guard rail, or mailbox) is misclassified as a stopped vehicle. To address the safety concerns of false activation, a current commercial system first limits automatic brake authority to maximal 3 m/s^2, which allows the driver to intervene by pressing the throttle pedal. Second, a fusion system of multiple sensors is employed to cross validate whether the lead target is a valid imminent threat. Third, a tracking strategy is used to monitor whether the lead target was moving in previous time frames.

Another challenge is latency of the system to appropriately react to a front cut-in vehicle. ACC is constantly monitoring targets within its sensing field of view. Any new vehicle coming to the field of view is instantly passed on to the process of threat assessment down the line. Because above-mentioned target validation should reduce the rate of false activation, substantial processing time (two or three time frames) is needed to accumulate sufficient confidence to react on the new target. The FSRACC system latency consists of two parts: the latency between a new target cutting in front and ACC reaction and the latency between issuing braking command and the vehicle response. Sometimes the system latency is so long that the driver may intervene by abrupt braking. This leads to a disruption of the operation of ACC and lower customer acceptability of the system.

In the test vehicle, a commercial ACC system and an enhanced FSRACC system with fusion were implemented. The commercial ACC system consists of two units: a radar and an ACC control module (ACM). The radar is a 77 GHz long-range radar (LRR) that can measure up to 20 objects within its observation range. The ACM hosts the ACC application code, including object selection and object validation module. The ACM reports the measurement of the most critical object.

In addition to objects from LRR, the fusion system (cf. Fig. 9.40) utilizes the objects from the vision system (VIS) and from SRR. Like the ACM, the fusion system includes a module that selects and validates the most critical in-path object.

Fig. 9.42(a) shows a typical scenario for ACC which begins with the subject vehicle (SV) traveling on a curved road at speed 48 km/h. Behind the SV in the adjacent lane is a single principal other vehicle (POV) traveling at a higher speed than that of the SV. The POV passes the SV and then changes to the same lane as the SV at a distance approximately 10 to 20 m. The POV keeps the lane for a while, switches back to its original lane in front of the SV, and then slows down, letting the SV pass.

The longitudinal measurements: displacement (x) and velocity (v_x) are presented in Fig. 9.42(c). The SV switched lane at a range of about 15 m. The plots (d) of both figures show the results of object validation of both systems: the commercial system and the proposed fusion system. The red dash-dot curves represent the Boolean variables that determine whether the object is selected as the closest in-path vehicle (CIPV) for the FSRACC.

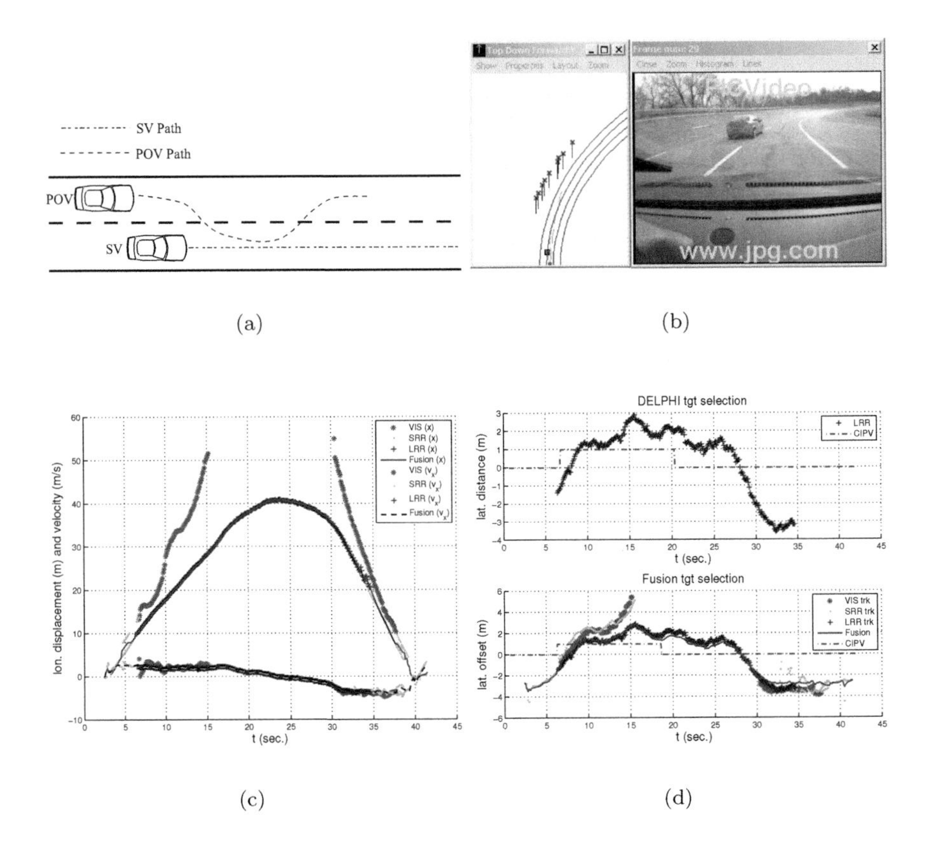

FIGURE 9.42
The result of a run of cut-in scenario on a curved road.

The scenario in Fig. 9.42 is a good example to show that the proposed fusion system can track the same object across different sensors. As shown in plots (c) and (d), in the process of the POV's lane-change (LC) maneuver, the target vehicle was

1. Detected by SRR (measurements denoted by green dots) at $t = 2$ seconds.
2. Detected by VIS (measurements denoted by symbol *) at $t = 7$ seconds.
3. Detected by LRR (measurements denoted by cross +) at $t = 7$ seconds.
4. LRR lost the track at $t = 35$ seconds.
5. VIS lost the track at $t = 15$ seconds, re-captured the track at t = 30 seconds, and finally lost the track at t = 37 seconds.

6. SRR lost the track at $t = 42$ seconds.

During the whole LC maneuver, the fusion system transits from one-sensor, two-sensor, three-sensor, two-sensor, and finally to one-sensor configurations smoothly without significant discontinuity in estimation.

The result shown in Fig. 9.42 demonstrates that the fusion system can validate the object significantly earlier than the LRR-based system. (b) is the screen snapshot of a logging camera when the LC occurs. The red squares in the top-down view window denote fused targets. The top and bottom curves in plot (c) show longitudinal measurements: x and v_x, respectively. The curves in (d) illustrate lateral measurements (y) and target selection flag.

The plots (c) and (d) in Fig. 9.42 illustrate an interesting limitation of mono-camera-based vision systems. As shown in the video window of plot (b), both vehicles were approaching a banked road with the POV driving in the outside adjacent lane. Elevation changes caused by the bank angle confused the vision system to believe the target was much farther away than it actually was. In a mono-camera system, the ground is assumed to be uniformly flat, and the range estimate is computed from the row coordinate (image plane) of the low edge of the vehicle boundary box.

9.7.2 Forward Collision Warning and Braking

Figure 9.43(a) shows the desired behavior for forward collision warning and braking (FCWB) in three phases of FCWB: collision warning, collision warning with brake support, and collision mitigation brake. In the first phase, FCWB detects when the vehicle ahead is slowing or stopped and warns the driver of the risk of a possible imminent crash. The system monitors the relative speed and following distance to the vehicle in front. When a vehicle gets too close to the vehicle in front, a signal (audible, visual, and/or haptic) warns the driver.

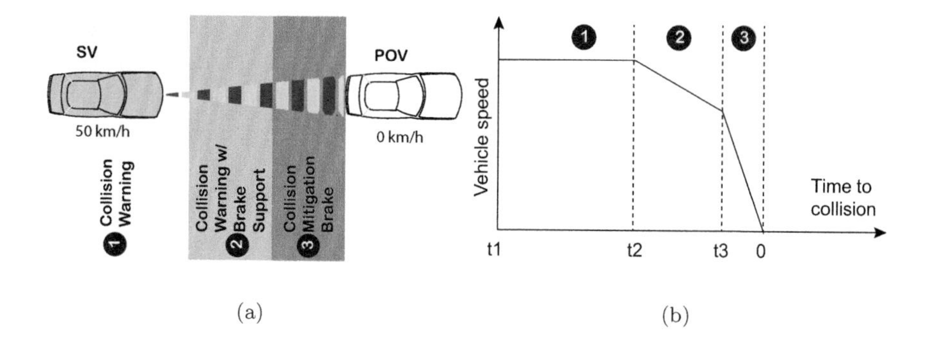

FIGURE 9.43
Forward collision warning and braking.

The system offers collision warning with brake support in the second phase. If the driver does not react after the collision warning has been given, the brake support function prepares the brake system to react quickly, and the brakes are applied slightly. A light jolt may be experienced.

In the event of an imminent crash if the driver has not responded by braking, the system applies harsh braking in the last phase automatically to help reduce the impact of the crash. The system may also activate the seat belt pre-tensioners, pre-charge the airbag systems in order to prepare for the imminent crash. The system may stop the vehicle completely to avoid the crash and issue a brake hold command until the driver intervenes.

In plot (b) we show FCWB behavior in a curve between time to collision (TTC) and SV's vehicle speed. The time 0 denotes the calculated impact time. Before the impact time at t_1 the system starts to provide a warning. If the driver does not react until time t_2, a slight brake is applied to the vehicle. Finally a full brake is issued if the driver still does not react and the time to collision is less than or equal to t_3.

Here we show the result of the fusion system in Fig. 9.40 in a few test runs for FCWB scenarios. We use a barrel in replacement of POV approaches. The subject vehicle (SV) initially is stopped, then is accelerated to a desired speed, and the speed maintained until FCWB is activated to avoid the imminent collision. As shown in Fig. 9.44, in scenes (a), (b), and (c) the SV approaches the barrel directly from left side and from right side, respectively. The blue squares and circles are the fused tracks and measurements from UWB radars in the left window, respectively.

This test determines whether the required collision countermeasures occur at a range that is consistent with FCWB warning and braking requirement.

In the scene (a) in Fig. 9.44, the SV directly approached a stationary barrel placed in the front. The longitudinal fusion estimates as well as the matched raw sensor measurements are shown in plot (a) in Fig. 9.45. The vehicular control output and response (i.e., request acceleration, brake on hold, and actual vehicle acceleration) are shown in plot (b). The barrel was acquired by both front SRR radars at time 5 seconds, the relative range was decreasing at a constant rate (-2.5 m/s) until the FCWB activated at time 10 seconds and stopped the vehicle at time 11 seconds. The vehicle automatic brake system was activated by requested acceleration (-2.5 m/s^2) at time 10 seconds, as shown in plot (b) of Fig. 9.45. The brake on-hold signal was issued after the vehicle was stopped and was released when the driver override event was detected (e.g., brake pedal pressed).

Scene (b) in Fig. 9.44 demonstrates FCWB activation when the SV approached the barrel in a curve from its left side. Plots (c) and (d) in Fig. 9.45 show the timing of object acquisition and control processes, respectively. The right front SRR detected the barrel first at time 3 seconds and was confirmed by the left SRR at time 4 seconds. The brake was triggered at time 6.5 seconds and then set to on-hold later.

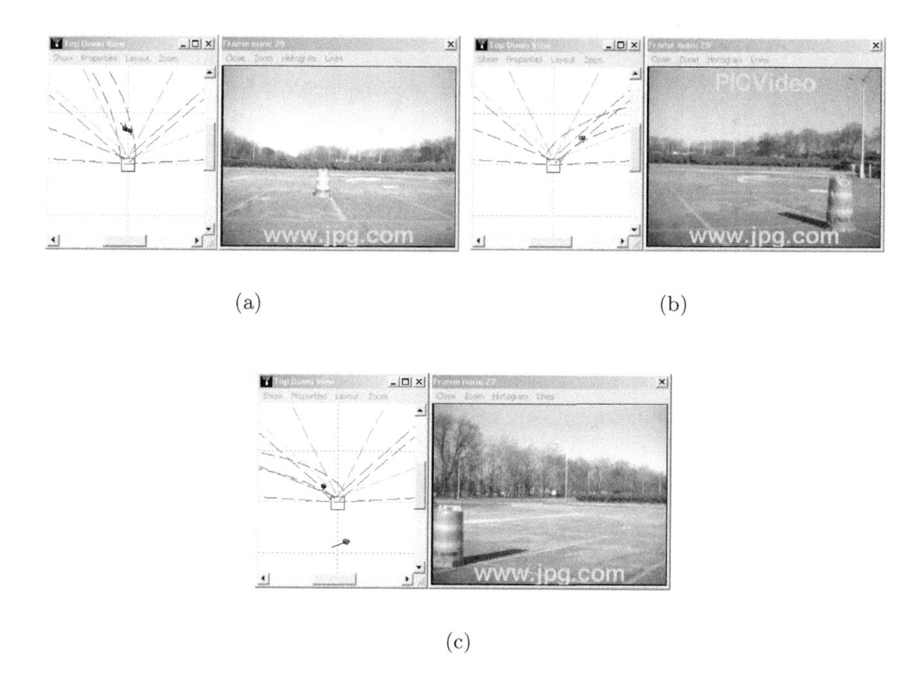

(a) (b)

(c)

FIGURE 9.44
Test scenarios for FCWB.

Similarly, the last scene (c) in Fig. 9.44 demonstrates FCWB activation but on the opposite side (right side). The result is shown in plots (e) and (f) in Fig. 9.45.

9.8 Radars and the Urban Grand Challenge

The ultimate application of automotive radar technology is shown in this section on a fully autonomous vehicle.

The Grand Challenge competitions, sponsored by the Pentagon's Defense Advanced Research Projects Agency (DARPA), were aimed at promoting the development of robotics and autonomous vehicles. A series of competitions from 2004 to 2007 were established. In 2005 the robotic race took place on a 132 mile course in Nevada consisting of desert roads and trails. In 2007 the Urban Challenge pushed robotic technology to determine whether the sensing, perception, behaviors, and computation were capable of handling city environment and traffic. Such vehicles would have obvious applications in advanced driver assistance systems (ADAS) showing up on show room floors in both vehicle as well as military applications.

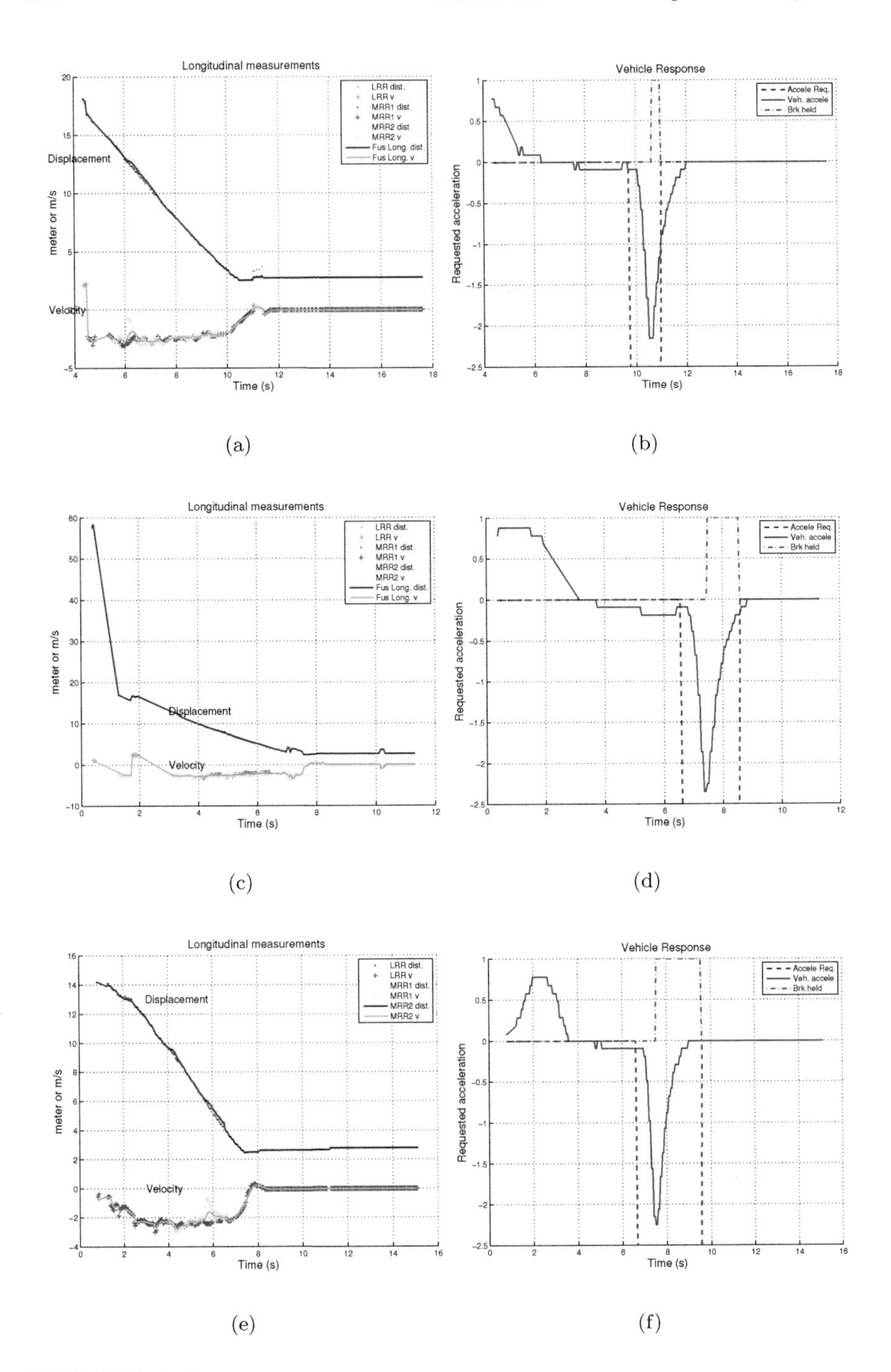

FIGURE 9.45
Results of FCWB testing runs for the scenes (a), (b), and (c) in Fig. 9.44. The left column shows results of the object acquisition while the right column shows the vehicular control signals.

The 2007 race pitted vehicles running completely autonomously (no outside communication or help) through an urban environment and series of tests. The vehicle had to be enabled to:

- Detect and track other vehicles at long range.
- Find a spot and park in a parking lot.
- Obey intersection precedence rules.
- Follow vehicles at a safe distance.
- React to dynamic conditions like traffic, blocked roads, or broken-down vehicles.

The race course, broken into three main areas, is approximately 55 miles with many formidable teams competing. In total, 11 teams were selected to run the final race and six arrived at the finish line. The 2007 DARPA Urban Grand Challenge highlighted the use of radars and lidars. The winning vehicle, from Tartanracing [35] (Fig. 9.46) used a total of eight UWB short-range radars (on the corners of the vehicle), five FMCW long-range radars (four on the front and one in the rear, and at least 10 lidars. The radars were critical in detecting moving traffic as well as augmenting and confirming stopped vehicles and objects along the route. The UWB radars were used to detect objects near (approximately 30 m) the vehicle while the FMCW radars were used for ranges up to 200 m.

There is a lot of detail that is available to readers and the authors assumed one could use search engine to locate the thousands of details available to the

FIGURE 9.46
Winning Tartanracing vehicle for the DARPA Urban Grand Challenge.

public. The reader is encouraged to follow through. A good place to start is Reference [36].

9.9 Summary

In this chapter, we have considered the basic concepts and technologies for radars that have been and will be deployed in automotive applications. The first section was devoted to high-level requirements in automotive domains. After introducing the design challenges, we discussed several key components of an automotive radar, such as antenna, analog front end, and radar processor. We developed and explored several waveforms for target range and range rate measurements. These include linear frequency modulation (LFM), frequency shift keying (FSK), and pulse compression LFM. We compared systems for their pros and cons, and thus showed pulse compression LFM's superiority in range–Doppler resolution.

We began the discussion of several signal processing algorithms for ranging and direction finding. We illustrated the relation among the linear matched filter, discrete Fourier transform, and ambiguity function. We showed that the probability of detection and the accuracy of estimation depends only on signal-to-noise ratio and shape of the ambiguity function.

Besides the radar technology, we also surveyed other commonly used technologies in automotive applications, such as ultrasonic, lidar, stereo vision, and monocular vision. The strengths and weaknesses of these sensing technologies were discussed in contrast with radar. After discussion of sensors of multiple modalities with complementary characteristics, we introduced the topic of fusion in order to create an accurate surrounding perception for improved decision making such as reduction of rates of false negatives and positives.

We presented adaptive cruise control (ACC) and forward collision warning and braking (FCWB) using the proposed fusion system as two case studies. Finally we outlined the usage of radars in the winning vehicle in the DARPA Urban Grand Challenge.

Bibliography

[1] http://auto.howstuffworks.com/cadillac-cyclone3.htm

[2] http://www.safecarguide.com/exp/statistics/statistics.htm

[3] Continental Automotive. http://www.atzonline.com/cms/images/radar_090602.jpg

[4] V. Viikari, T. Varpula, and M. Kantanen, "24 GHz Automotive Radar for Detecting Low Friction spot due to Water, Ice or Snow on Asphalt," URSI, 2008.

[5] K. Strohm, H.-L. Bloecher, R. Schneider, J. Wenger, "Development of Future Short Range Radar Technology," European Radar Conference EuRAD 2005, Conference Proceedings, Paris, France, Oct. 3–7, 2005, pp. 165–168.

[6] A. J. Joseph, D. L. Harame, B. Jagannathan, D. Coolbaugh, D. Ahlgren, J. Magerlein, L. Lanzerotti, N. Feilchenfeld, S. Onge, J. Dunn, and E. Nowak, "Status and Direction of Communication Technologies SiGe BiCMOS and RFCMOS," *Proc. of the IEEE*, Vol. 93, No. 9, Sept. 2005, pp. 1539–1558.

[7] Federal Communications Commission (FCC) Rules and Regulations, Title 47 of the Code of Federal Regulations (CFR), Part 15.

[8] "Commission Implementing Decision of 29 July 2011 amending Decision 2005/50/EC on the harmonisation of the 24 GHz range radio spectrum band for the time-limited use by automotive short-range radar equipment in the Community," *Official Journal of the European Union*, EN, L198/71, 30.7.2011.

[9] "Short range radar equipment operating in the 24 GHz range; Part 1: Technical requirements and methods of measurement," Draft ETSI EN 302 288-1 V1.5.1 (2011-05), European Standard.

[10] "Ultra-Wideband Short-Range Radars for Automotive Applications," Spectrum Planning & Engineering, Radiofrequency Planning Branch Document SP 205, Australian Communications and Media Authority November 2005.

[11] *Radio Law Handbook, Third Edition,* Japan Automobile Importers Association (JAIA) Environmental.

[12] M. Murad, J. Nickolaou, G. Raz, J. S. Colburn, K. Geary, "Next Generation Short Range Radar (SRR) for Automotive Applications," 2012 RadarConn.

[13] W. Menzel. 50 Years of Millimeter-Waves: A Journey of Development. Microwave Journal. August 18, 2008. 51(8) 28-32. `http://www.microwavejournal.com/articles/print/6708-50-years-of-millimeter-waves-a-journey-of-development`

[14] M. Skolnik. *Introduction to Radar Systems*, 3rd ed., McGraw Hill, New York, 2001.

[15] M. Richards. *Fundamentals of Radar Signal Processing*, McGraw Hill, New York, 2005.

[16] Eugin Hyun and Jong-Hun Lee. "A method for multi-target range and velocity detection in automotive FMCW radar." In *Proceedings of the 12th International IEEE Conference on Intelligent Transportation Systems*, pages 7–11, St. Louis, US, October 2007.

[17] M. Meinecke and H. Rohling. "Combination of LFMCW and FSK modulation principles for automotive radar systems." In *Proceedings of German Radar Symposium GRS200*, Berlin, October 2000.

[18] A. Bazzi, T. Chonavel, C. Karnfelt, A. Peden, and F. Bodereau. "Strategies for FMCW radars." In *Proceedings of International Conference on Intelligent Transport Systems Telecommunications.* pp 108–111, October 20–22, 2009.

[19] B. Ristic and B. Boashash. Comments on "The Cramer-Rao lower bound for signals with constant amplitude and polynomial phase." *IEEE Trans. Sig. Proces.* 46(6):1708–1709, 1998.

[20] H. Cramer. *Mathematical Methods of Statistics*, Princeton University Press. 1951.

[21] B. Friedlander. Wireless direction-finding fundamentals. In T. Tuncer and B. Friedlander, *Classic and Modern Direction-of-Arrival Estimation.* Academic Press. 2009.

[22] D. Rhodes. *Introduction to Monopulse*, Artech House, Norwood, MA, 1980.

[23] J. C. Becker. "Fusion of heterogeneous sensors for the guidance of an autonomous vehicle." In *The Third International Conference on Information Fusion*, pages 11–18, Paris, France, July 10–13, 2000.

[24] H. Gunes and M. Piccardi. "Affect recognition from face and body: Early fusion vs. late fusion." In *IEEE International Conference on Systems, Man and Cybernetics*, Vol. 4, pp. 3437–3443, Oct. 10–12, 2005.

[25] D. L. Hall. *Mathematical Techniques in Multisensor Data Fusion.* Artech House, Norwood, MA, 1992.

[26] N. Kaempchen, M. Buehler, and K. Dietmayer. "Feature-level fusion for free-form object tracking using laserscanner and video." In *IEEE Intelligent Vehicles Symposium*, pages 453–458, Las Vegas, USA, June 6–8, 2005.

[27] R. Labayrade, C. Royere, and D. Aubert. "A collision mitigation system using laser scanner and stereovision fusion and its assessment." In *IEEE Intelligent Vehicles Symposium*, pages 441–446, Las Vegas, USA, June 6–8, 2005.

[28] R. C. Luo, C. Yih, and K. L. Su. "Multisensor fusion and integration: Approaches, applications, and future research directions." *IEEE Sensors Journal*, 2(2):107–119, 2002.

[29] R. Mobus and U. Kolbe. "Multi-target multi-object tracking, sensor fusion of radar and infrared." In *IEEE Intelligent Vehicles Symposium*, pages 732–737, Parma, Italy, June 14–17, 2004.

[30] K. A. Redmill, J. I. Martin, and Umit Ozguner. "Sensing and sensor fusion for the 2005 Desert Buckeyes DARPA Grand Challenge offroad autonomous vehicle." In *IEEE Intelligent Vehicles Symposium*, pages 528–533, Tokyo, Japan, June 13–15, 2006.

[31] Y. B. Shalom and E. Tse. "Tracking in a cluttered environment with probabilistic data association." *Automatica*, II(9):451–460, 1975.

[32] M. Skutek, D. T. Linzmeier, N. Appenrodt, and G. Wanielik. "A pre-crash system based on sensor data fusion of laser scanner and short range radars." In *2005 8th International Conference on Information Fusion*. Vol. 2, July 25–28, 2005.

[33] D. Smith and S. Singh. "Approaches to multisensor data fusion in target tracking: A survey." *IEEE Transactions on Knowledge and Data Engineering*. 18(12):1696–1710, 2006.

[34] J. Weng, T. Huang, and N. Ahuja. *Motion and Structure from Image Sequences*. Springer-Verlag, Berlin, 1993.

[35] `www.tartanracing.org`

[36] `http://archive.darpa.mil/grandchallenge/index.asp`

[37] S. Zeng, "A target tracking system using sensors of multiple modalities." *Int. J. Vehicle Autonomous Sys.* 11(4):384–404, 2013.

10

Through-Wall Radar

Through-wall imaging is a potentially game-changing military technology that could be used to locate enemy combatants inside of a building in an urban combat zone. Unfortunately this technology continues to be a topic of research rather than an operational capability. There are many technologies being researched to image through walls but radar is most promising because radio waves routinely propagate through walls to great effect. Examples include an AM or FM broadcast receiver that works indoors. Cell phones work indoors. Wireless routers deployed indoors often provide good coverage outdoors.

A through-wall radar system directs its transmit and receive antennas toward a wall and image or detect what is on the other side of that wall (Fig. 10.1). A small fraction of transmitted field incident on the wall passes through. Most of the transmit field is scattered back toward the radar and some of it is absorbed in the wall itself. Energy that makes it through the wall scatters off the target scene behind the wall. The scattered field is incident on the back side of the wall where it is further reduced by reflection and absorption.

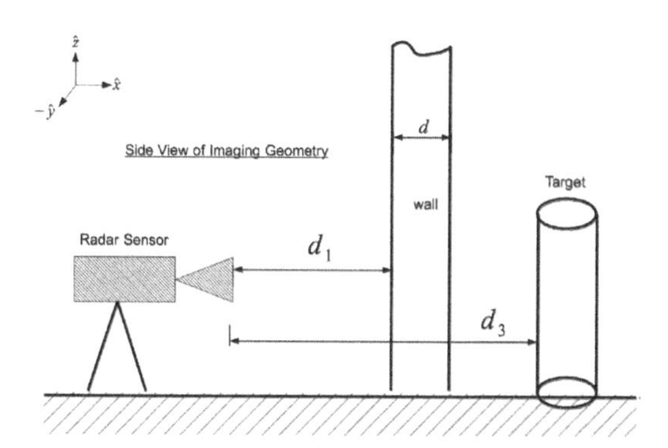

FIGURE 10.1
A through-wall radar system works by directing its transmit and receive antennas at a wall and imaging or detecting what is on the other side of that wall.

Finally, a very small quantity of the scattered field that makes it back through the wall is collected by the radar receive antenna.

At 3 GHz, two-way wall losses for 4 and 8 inch thick solid concrete walls are typically 45 dB and 90 dB respectively [1] and [2]. The lower the frequency of the radar, the more easily it penetrates the wall because wall losses are proportional to frequency. Unfortunately the lower the frequency of the radar, the larger it must become to provide sufficient cross range resolution so that targets can be differentiated from each other, easily detected, counted, and tracked.

In addition to this, most microwave energy radiated at a wall is scattered back toward the radar, causing the radar's receiver to be saturated. Target scenes are generally very short, where the radar might be directly against the wall or a short stand-off range from the wall (20 to 30 feet). For these reasons it is extremely difficult to image through walls using standard radar architectures.

A great deal of through-wall radar development has focused on UWB short-pulse radar systems, where the air–wall boundary can be range-gated out in the time domain to prevent the receiver from being saturated. Examples operating in the 1 to 3 GHz frequency range are treated in [3]–[7]. In order to achieve the average power necessary for reasonable signal-to-noise ratios (SNRs) these radar systems must operate at a high peak power, or alternatively, at a low peak power with a high pulse rate frequency (PRF) using coherent integration of numerous scattered pulses. For this reason, UWB short-pulse radar systems rely on the latest ADC technology to acquire the wide instantaneous bandwidth scattered impulses.

Most through-wall radar systems use some method of beamforming (either SAR or array) to localize targets. Many through-wall radar systems place their antenna elements directly on or in close proximity to the wall in order to reduce air-wall path loss. In this configuration the effects of Snell's law reduce the performance of free space beamforming algorithms by distorting wave propagation through the wall; therefore, much research focusing on through-wall beamforming algorithms has been to develop methods to counter the Snell's law effects [8]–[13]. Other radar systems operate at stand-off ranges using greater average power and therefore do not have to contend with these effects.

Switched array techniques are useful in through-wall imaging because the targets behind the wall (usually human targets) do not move fast enough to smear a radar image. Switched arrays have been used for short-range free space radar imaging [14]–[16] and for through-wall applications [3], [4], [5], [6], [7], [17]–[21], and [22].

In this chapter a simple radar range equation for through-wall radar systems will be shown (Sec. 10.1). A through-wall model will be developed that simulates range profiles, 2D rail SAR imagery, and 2D switched array imagery (Sec. 10.2). And finally, three examples of through-wall imaging sensors will be discussed (Sec. 10.3).

10.1 Radar Range Equation for Through-Wall Radar

The key difference between this and other radar range equations presented in this book is the inclusion of the two-way propagation loss through the wall (L_{wall}) in the denominator. In addition to this, the equation is generalized to be applicable to both UWB impulse and FMCW radars. An approximation for maximum range of a through-wall radar system for both FMCW and UWB impulse radar is

$$R_{max}^4 = \frac{NP_tG_{tx}A_{rx}\rho_{rx}\sigma e^{(2\alpha R_{max})}}{(4\pi)^2kT_oF_nB_n(SNR)_1L_sL_{wall}}, \tag{10.1}$$

where:

R_{max} = maximum range of radar system (m)
P_t = RMS transmit power during transmit pulse (watts)
G_{tx} = transmit antenna gain
A_{rx} = receive antenna effective aperture (m^2)
ρ_{rx} = receiver antenna efficiency
σ = radar cross section (m^2) for target of interest
L_s = miscellaneous system losses
α = attenuation constant of propagation medium before and after the wall
F_n = receiver noise figure (derived from procedure outlined in Sec 1.1.5.4)
k = $1.38 \cdot 10^{-23}$ (joul/deg) Boltzmann's constant
T_o = 290^oK standard temperature
B_n = system noise bandwidth (Hz)
$(SNR)_1$ = single-pulse signal-to-noise ratio requirement
N = number of range profiles used in synthesizing an aperture or array
L_{wall} = two-way wall loss

The two-way wall loss L_{wall} is the ratio of total power passed through the wall divided by total power radiated at the wall squared. This ratio is squared to account for the two-way wall loss rather than just the one-way wall loss. Examples of using this parameter are shown later in this chapter.

For a through-wall radar with only 1D (non-imaging) capabilities including ranging and Doppler, let $N = 1$. For a through-wall radar that is used as a rail SAR imaging device or a phased array, N is the number of range profiles acquired.

For an impulse radar, the noise bandwidth is simply the -3 dB roll-off frequency (f_{-3db}) of the anti-aliasing filter. For a well-designed impulse radar where the pulse width matches the radar receiver bandwidth, the -3 dB roll-off frequency should correspond to the bandwidth of the transmitted impulse; therefore the noise bandwidth $B_n = 1/T_p$.

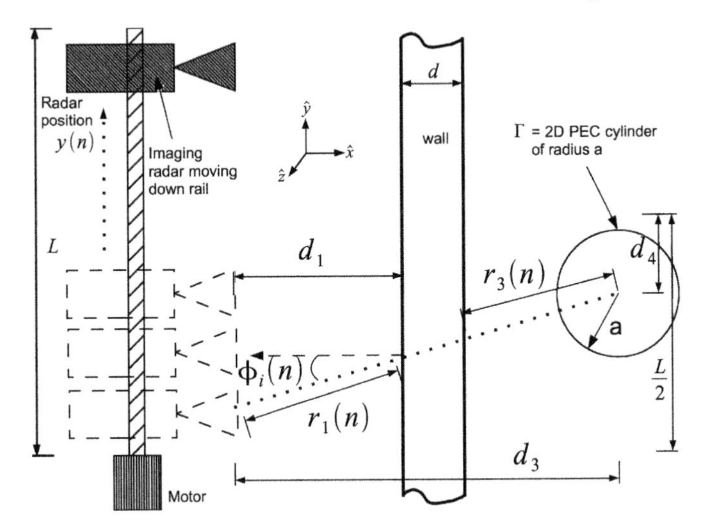

FIGURE 10.2
Through-wall imaging geometry.

For an FMCW radar, the noise bandwidth is inversely proportional to the discrete sample length $B_n = 1/t_{sample}$. For a direct conversion radar (this applies to all FMCW radar examples in this book), the noise bandwidth is twice as wide as a radar following an image rejection architecture where $B_n = 2/t_{sample}$.

10.2 Through-Wall Models

Through-wall imaging is challenging and therefore requires the use of models to explore various radar architecture trades. Three types of through-wall models will be shown, including one that provides simple 1D range profiles (Sec. 10.2.1), another that simulates linear rail SAR data (Sec. 10.2.2), and one that simulates a switched-antenna array (Sec. 10.2.3).

10.2.1 1D Model for Simulating Range Profiles

Before a complete full image through-wall imaging model can be shown it is important to first discuss how to model a single range profile.

The through-wall rail SAR geometry is shown in Fig. 10.2, where a wall of thickness d is placed between the rail SAR and a perfect electric conductor (PEC) cylinder of radius a. The radar system is located r_1 from the front of the wall, and the cylinder is located r_3 behind the wall. When simulating a range profile of this target scene it is assumed that the radar is located at a fixed location on the rail such that the angle of incidence is ϕ_i.

The wall model is based on wave matrix theory from [24], where the normalized impedance of the dielectric with a plane wave incident at an angle of ϕ_i from the normal is

$$Z = \frac{\cos\phi_i}{\sqrt{(\epsilon_r + \frac{\sigma}{j\omega\epsilon_o}) - \sin^2\phi_i}} \tag{10.2}$$

where, for a solid concrete wall, $\epsilon_r = 5$ is the relative permittivity and the conductivity σ is assumed to be varying linearly with frequency from 0.142 to 0.186 S/m over the 1.926 to 4.069 GHz frequency range of the transmit chirp [1]. The instantaneous radial frequency $\omega = 2\pi f$ radians/s and f is the frequency of the radar device in Hz.

The phase delay of the wave inside of the dielectric wall d for an oblique incidence angle ϕ_i is

$$\theta = k_o d\sqrt{(\epsilon_r + \frac{\sigma}{j\omega\epsilon_o}) - \sin^2\phi_i}\ , \tag{10.3}$$

where the free space wave number is

$$k_o = \omega\sqrt{\mu_o\epsilon_o}. \tag{10.4}$$

It is assumed that we will be using vertically polarized antennas where the incident wave is propagating in the $\hat{x}$ direction with the electric field component in the $\hat{z}$ direction and the magnetic field component in the $-\hat{y}$ direction. Therefore the polarization is transverse magnetic to the z axis (TM^z) [25] and thus the wave amplitude coefficient c_1 at the dielectric boundary is

$$c_1 = E_o e^{-jk_o r_1} \tag{10.5}$$

where r_1 is the distance from the radar system to the surface of the dielectric wall.

The simulated range profile is calculated by solving the wave matrix equations for the complex amplitude b_1 of the field traveling in the normal direction $\hat{n}$ at the air–wall interface, which is given by

$$b_1 = \frac{c_3}{(1+\frac{Z-1}{Z+1})(1+\frac{1-Z}{Z+1})}\Big[\frac{Z-1}{Z+1}e^{j\theta} + \frac{1-Z}{Z+1}e^{-j\theta} + \Gamma\Big(\frac{Z-1}{Z+1}\frac{1-Z}{Z+1}e^{j\theta} + e^{-j\theta}\Big)\Big] \tag{10.6}$$

where the complex amplitudes (c_1 and c_3) of the field traveling in the $-\hat{n}$ direction at the interfaces of the wall and cylinder, respectively, are related by

$$c_3 = \frac{c_1(1+\frac{Z-1}{Z+1})(1+\frac{1-Z}{Z+1})}{e^{j\theta} + \frac{Z-1}{Z+1}\frac{1-Z}{Z+1}e^{-j\theta} + \Gamma(\frac{1-Z}{Z+1}e^{j\theta} + \frac{Z-1}{Z+1}e^{-j\theta})}. \tag{10.7}$$

TABLE 10.1
Through-wall 1D range profile example target scene parameters.

$$\begin{aligned}
BW &= f_{stop} - f_{start} \text{ transmit bandwidth} \\
f_{start} &= 1.926 \text{ GHz start frequency of chirp} \\
f_{stop} &= 4.069 \text{ GHz stop frequency of chirp} \\
d &= 10 \text{ cm thickness of the wall} \\
d_1 &= 6.1 \text{ m distance from rail center to wall} \\
d_3 &= 9.1 \text{ m distance from rail center to PEC cylinder} \\
a &= 7.62 \text{ cm radius of a PEC cylinder behind the wall} \\
\phi_i &= 0 \text{ incidence angle} \\
r_1 &= d_1 \text{ at } \phi_i = 0 \text{ radar system distance from the wall} \\
r_3 &= d_3 \text{ at } \phi_i = 0 \text{ distance from back of the wall to the PEC cylinder} \\
T_p &= \text{arbitrary, in 1000 steps}
\end{aligned}$$

The cylinder oriented vertically in the $\hat{z}$ direction, and thus the scattering solution Γ of a 2D PEC cylinder from [25] is given by

$$\Gamma = -e^{-j2k_o r_3} \sum_{n=0}^{\infty} (-j)^n \varepsilon_n \frac{J_n(k_o a)}{H_n^{(2)}(k_o a)} H_n^{(2)}(k_o r_3(n)) \cos n\phi, \tag{10.8}$$

where

$$\varepsilon_n = \begin{cases} 1 & \text{for } n = 0 \\ 2 & \text{for } n \neq 0 \end{cases}.$$

Since the radar system is effectively mono-static, the bi-static observation angle ϕ is $-\pi$.

The received field, a scattered plane wave from the dielectric surface, is represented by

$$E_s = b_1 e^{-jk_o r_1}. \tag{10.9}$$

To apply this model to an S-band through-wall imaging scenario, the IDFT of E_s was applied to a number of test frequencies that emulate an S-band LFM transmitted pulse from 1.929 GHz to 4.069 GHz in 1000 steps following the target scene parameters outlined (Table 10.1). The incident wave amplitude $E_o = 1$.

The simulated range profile shows the locations of the front of the wall at approximately 41 ns and the front of the cylinder at approximately 62 ns (Fig. 10.3). The scattered return off of the wall–air interface is visible at approximately 42.5 ns, where the conductivity of the wall attenuates the return off the back side of the wall. The scattered return from the wall has the greatest magnitude in this range profile, where the cylinder behind the wall is approximately 35.8 dB below the wall.

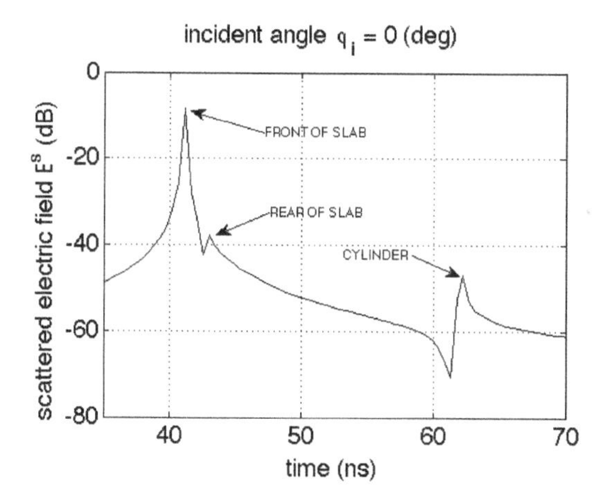

FIGURE 10.3
Simulated range profile of a 10 cm thick lossy-dielectric wall (slab) in front of a 7.62 cm radius cylinder at normal incidence.

10.2.2 2D Model for Simulating Rail SAR Imagery

The 1D model from Sec. 10.2.1 is expanded to provide a simulated SAR image data set of a cylinder behind a wall. The geometry is shown in Fig. 10.2. The SAR is composed of a radar sensor mounted on a linear rail of length L. The antenna is directed toward the target scene which is made up of a dielectric wall and a 2D PEC cylinder at ranges d_1 and d_3 from the antenna. The cylinder location along the y axis is defined by the offset distance d_4 from the rail center $\frac{L}{2}$. The 2D PEC cylinder has a radius a and is located behind the wall. The wall has a thickness of d. The radar sensor moves down the linear rail, acquiring evenly spaced LFM range profiles at incremental locations $y(n)$. The incident angle ϕ_i changes with respect to the radar at position $y(n)$ on the rail relative to the 2D PEC cylinder:

$$\phi_i(n) = \cos^{-1}\left[\frac{d_3}{\sqrt{\left(y(n) - \frac{L}{2} - d_4\right)^2 + d_3^2}}\right], \tag{10.10}$$

where $r_1(n)$ is the distance from the radar antenna to the surface of the dielectric wall in the direction of the cylinder

$$r_1(n) = \frac{d_1}{\cos\phi_i(n)} \tag{10.11}$$

and $r_3(n)$ is the distance from the opposite side of the wall to the 2D PEC cylinder

$$r_3(n) = \frac{d_3 - d_1 - d}{\cos\phi_i(n)}. \tag{10.12}$$

The quantity $y(n)$ depends on the variables $\phi_i(n)$, $r_1(n)$, and $r_3(n)$, which on direct substitution into Equations (10.2) through (10.5) yield the $y(n)$-dependent scattered field equation represented by Equation (10.9).

These calculations are represented by the frequency and rail position-dependent scattered field matrix $E_s\big(y(n), \omega(t)\big)$, where $y(n)$ is the cross range radar position (in meters) on the linear rail shown in Figure 10.5 and $\omega(t)$ is the instantaneous radial frequency at time t for an LFM modulated transmit signal:

$$\omega(t) = 2\pi\left(c_r t + f_c - \frac{BW}{2}\right). \tag{10.13}$$

In this c_r is the chirp rate in Hz/s, f_c is the radar center frequency, and BW is the chirp bandwidth. For the simulated imagery shown in this chapter, t spans 0 to pulse time T_p in 256 steps.

Coherent background subtraction is used in order to image the cylinder behind a lossy wall. One scattered data set was simulated without the cylinder by letting $\Gamma = 0$ represented by E_{sBack} and another was simulated with the cylinder present where Γ is represented by Equation (10.8). The difference between these two data sets is the background subtracted image data set

$$E_{sTargets}\big(y(n), \omega(t)\big) = E_{sScene}\big(y(n), \omega(t)\big) - E_{sBack}\big(y(n), \omega(t)\big). \tag{10.14}$$

Using the target scene geometry and radar parameters outlined (Table 10.2), the SAR algorithm described in Chapter 4 was used to process images of this simulated data for the cylinder in free space and behind the wall (Fig. 10.4). The resulting relative magnitude of the cylinder behind the dielectric wall is -22 dB relative to no wall, the down range location is -922 cm, the down range extent is approximately 8.1 cm, and the cross range extent is approximately 18.4 cm. The resulting down range location of the cylinder without the wall is approximately -904 cm, the down range extent is 8.1 cm, and the cross range extent is approximately 17.1 cm. The presence of the lossy wall does not significantly distort the SAR image.

These results show that the wall causes the cylinder's image to be slightly offset in down range position. The cylinder image is not distorted noticeably because there is no change in down range extent and only a 1.3 cm increase in cross range extent. The return magnitude of the cylinder is lower (-22 dB) when located behind the wall.

10.2.3 2D Model for Switched or Multiple Input Multiple Output Arrays

The 2D rail SAR model from Sec. 10.2.2 is expanded here to account for a switched antenna array imaging system. Although this model applies to a

TABLE 10.2
Through-wall rail SAR model example parameters.

$BW =$ $f_{stop} - f_{start}$ transmit bandwidth (Hz)
$f_{start} =$ 1.926 GHz start frequency of chirp
$f_{stop} =$ 4.069 GHz stop frequency of chirp
$f_c =$ $(f_{start} + f_{stop})/2$ center frequency of the radar (Hz)
$T_p =$ 10 ms pulse length in 256 time steps
$c_r =$ $BW/T_p = 214$ GHz/s chirp rate
$L =$ 2.44 m length of linear rail SAR
$d =$ 10 cm thickness of the wall
$a =$ 7.62 cm radius of a PEC cylinder behind the wall
$d_1 =$ 6.1 m distance from linear rail SAR to wall
$d_3 =$ 9.1 m distance from linear rail SAR to cylinder behind wall
$d_4 =$ 0 position of cylinder relative to linear rail center
$n =$ 1 to 48 locations of $y(n)$ over which the SAR will acquire data

specific switched array radar system it can be scaled to accommodate almost any switched array geometry. Specifically, we will consider the S-band switched array radar imaging system described in Sec. 6.5.

Similar to the rail SAR through-wall geometry, through-wall geometry is shown in Fig. 10.5, where a wall of thickness d is placed between the antenna array and a perfect electric conductor (PEC) cylinder of radius a. The antenna array length is L. The array is directed toward the target scene which is made up of a dielectric wall and a 2D PEC cylinder at ranges d_1 and d_3 from the array. The cylinder location along the y axis is defined by the offset distance d_4 from the array center $L/2$.

According to bi-static radar theory, a separate transmitter and receiver function like a mono-static radar with a phase center located on the baseline between the transmitter and receiver, where the angle bisector of the triangle made of the transmitter, receiver, and target intersect [26], [27]. For this phased array, three points make up each triangle: the transmitter element, the receive element, and the point target p (Fig. 10.6). Each transmitter-to-receiver baseline is represented by $\vec{pc}(n)$, where $n = 1$ to 44 representing each bi-static baseline. The location of the phase centers along $\vec{pc}(n)$ is the position vector $\vec{pos}(n)$, which depends on the length of $\vec{rrx}(n)$ and $\vec{rtx}(n)$ in the direction of $\vec{pc}(n)$ as determined by the angle bisector theorem

$$\vec{pos}(n) = \frac{\vec{pc}(n)}{\left(\frac{|\vec{rrx}(n)|}{|\vec{rtx}(n)|} + 1\right)}. \tag{10.15}$$

The origin of the coordinate system with respect to the antenna array is located at ANT1 in Fig. 6.13 (the x-axis is down range away from the array, the y-axis is the length of the array, and the z-axis is the height of the array). Equation (10.15) shows that the effective mono-static element spacing

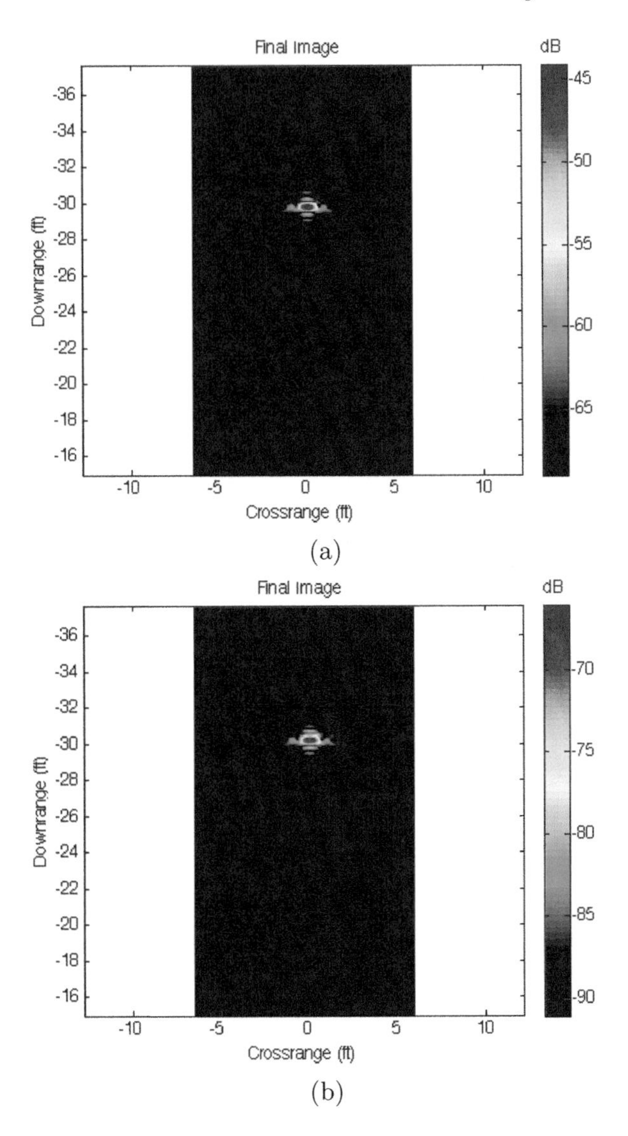

FIGURE 10.4
Simulated SAR imagery of a 2D cylinder with radius $a = 7.62$ cm in free space (a) and behind a 10 cm thick lossy-dielectric wall (b).

is not precisely uniform, given the physical layout of the baselines and the short-range geometry of the target scene that this radar system is designed to image (typically between 4.5 and 20 m down range); however the simulated and measured free space results show that errors are negligible compared to

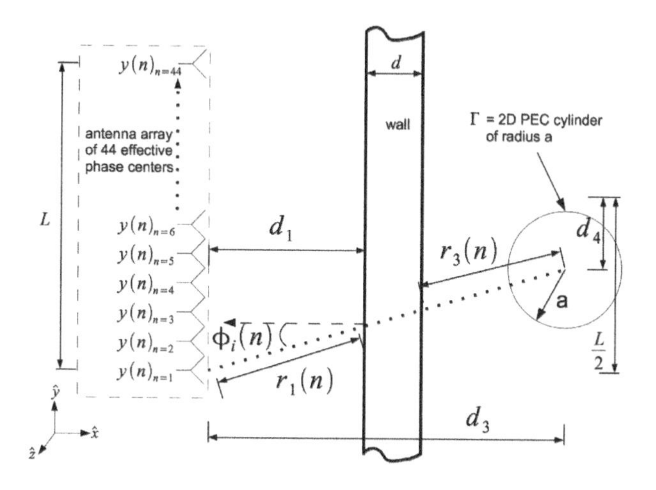

FIGURE 10.5
Through-wall imaging geometry.

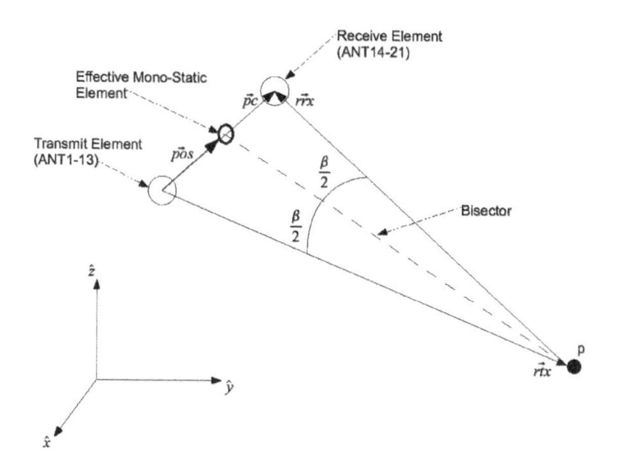

FIGURE 10.6
Location of the effective mono-static element.

the 10 cm wavelength of the 3 GHz radar center frequency [32]. In effect, this bi-static array synthesizes a 44 element $\lambda/2$-spaced linear array.

The incident angle $\phi_i(n)$ between the array phase centers and the wall is dependent upon the location of the phase center position $y(n)$ on the array relative to the 2D PEC cylinder equation (10.10), where $y(n) = pos(n)\hat{y}$, $r_1(n)$

is the distance from the phase center to the surface of the dielectric wall in the direction of the cylinder represented by Equation (10.11) and $r_3(n)$ is the distance from the opposite side of the wall to the 2D PEC cylinder represented by Equation (10.12). For the simulated imagery shown in this section, t spans 0 to T_p in 512 steps, where T_p is the pulse time.

The array uses vertically polarized antennas where the incident plane wave is propagating in the $\hat{x}$ direction with the electric field component in the $\hat{z}$ direction and the magnetic field component in the $-\hat{y}$ direction. Therefore the polarization is transverse magnetic to the z axis (TM^z) [25], and thus the wave amplitude coefficient $c_1(n)$ (Equation (10.5)) at the dielectric boundary is represented by Equation (10.5).

For each phase center n, a simulated range profile is solved for the complex amplitude $b_1(n)$ of the field traveling in the normal direction $\hat{n}$ at the air-wall interface, which is given by Equation (10.6) where the complex amplitudes ($c_1(n)$ and $c_3(n)$) of the field traveling in the $-\hat{n}$ direction at the interfaces of the wall and cylinder, respectively, are related by Equation (10.7).

The cylinder is oriented vertically in the $\hat{z}$ direction, and thus the scattering solution Γ of a 2D PEC cylinder is given by Equation (10.8). Since the radar system is effectively mono-static, the bi-static observation angle ϕ is $-\pi$.

The received field, a scattered plane wave from the dielectric surface, is represented by Equation (10.9), where we let the incident wave amplitude $E_o = 1$. A complete set of 44 range profiles acquired for every phase center across the array is represented by the frequency and phase center position-dependent scattered field matrix $E_s\big(y(n), \omega(t)\big)$.

The scattered return from the wall is significantly higher than the cylinder, and the range sidelobes of the wall mask the image of the cylinder. For this reason, coherent background subtraction is used when modeling the cylinder behind the wall so that the image of the cylinder can be observed in the data. This is accomplished by simulating a data set without the cylinder by letting $\Gamma = 0$, represented by E_{sBack}, and another with the cylinder present represented by E_{sScene} where Γ is provided by Equation (10.8). Coherent background subtraction is accomplished by taking the difference between these two simulated data sets (Equation (10.14)).

For all simulated imagery using this model the parameters shown in Table 10.3 will be used. The SAR imaging algorithm from Chapter 4 was used to process simulated imagery.

10.3 Examples of Through-Wall Imaging Systems

Three examples of through-wall radar systems are shown, all of which are at S-band. S-band is the very highest frequency that might be used for through-wall radar; one might call it the 'knee' of the curve for through-wall attenuation.

TABLE 10.3
Through-wall switched array model parameters.

$BW =$	$f_{stop} - f_{start}$ transmit bandwidth
$f_{start} =$	1.926 GHz start frequency of chirp
$f_{stop} =$	4.069 GHz stop frequency of chirp
$f_c =$	$(f_{start} + f_{stop})/2$ center frequency of the radar
$T_p =$	2.5 ms pulse length in 512 time steps
$c_r =$	$BW/T_p = 857$ GHz/s chirp rate
$L =$	2.24 m length of linear rail SAR
$d =$	10 cm thickness of the wall
$a =$	7.62 cm radius of a PEC cylinder behind the wall
$d_1 =$	distance from linear rail SAR to wall
$d_3 =$	distance from linear rail SAR to cylinder behind wall
$d_4 =$	position of cylinder relative to linear rail center
$n =$	1 to 44 locations of $y(n)$ over which the SAR will acquire data

It is a compromise between cross range resolution for a given aperture size and wall loss at a given frequency [1] and [2]. The trade is simple, better wall penetration achieved at lower frequencies but the size of the radar must be proportionally larger. S-band facilitates a reasonable from factor while also providing excellent imaging performance and good sensitivity.

FMCW was chosen as the radar architecture to take advantage of the high SNR achieved using long-duration LFM waveforms with low transmit power. Unfortunately, a conventional FMCW radar (Secs. 3.3.1 and 3.3.2) has limited capability for through-wall applications because, as was shown above, the greatest signal return from a through-wall target scene is the wall itself. The wall return sets the upper bound of the radar digitizer's dynamic range, and depending on radar frequency, wall thickness and type of material, makes it difficult to image a relatively low RCS (in comparison to the wall) target behind the wall.

For these reasons, a range-gated FMCW architecture (Sec. 3.3.3) was chosen so that long duration (1 to 10 ms) LFM waveforms could be used in small through-wall target scenes while also maintaining the ability to gate-out the wall response. Specifics of this range gate architecture were discussed (Sec. 3.3.3.1) and its effectiveness modeled for a 7.6 cm radius PEC cylinder behind a 10 cm thick solid concrete wall (Figs. 3.19 through 3.21).

Radars discussed in this section are capable of imaging through solid concrete and other walls with only milliwatts of transmit power. A rail SAR will be shown (Sec. 10.3.1) along with a switched array (Sec. 10.3.2) and a high-performance real-time through-wall imaging system (Sec. 10.3.3).

TABLE 10.4
Range gated S-band FMCW rail SAR through-wall imaging system specifications.

$P_{ave} =$ $10 \cdot 10^{-3}$ (watts)
$G_{tx} =$ 12 dBi antenna gain (estimated)
$G_{rx} =$ 12 dBi antenna gain (estimated)
$A_{rx} =$ $G_{rx}\lambda_c^2/(4\pi)$ (m^2)
$\lambda_c =$ c/f_c (m) wavelength of carrier frequency
$f_c =$ 3 GHz center frequency of radar (Hz)
$\rho_{rx} =$ 1 because antenna efficiency is accounted for in antenna gain
$\sigma =$ 0 (m^2) for automobile at 10 GHz
$L_s =$ 6 dB miscellaneous system losses
$\alpha =$ 0 attenuation constant of propagation medium
$F_n =$ 3.5 dB receiver noise figure
$B_n =$ $2/t_{sample}$ system noise bandwidth (Hz) where $t_{sample} = 10$ ms
$(SNR)_1 =$ 13.4 dB
$N =$ 48 number of range profiles used in synthesizing the aperture for 2 inch spacing across the rail
$L_{wall} =$ 45 dB two-way wall loss for a 10 cm thick solid concrete wall

10.3.1 S-Band Range Gated FMCW Rail SAR

A rail SAR through-wall imaging sensor [28] and [29] will be discussed in this section, where its maximum range through a wall is estimated and results are compared to the model.

10.3.1.1 Implementation

Implementation details were described previously (Sec. 5.3.2).

10.3.1.2 Expected Performance

Substituting specifications (Table 10.4) for a 10 cm thick solid concrete wall into the radar range equation (10.1), the maximum range is estimated to be 105 m for a 0 dBsm human target using the MATLAB® script [23].

The IDFT is applied over 10 ms up-chirps with a direct conversion receive architecture resulting in a 200 Hz effective noise bandwidth. The number of range profiles required to process a SAR image is 48 ($= N$).

This radar chirps from 1.926 to 2.069 GHz. Expected range resolution is 6.2 cm with no weighting ($K_r = 0.89$). The cross range resolution is expected to be 16.8 cm when the targets are located 9.1 m down range, centered with respect to the rail with a length of 2.4 m, with no weighting ($K_r = 0.89$). This can be estimated using the same MATLAB script [23].

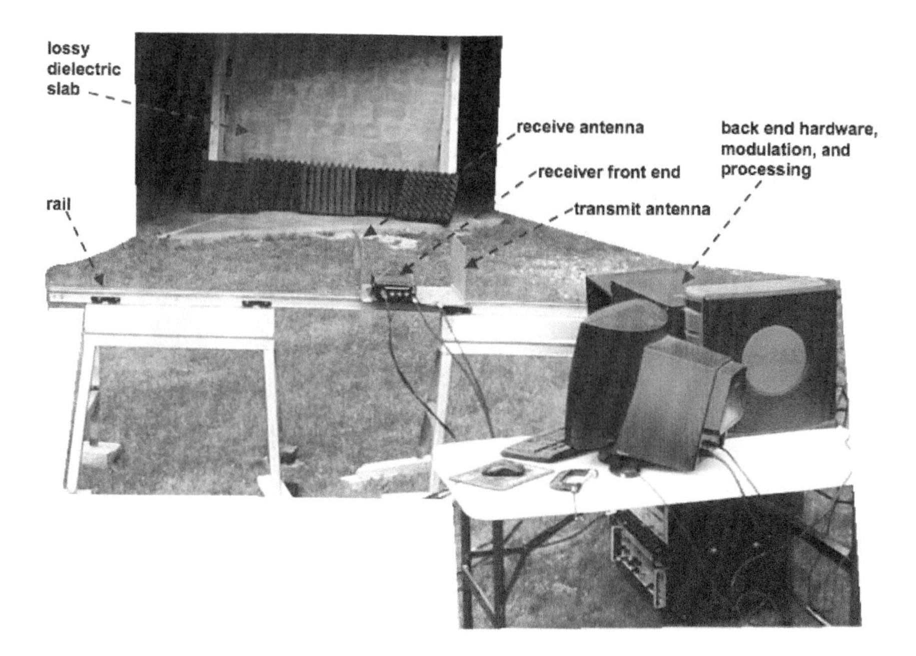

FIGURE 10.7
Through-wall imaging with the S-band range-gated rail SAR made of solid concrete blocks.

10.3.1.3 Results

This SAR was deployed in a through-wall imaging configuration where a 4 inch thick solid concrete wall was placed 30 feet down range from the SAR and targets were placed behind the wall up to 10 feet further down range (Fig. 10.7).

Using the model discussed (Sec. 10.2.2) a 6 inch diameter cylinder was placed behind a wall and simulated (Fig. 10.8a). This model was compared to measured results of an actual 6 inch diameter cylinder placed behind the wall (Fig. 10.8b). Measurements and simulation were in agreement, showing the efficacy of the range gate circuit for imaging through walls and not causing significant increases in image distortion. This measured data can be downloaded and processed [30].

To measure the SAR's sensitivity when imaging through a concrete wall a target scene of three 6 inch tall carriage bolt point targets was imaged (Fig. 10.9). Each bolt is clearly shown in this image and the fact that these bolts can be imaged through a concrete wall demonstrates the SAR's sensitivity.

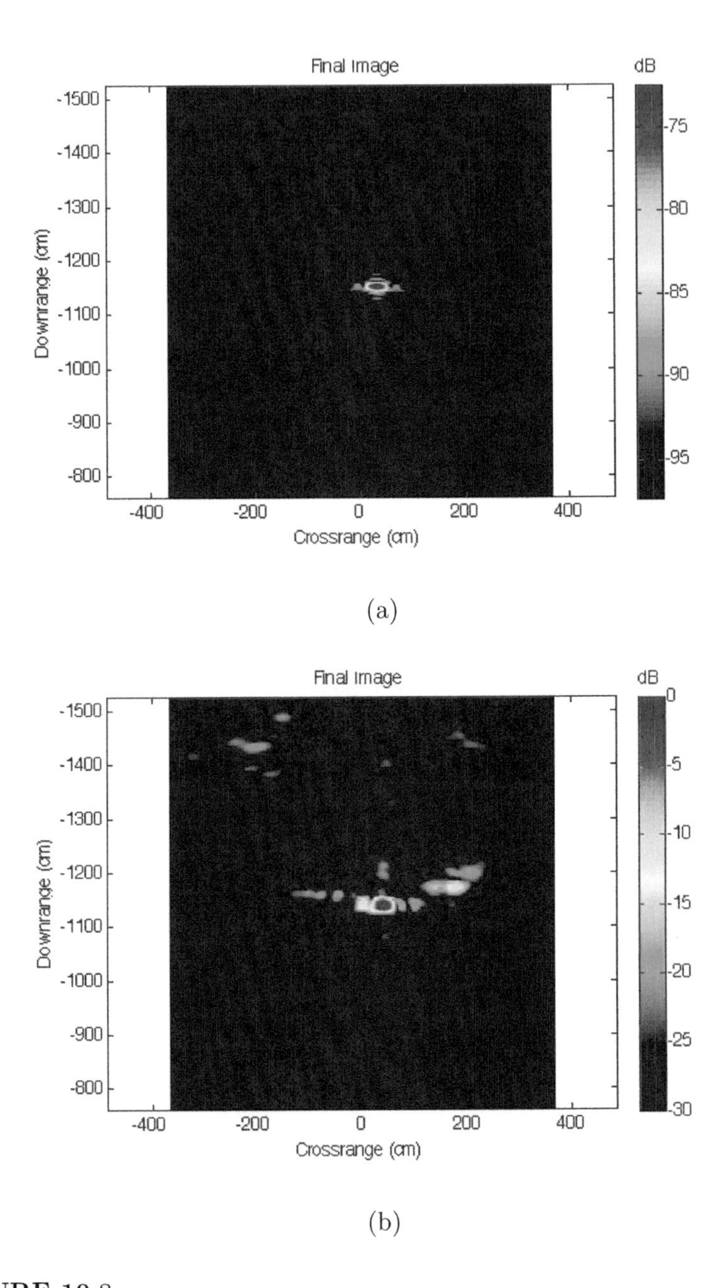

FIGURE 10.8
Image of a 6 inch diameter cylinder through a 4 inch thick solid concrete wall simulated (a) and measured (b).

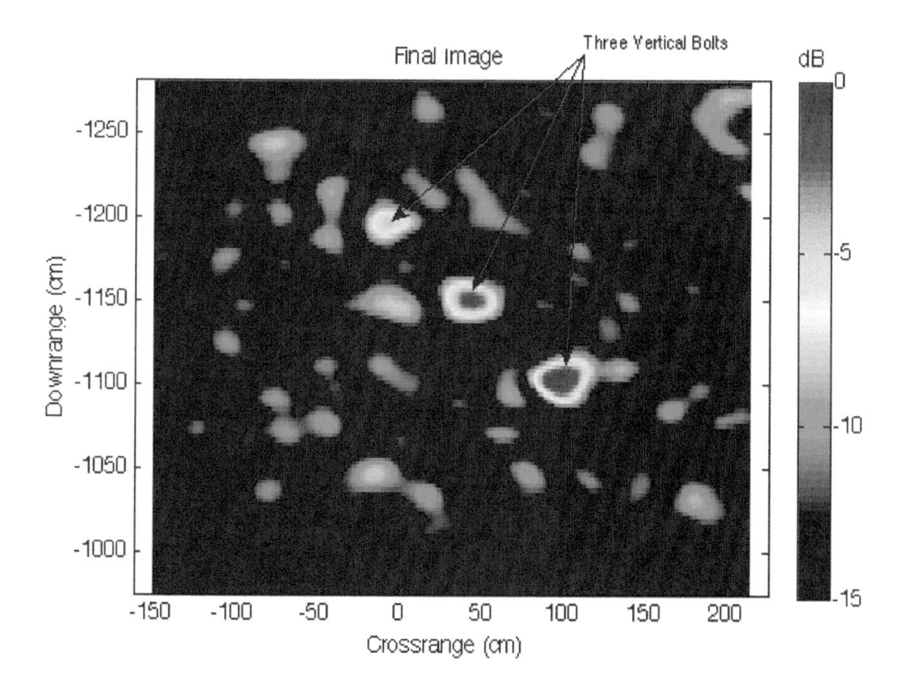

FIGURE 10.9
Row of 6 inch tall carriage bolts through a 4 inch solid concrete wall.

The expected range and cross range resolution for the bolt at 1114 cm are 6.2 cm and 20.5 cm using the MATLAB script [23]. The measured range and cross range resolution for the first bolt at 1114 cm are 8.9 cm and 21.5 cm, showing that the range resolution performance is very close to what can be achieved at best possible theoretical conditions. This measured data can be downloaded and processed [31].

10.3.2 S-Band Switched Array

In the previous section a rail SAR was shown capable of imaging through walls but unfortunately it requires about 20 minutes to acquire image data. This time frame is not valuable in practical applications. By connecting the front ends from the FMCW rail SAR to a switched antenna array, the time required to collect an image is reduced to approximately 2 seconds [28], [32]–[33]. A combination of the range-gated architecture and the switched array provides for a near real-time through-wall imaging system.

TABLE 10.5
S-Band switched array specifications.

$P_{ave} =$	$1 \cdot 10^{-3}$ (watts)
$G_{tx} =$	12 dBi antenna gain (estimated)
$G_{rx} =$	12 dBi antenna gain (estimated)
$A_{rx} =$	$G_{rx}\lambda_c^2/(4\pi)$ (m^2)
$\lambda_c =$	c/f_c (m) wavelength of carrier frequency
$f_c =$	3 GHz center frequency of radar
$\rho_{rx} =$	1 because antenna efficiency is accounted for in antenna gain
$\sigma =$	0 (m^2) for automobile at 10 GHz
$L_s =$	6 dB miscellaneous system losses
$\alpha =$	0 attenuation constant of propagation medium
$F_n =$	3.5 dB receiver noise figure
$B_n =$	$2/t_{sample}$ system noise bandwidth (Hz) where $t_{sample} = 10$ ms
$(SNR)_1 =$	13.4 dB
$N =$	44 number of range profiles used in synthesizing the aperture for 2 inch spacing across the rail
$L_{wall} =$	45 dB two-way wall loss for a 10 cm thick solid concrete wall

10.3.2.1 Implementation

Implementation details were shown previously (Sec. 6.5).

10.3.2.2 Expected Performance

Substituting specifications (Table 10.5) for a 10 cm thick solid concrete wall into the radar range equation (10.1), the maximum range is estimated to be 40 m for a 0 dBsm human target using the MATLAB script [23]

The IDFT is applied over 2.5 ms up-chirps with a direct conversion receive architecture resulting in a 800 Hz effective noise bandwidth. The number of range profiles required to process a SAR image is 44 ($= N$).

This radar chirps from 1.926 to 2.069 GHz. Expected range resolution is 6.2 cm with no weighting ($K_r = 0.89$). The cross range resolution is expected to be 16.8 cm when the targets are located 9.1 m down range, centered with respect to the rail with a length of 2.4 m, with no weighting ($K_r = 0.89$). This can be estimated using the same MATLAB script [23].

10.3.2.3 Results

Imaging through a 4 inch thick solid concrete wall was achieved for a variety of moving and stationary targets. In all scenarios, background subtraction was used to reduce in-scene clutter (such as support beams and other infrastructure) because all measurements were conducted within the author's garage. It was shown in Fig. 3.21 that background subtraction is not necessary to image through a concrete wall when using the range-gated FMCW architecture but in practice it helps to reduce clutter. The results in the following experiments will show that the locations of the targets behind the wall were clearly discernible.

To quantify the performance of this radar system, the same target scene is both simulated and measured in the laboratory. This target scene consists of a cylinder with a radius of 15.2 cm located down range at 907 cm, cross range at 25 cm, and behind a wall located 6.1 m down range made of solid concrete with a thickness of 10 cm. The simulated image of this target scene using the model discussed (Sec. 10.2.3) is shown in Figure 10.10a. For comparison, the measured image of this target scene is shown in Figure 10.10b. The measured cylinder is about 20 dB above the noise floor, where the noise floor is clearly shown in the image. The point spread functions of both simulated and measured imagery are similar, showing the validity of the model except for the close-in cross range sidelobes which are slightly elevated.

Down range and cross range cuts are plotted (Fig. 10.11) of the simulated and measured images (Fig. 10.10). The measured and simulated images compare well in down range, except that the first sidelobe closest to the radar appears to be elevated in the measurement. Measured and simulated cross range main lobes are in close agreement, however the first sidelobes are approximately 3 dB above the simulated result. The differences are likely due to transmit leakage into adjacent elements through one of several paths: the relatively low measured isolation of the antenna switches at 4 GHz (35 dB), feed line coupling because all feed lines are bundled tightly into two harnesses (transmit and receive), and mutual coupling of antenna elements. All aforementioned affects would shift the assumed bi-static phase centers, thereby increasing the cross range sidelobes.

It was observed that theoretical best possible resolution in free space is nearly identical to the measured range resolution when imaging through a lossy wall with the system placed at a stand-off range of d_3 (= 6.1 m).

The expected down range resolution is 6.2 cm. The down range resolution derived from the measurements is 9.4 cm. This result shows that when imaging through a lossy wall of concrete at a stand-off range the switched antenna array radar is performing close to the best possible range resolution and that the wall has little effect on range resolution when the radar is located at a stand-off range.

The expected cross range resolution for the cylinder located at 907 cm down range and 25 cm cross range is 18.2 cm. The measured cross range is 12.6 cm, which appears to be better than free space but this is likely due to element coupling (as mentioned above) or array calibration causing subtle cancellation close to the main lobe. Regardless of this, the measured cross range sidelobes show that the switched antenna array radar is performing close to the smallest theoretical cross range resolution possible when imaging through a lossy wall at a stand-off range.

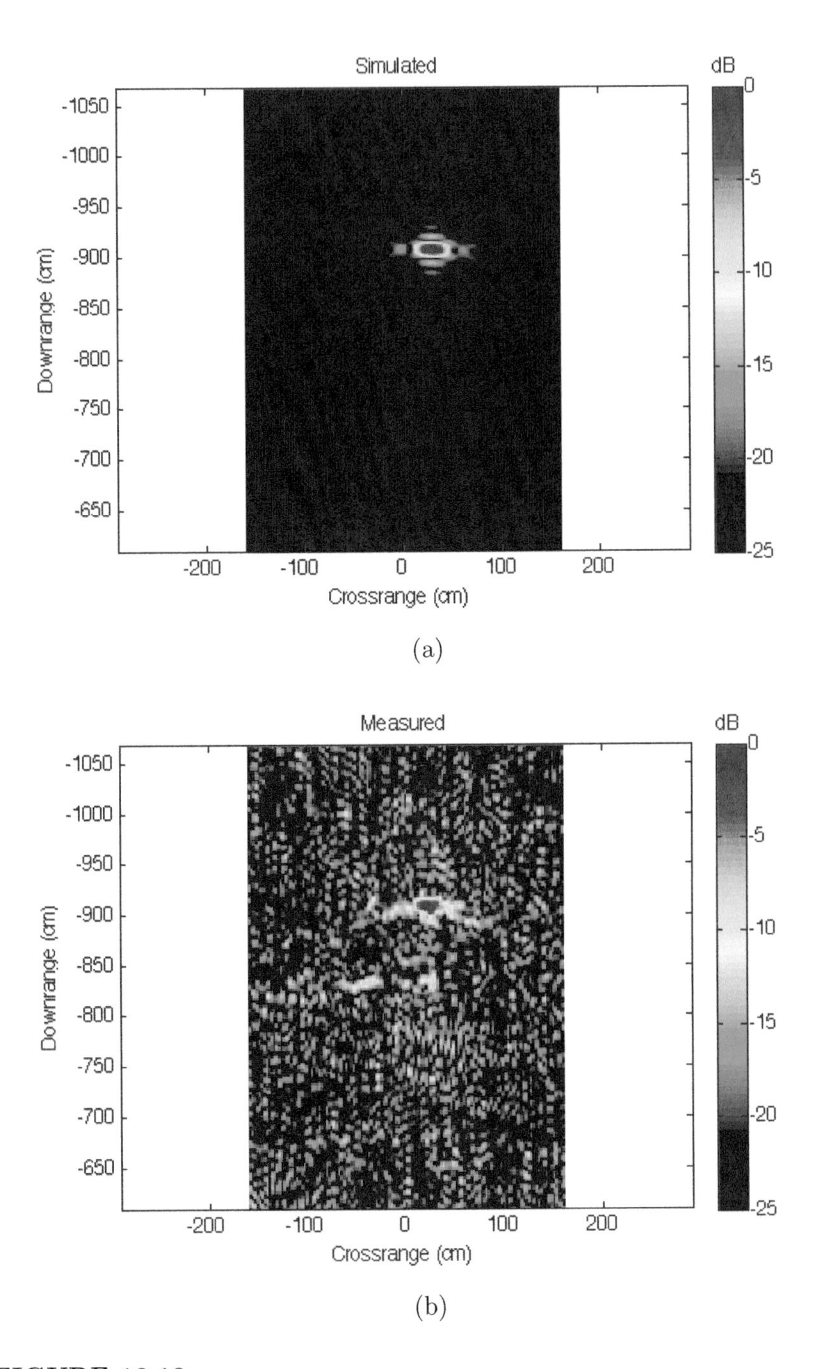

FIGURE 10.10
Simulated (a) and measured (b) cylinder (radius $a = 15.2$ cm) through a lossy wall (solid concrete with thickness $d = 10$ cm).

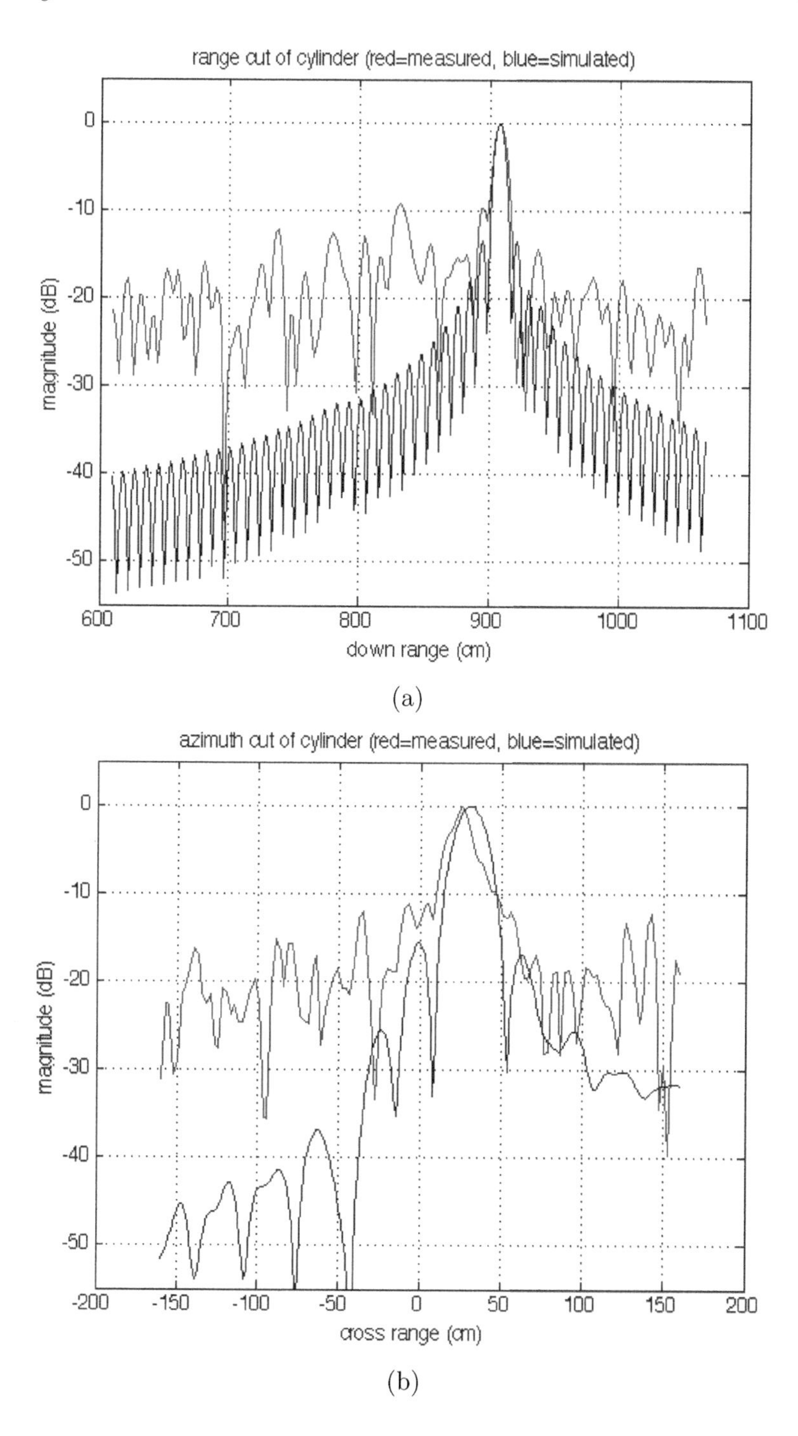

FIGURE 10.11
Measured and simulated down range (a) and cross range (b) sidelobes.

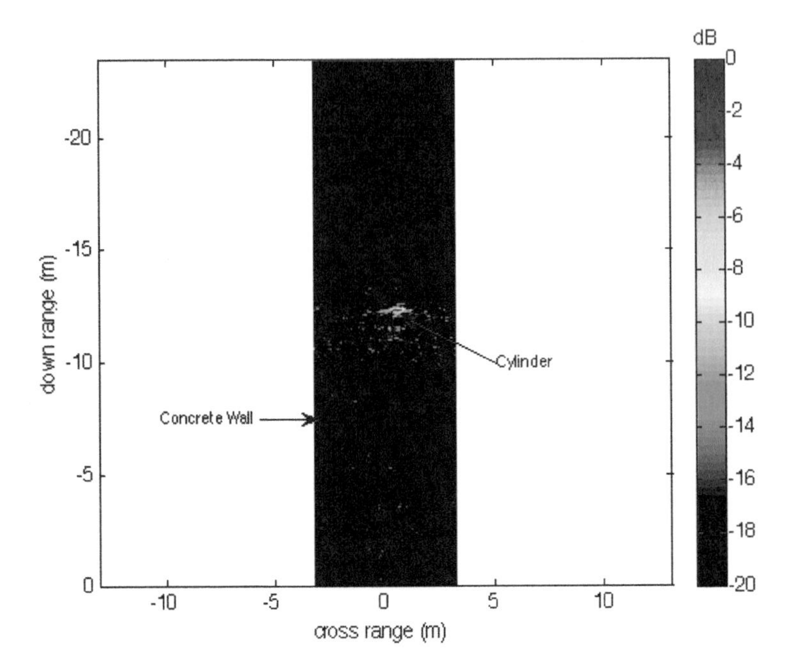

FIGURE 10.12
Large-target imagery example: a 30.5 cm diameter cylinder is imaged through a 10 cm thick concrete wall.

A larger RCS cylinder with a diameter of 30.5 cm was imaged through the wall (Fig. 10.12). The full extent of the radar image is clearly showing the target position. But more importantly, the lossy wall located 610 cm down range in front of the cylinder is not shown, demonstrating the effectiveness of both the range-gated FMCW radar architecture and coherent background subtraction for eliminating the unwanted returns.

Three 12 oz aluminum soda cans were imaged through the wall (Fig. 10.13). Although the RCSs of these targets are significantly smaller than those of both the 15.2 cm and 30.5 cm diameter cylinders, the location of each is clearly shown, demonstrating this radar's sensitivity.

Through-wall radar demonstrations are shown for a 6 inch diameter metal cylinder and a 12 oz soda can [34] and [35].

10.3.3 Real-Time Through-Wall Radar Imaging System

A higher speed, greater transmit power, and better sensitivity radar was developed at MIT Lincoln Laboratory following the architecture and array layouts previously discussed. This device is capable of imaging at a rate greater than

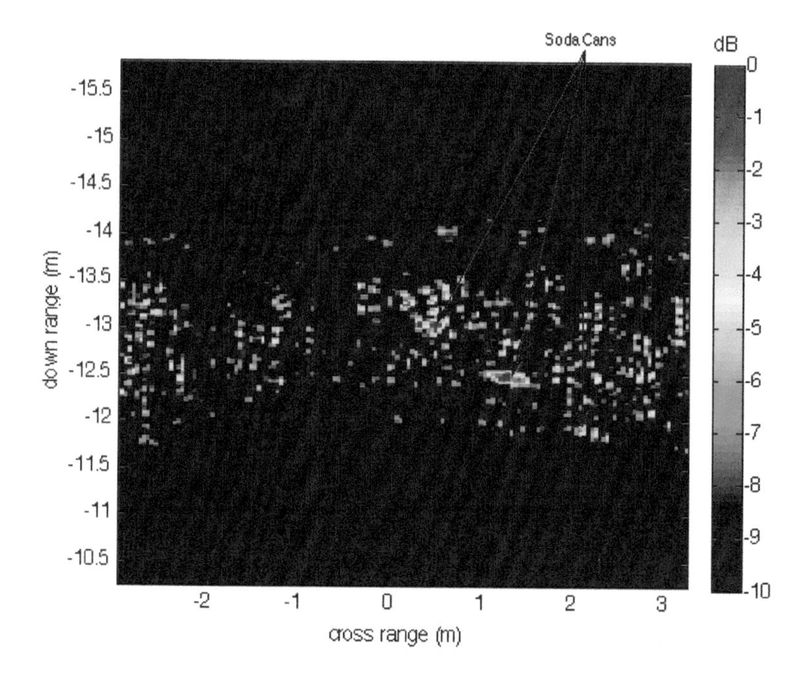

FIGURE 10.13
Small-target imagery example: three 12 oz soda cans are imaged through a 10 cm concrete wall.

10 Hz, providing real-time imagery of what is behind a wall. It uses frame-to-frame CCD revealing the location of any moving targets behind the wall even if they are standing still [36]–[41]. This system will be shown to be effective at imaging human targets through 10 cm, 20 cm thick solid concrete and masonry block (cinder block) walls.

10.3.3.1 Implementation

This through-wall radar system uses the architecture and antenna array described previously (Sec. 10.3.2). It shares the same array layout, imaging algorithm, switch and sequencing. A photo is shown (Fig. 10.14).

But this radar differs from the previous work in that it uses a real-time data acquisition system and imaging algorithm that provides a 10 Hz frame rate [37]. It uses a more sophisticated antenna element that provides greater gain and efficiency [41]. It transmits 500 mW of average power, uses 1 dB noise figure LNAs, and has significantly less switch loss [40]. Additionally, this radar uses detection and tracking algorithms so that the end user is not required to look at 'blobs' on the screen, rather a 'trail of breadcrumbs' can be displayed that is representative of the targets' current and previous locations [39].

FIGURE 10.14
The MIT through-wall radar imaging system.

10.3.3.2 Expected Performance

This radar transmits an average power of 500 mW, chirps from approximately 2 to 4 GHz, the cascaded noise figure is approximately 1 dB, the antenna gain is 15 dB, and it uses 44 phase centers to acquire a radar image. Using the radar range equation for through-wall radar systems (10.1), the estimated maximum range through a 20 cm thick solid concrete wall with a two-way attenuation of 90 dB is 20 m. Results for different types and stack-ups of wall material can be scaled from this estimate.

10.3.3.3 Results

Numerous real-time radar videos were acquired of moving and stationary targets. The focus of this work was to develop a system for locating human targets (moving or stationary) inside a small building or structure and for this reason human target results will be discussed here.

Single-frame imagery of human targets in free space (no wall), behind a 10 cm (4 inch) solid concrete, cinder block, and 20 cm (8 inch) solid concrete walls are shown (Fig. 10.15). In all cases the locations of the human targets are clearly shown. Compared to free space, the imagery through the 10 and 20 cm concrete walls shows a slight increase in clutter. Compared to free space, there is noticeably more clutter when imaging through the cinder block wall.

All data was processed using CCD, where the current frame was coherently subtracted from the previous one. This radar runs at a high frame rate of 10

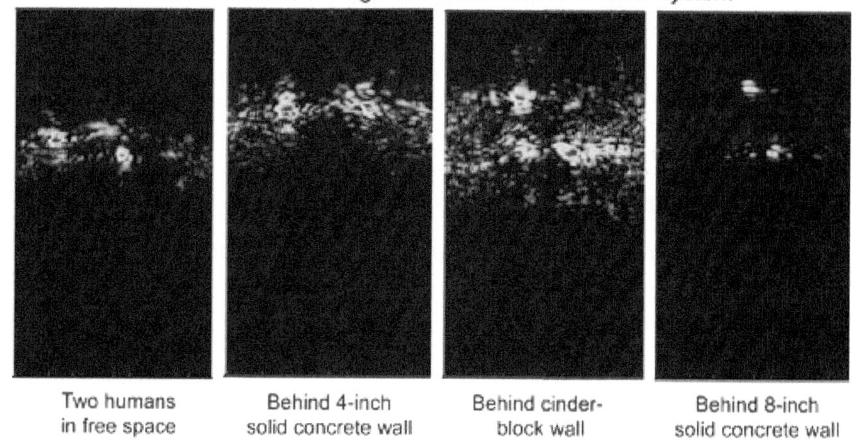

FIGURE 10.15
Imagery through three types of wall and in free space using the MIT through-wall radar imaging system.

TABLE 10.6
Measured performance summary of the real-time through-wall radar system on various types of walls with human targets that are moving and standing still.

Wall Type	1 Target Walking	2 Targets Walking	Standing Still	Sitting Still	Standing and Holding Breath	Sitting and Holding Breath
Free space (no wall)	Good	Good	Good	Good	Good	Good
10 cm Concrete	Good	Good	Good	Good	Good	Poor
Cinder block	Good	Moderate	Good	Moderate	Good	Poor
20 cm Concrete	Good	Good	Moderate	Poor	Poor	Poor

Hz; combining this with CCD allows for very minute phase changes to be shown in the resulting imagery. With this it was observed that even when a human target is standing still it could be located through 10 cm, 20 cm solid concrete walls and cinder block walls (Table 10.6).

A detection algorithm was also developed that plots a 'trail of breadcrumbs' target location history of all moving targets and an overall target count [39]. This algorithm greatly simplifies the use of this radar system by reducing the training requirements for end users.

10.4 Summary

Through-wall radar is a potentially game-changing technology for the urban war fighter. It provides the ability to precisely locate moving targets inside of a building. The basic principles of through-wall radar were discussed followed by a method of estimating performance. Field use of through-wall radar systems remains a challenge because of their size, which is proportional to performance but at odds with the end users' requirements of small and portable. Three S-band through-wall radar systems were shown, including a rail SAR, switched array, and a real-time switched array system. Each radar was capable of imaging through concrete walls. The real-time system used frame-to-frame CCD and was shown to have the remarkable capability of locating human targets moving or standing still.

Bibliography

[1] U.B. Halabe, K. Maser, and E. Kausel, "Propagation characteristics of electromagnetic waves in concrete," Technical Report AD-A207387, Massachusetts Institute of Technology Department of Civil Engineering, March 1989.

[2] P. R. Hirschler-Marchand, "Penetration losses in construction materials and buildings," MIT Lincoln Laboratory Project Report TrACC-1 Rev. 1, 19 July 2006.

[3] M. A. Barnes, S. Nag, and T. Payment, "Covert situational awareness with handheld ultra-wideband short pulse radar," Radar Sensor Technology VI, *Proceedings of SPIE*, Vol. 4374, 2001.

[4] S. Nag, M. A. Barnes, T. Payment, and G. W. Holladay, "An ultra-wideband through-wall radar for detecting the motion of people in real time," Radar Sensor Technology and Data Visualization, *Proceedings of SPIE*, Vol. 4744, 2002.

[5] R. Benjamin, I. J. Craddock, E. McCutcheon, and R. Nilavalan, "Through-wall imaging using real-aperture radar," Sensors, and Com-

mand, Control, Communications, and Intelligence (C3I) Technologies for Homeland Security and Homeland Defense IV, *Proceedings of SPIE*, Vol. 5778, 2005.

[6] A. Berri and R. Daisy, "High-resolution through-wall imaging," Sensors, and Command, Control, Communications, and Intelligence (C3I) Technologies for Homeland Security and Homeland Defense V, *Proceedings of SPIE*, Vol. 6201J, 2006.

[7] W. Zhiguo and L. Xi, F. Yuanchun, "Moving target position with through-wall radar," *Proceedings of the CIE International Conference on Radar*, 2006.

[8] F. Ahmad, M. G. Amin, S. A. Kassam, and G. J. Frazer, "A wideband, synthetic aperture beamformer for through-the-wall imaging," *IEEE International Symposium on Phased Array Systems and Technology*, 2003, pp. 187–192.

[9] F. Ahmad, M. G. Amin, and S. A. Kassam, "Through-the-wall wideband synthetic aperture beamformer," *IEEE Antennas and Propagation Society International Symposium*, 2004, Vol. 3, pp. 3059–3062.

[10] F. Ahmad, M. G. Amin, and S. A. Kassam, "Synthetic aperture beamformer for imaging through a dielectric wall," *IEEE Transactions on Aerospace and Electronic Systems*, Vol. 41, No. 1, 2005, pp. 271–283.

[11] M. Lin, Z. Zhongzhao, and T. Xuezhi, "A novel through-wall imaging method using ultra wideband pulse system," IIH-MSP International Conference on Intelligent Information Hiding and Multimedia Signal Processing, 2006, pp. 147–150.

[12] M. Dehmollaian and K. Sarabandi, "Refocusing through building walls using synthetic aperture radar," *IEEE Transactions on Geoscience and Remote Sensing*, Vol. 46, No. 6, 2008, pp. 1589–1599.

[13] M. G. Amin and F. Ahmad, "Wideband synthetic aperture beamforming for through-the-wall imaging," *IEEE Signal Processing Magazine*, 2008, pp. 110–113.

[14] P. J. F. Swart, J. Schier, A. J. van Gemund, W. F. van der Zwan, J. P. Karelse, G. L. Reijns, P. van Genderen, L. P. Ligthart, and H. T. Steenstra, "The Colorado multistatic FMCW radar system," *IEEE European Microwave Conference*, 1998, Vol. 2, pp. 449–454.

[15] V. Katkovnik, M. S. Lee, and Y. H. Kim, "High-resolution signal processing for a switch antenna array FMCW radar with a single channel receiver," *IEEE Proceedings of Sensor Array and Multichannel Signal Processing*, 2002, pp. 543–547.

[16] M. S. Lee, V. Katkovnik, and Y. H. Kim, "System modeling and signal processing for a switch antenna array," *IEEE Transactions on Signal Processing*, Vol. 52, No. 6, 2004, pp. 1513–1523.

[17] A. R. Hunt, "Image formation through walls using a distributed radar sensor array," *Proceedings of the 32nd IEEE Applied Imagery Pattern Recognition Workshop*, 2003, pp. 232–237.

[18] M. Mahfouz, A. Fathy, Y. Yang, E. E. Ali, and A. Badawi, "See-through-wall imaging using ultra wideband pulse systems," *IEEE Proceedings of the 34th Applied Imagery and Pattern Recognition Workshop*, 2005.

[19] Y. Yang, C. Zhang, S. Lin, and A. E. Fathy "Developement of an ultra wideband Vivaldi antenna array," *IEEE Antennas and Propagation Society International Symposium*, Vol. 1A, 2005, pp. 606–609.

[20] H. Burchett, "Advances in through-wall radar for search, rescue and security applications," *The Institution of Engineering and Technology Conference on Crime and Security*, 2006, pp. 511–525.

[21] Y. Yang and A. Fathy, "Design and implementation of a low-cost real-time ultra-wide band see-through-wall imaging radar system," *IEEE/MTT-S International Microwave Symposium*, 2007, pp. 1476–1470.

[22] K. E. Browne, R. J. Burkholder and J. L. Volakis, "Through-Wall Opportunistic Sensing System Utilizing a Low-Cost Flat-Panel Array," *IEEE Transactions on Antennas and Propagation*, 59(3): 2011, 859–868.

[23] radar_range_eq_thruwall.m, `http://glcharvat.com/shortrange/through-wall-radar/`

[24] R.E. Collin, *Field Theory of Guided Waves*, 2nd ed., IEEE Press, Piscataway, NJ, 1991.

[25] C.A. Balanis, *Advanced Engineering Electromagnetics*, John Wiley & Sons, New York, NY, 1989.

[26] N. J. Willis, *Bistatic Radar*, Scitech Publishing, Inc., Raleigh, NC, 1995.

[27] S. Gabig, K. Wilson, P. Collins, J. Terzuoli, A. J. , G. Nesti, and J. Fortuny, "Validation of near-field monostatic to bistatic equivalence theorem," *Proceedings of IEEE International Geoscience and Remote Sensing Symposium*, 2000, Vol. 3, pp. 1012–1014.

[28] G. L. Charvat, "A Low-Power Radar Imaging System," Ph.D. dissertation, Department of Electrical and Computer Engineering, Michigan State University, East Lansing, MI, 2007.

[29] G. L. Charvat, L. C. Kempel, E. J. Rothwell, C. Coleman, and E. L. Mokole, "A through-dielectric radar imaging system," *IEEE Transactions on Antennas and Propagation*, Vol. 58, No. 8, 2010, pp. 2594–2603.

[30] S_band_rg_thruwall_6in_cylinder.zip, `http://glcharvat.com/shortrange/through-wall-radar/`

[31] S_band_rg_thruwall_pt_targets.zip, `http://glcharvat.com/shortrange/through-wall-radar/`

[32] G. L. Charvat, L. C. Kempel, E. J. Rothwell, C. Coleman, and E. L. Mokole, "An ultrawideband (UWB) switched-antenna-array radar imaging system," 2010 International Symposium on Phased Array Systems and Technology, October 12-15, Waltham, MA.

[33] G. L. Charvat, L. C. Kempel, E. J. Rothwell, C. Coleman, and E. L. Mokole, "A through-dielectric ultrawideband (UWB) switched-antenna-array radar imaging system," *IEEE Transactions on Antennas and Propagation*, Vol. 60, No. 11, 2012, pp. 5495–5500.

[34] Radar imaging a cylinder through a 4" solid concrete wall, `http://glcharvat.com/shortrange/through-wall-radar/`

[35] Radar imaging a 12 oz soda can through a 4" solid concrete wall, `http://glcharvat.com/shortrange/through-wall-radar/`

[36] G. L. Charvat, T. S. Ralston, J. E. Peabody, "A through-wall real-time MIMO radar sensor for use at stand-off ranges," *Proceedings of the Tri-Service Radar Symposium*, 2010.

[37] T. S. Ralston, G. L. Charvat, and J. E. Peabody, "Real-time through-wall imaging using an ultrawideband multiple-input multiple-output (MIMO) phased array radar system," *Proceedings of the IEEE International Symposium on Phased Array Systems and Technology*, 2010, pp. 551–558.

[38] E. Finn, "Seeing through walls,," *MIT News*, October 18, 2011.

[39] G. L. Charvat, J. Goodwin, M. Tobias, J. Pozderac, and J. Peabody, "A real-time through-wall radar system using a time division multiplexed (TDM) multiple-input multiple-output (MIMO) antenna array; measured results and performance," Atlanta, GA: IEEE Radar Conference, 2012.

[40] G. L. Charvat, J. E. Peabody, J. Goodwin, and M. Tobias, "A real-time through-wall imaging system," *MIT Lincoln Laboratory Journal*, June 2012.

[41] J. S. Sandora, G. L. Charvat, "An Ultra-Wideband Vivaldi and Linear Hybrid Taper Antenna for use in a Near-Field Real-Time Phased Array Radar System," *Proceedings of the IEEE International Symposium on Phased Array Systems and Technology*, Waltham, MA, 2013.

Index

Note: Page numbers ending in "e" refer to equations. Page numbers ending in "f" refer to figures. Page numbers ending in "t" refer to tables.

Y

For Product Safety Concerns and Information please contact our EU representative GPSR@taylorandfrancis.com Taylor & Francis Verlag GmbH, Kaufingerstraße 24, 80331 München, Germany

Printed and bound by CPI Group (UK) Ltd, Croydon, CR0 4YY

01/05/2025

01858495-0001